LARRY C. BURTON

61552

9 95

ELEMENTS

OF

THERMOSTATISTICS

ELEMENTS OF THERMOSTATISTICS

D. TER HAAR
UNIVERSITY READER IN THEORETICAL PHYSICS
AND FELLOW OF MAGDALEN COLLEGE,
OXFORD

Second Edition

Holt, Rinehart and Winston
New York Chicago San Francisco
Toronto London

Library of Congress Catalog Card Number: 66–18800

28377–0216

Printed in the United States of America

TO

MY

WIFE

CONTENTS

CHAPTER 3/ THE PARTITION FUNCTION

CHAPTER 4/ BOSE-EINSTEIN AND FERMI-DIRAC STATISTICS

CHAPTER 5/ CLASSICAL ENSEMBLES

CHAPTER 6/ THE ENSEMBLES IN QUANTUM STATISTICS

CHAPTER 7/ RELATIVISTIC STATISTICS

CHAPTER 8/ FLUCTUATIONS AND IRREVERSIBLE PROCESSES

CHAPTER 9/ THE THIRD LAW OF THERMODYNAMICS

PREFACE TO THE SECOND EDITION

More than thirteen years have elapsed since the manuscript of the first edition was completed. For some time it has been clear that a second edition was needed. In the last fifteen years statistical mechanics has developed at an enormous pace, and a textbook on statistical mechanics should take these developments into account. When I started to contemplate what changes should be made, it soon became clear that, if the book were to cover even approximately the same ground as the first edition did, it would have to be greatly expanded. It was then decided to split the book into two parts, the first volume covering the basic theory; a second volume, advanced theory. As a result, the present volume is now probably more suitable as a textbook for advanced undergraduate courses than the first edition was. To improve its usefulness a variety of problems has been included. I am greatly indebted to R. Kubo and the North Holland Publishing Company for permission to incorporate some of the problems from Kubo's *Statistical Mechanics* and to the Oxford University Press for permission to incorporate problems from Oxford University Examination Papers. To some extent I have adopted Kubo's philosophy in that I feel that, to get the greatest advantage from the present volume, the student should work through most of the problems and so learn the subject by practicing it. A number of topics—especially in the theory of metals and semiconductors, which were treated in the text of the first edition, have now been relegated to the problems section, partly to give the student experience in using statistical methods and partly to make room for other topics.

I should like to express my thanks to the many readers, users, and reviewers who have given me the benefit of their comments and criticism. In many cases I have followed their advice, and I have found that the reactions of my own students especially have been invaluable in finding out obscure passages that needed clarification.

I have finally had the strength of my convictions and have used Kramers' terminology "Thermostatistics" in the title of the second edition.

A comparison between the first edition and the second edition will show that the present volume contains Chapters I to IV of the first edition as Chapters 1 to 4; Chapters V and VI of the first edition make up the present

Chapter 5, while Chapter VII becomes Chapter 6. Parts of Chapter IX of the first edition are incorporated in Chapter 4 and in a problem at the end of Chapter 5; Chapter X and XI of the old edition have been nearly completely incorporated in various problems; and some topics of Chapter XIII have also been incorporated in problems. A few sections of the first appendix have been put either in problems or in Chapters 5 and 6; Appendix II has become Chapter 8; Appendix III, Chapter 9; Appendix IV, Section 4.5; and Appendix VI, Chapter 7. The various parts of the mathematical appendix have been inserted at the appropriate spots in the earlier chapters. The main new topics covered in this book are the Saha equilibrium (Section 6.9), the Kramers-Casimir discussion of the third law (Section 9.2), fluctuations (Sections 7.1 and 7.2), and a more extensive discussion of the density matrix. A fair amount of new material is also to be found in the problems sections. Readers will probably also notice that the notation has been changed considerably to bring it in line with present-day practice.

Once again I appeal to readers and reviewers to let me have the benefit of their detailed criticism.

D. t. H.

Magdalen College, Oxford
May 1966

PREFACE TO THE FIRST EDITION

It is usual for an author to explain in a preface the reasons for writing the particular book which he is presenting to the public, to state who are the readers whom he has in mind for the book, to sketch the history of the writing of the book, and, last but not least, to express his thanks to all people who have been of assistance during the completion of the book.

The reason for writing another textbook on statistical mechanics was the feeling that there should be a textbook which combined in not too large a volume an outline of the main elements of statistical mechanics, starting from the Maxwell distribution and ending with quantum mechanical grand ensembles, with an account of a number of successful applications of these elements. Almost all existing textbooks stress only one or the other of these two aspects. It is hoped that readers will point out to me how far my own attempt has been successful, and I should at this point like to express the hope that reviewers will let me have the benefit of their detailed criticism.

The book is meant to be a textbook and is thus primarily intended for students. I have had in mind graduate students. This means that it should be used as a text for graduate lectures in the United States or for postgraduate lectures in the United Kingdom. It will probably be too advanced as a textbook for honors courses in British universities, although parts of it might be used as such and have been used as such by me. It is hoped that the book can also be used as a research tool and that it is possible to see from the applications how the theory might be applied to other subjects. For that reason I have tried to give as complete a bibliography as was feasible in the framework of a textbook. As the manuscript of the book was essentially completed at the beginning of 1952, references to papers published in 1951 or 1952 will not be complete.

A first rough outline of the book was sketched during the last war-winter in Leiden. A number of students were deprived of the regular lecture courses because Leiden University was closed by the occupying authorities, and the outline of the present book served as a substitute for the regular statistical mechanics course. The manuscript then rested until I wrote the first draft of Parts A and B at Purdue University during 1947–1948. There was another interval until 1950, when I came to St. Andrews, where the manuscript in its present form was started and finished.

It is a pleasant task to thank the many physicists who have given me their advice on parts of the manuscript. My thanks are especially due to Professors F. J. Belinfante, H. M. James, K. Lark-Horovitz, H. Margenau, R. E. Peierls, and F. E. Simon, who have helped me with their criticism and advice. If the approach is sometimes not very clear and if my English has sometimes a distinct foreign flavor, it cannot be blamed on Professor F. Y. Poynton, who has tried to make Parts A and B as far as possible easy reading for students, or on Professors E. S. Akeley and J. F. Allen, who have tried to weed out of the text all barbarisms. I should like to express to them my warmest gratitude. Finally I should like to express my great indebtedness to Professor H. A. Kramers. Anybody who is familiar with his lectures on statistical mechanics will immediately see how much this book owes to him. It is far from a platitude to say that it would never have been written but for Professor Kramers. Not only did he give me an outline of the contents of Parts A and B; in discussions and in lectures he has taught me the fundamental ideas of the subject. I can therefore with some justification claim that the method of treatment in Parts A and B goes straight back to Boltzmann, via Kramers and Ehrenfest. Large parts of the book are, indeed, nearly wholly based on a series of lectures given by Professor Kramers in Leiden during 1944–1945.

In conclusion, I should like to express my thanks to Professor K. Lark-Horovitz and Miss A. Scudder for their help in editing the manuscript.

D. t. H.

Department of Natural Philosophy
St. Andrews
January, 1954

INTRODUCTION

Thermostatistics, to use the term coined by Kramers for statistical mechanics, is a subject that can be fruitfully studied only if a great number of other subjects in physics are well understood. It has thus been assumed that the reader is well acquainted with classical mechanics, quantum theory, thermodynamics, and calculus. As an indication of the standard assumed, we give the following, rather arbitrary, list of textbooks.

GENERAL THEORETICAL PHYSICS

R. M. Eisberg, *Fundamentals of Modern Physics*, New York, 1961.
G. Joos, *Theoretical Physics*, New York, 1944.
R. B. Leighton, *Modern Physics*, New York, 1959.

CLASSICAL MECHANICS

D. ter Haar, *Elements of Hamiltonian Mechanics*, Amsterdam, 1961.
L. D. Landau and E. M. Lifshitz, *Mechanics*, Oxford, 1960.

QUANTUM THEORY

A. S. Davydov, *Quantum Mechanics*, Oxford, 1965.
H. A. Kramers, *Quantum Mechanics*, Amsterdam, 1957.
A. Messiah, *Quantum Mechanics*, Amsterdam, 1960.

THERMODYNAMICS

D. ter Haar and H. Wergeland, *Elements of Thermodynamics*, Reading, Mass., 1966.
L. D. Landau and E. M. Lifshitz, *Statistical Physics*, Oxford, 1958.

MATHEMATICAL PHYSICS

E. T. Whittaker and G. N. Watson, *Modern Analysis*, Cambridge, 1927.
H. and B. S. Jeffreys, *Mathematical Physics*, Cambridge, 1946.

Many subjects that belong to statistical mechanics have not been treated in the present book. From among the large number of textbooks now available we mention the following ones as treating topics not to be found in the present textbook or—more often—as treating the various topics in a different manner. We wish to emphasize that this list is at least as arbitrary as the list of books given above.

J. de Boer and G. E. Uhlenbeck (eds.), *Studies in Statistical Mechanics*, Amsterdam, Vol I, 1962 (continuing series).
R. H. Fowler, *Statistical Mechanics*, Cambridge, 1936.
R. H. Fowler and E. A. Guggenheim, *Statistical Thermodynamics*, Cambridge, 1939.
J. W. Gibbs, *Elementary Principles in Statistical Mechanics*, New Haven, 1902.
T. L. Hill, *Statistical Mechanics*, New York, 1956.
K. Huang, *Statistical Mechanics*, New York, 1963.
C. Kittel, *Elementary Statistical Physics*, New York, 1958.
L. D. Landau and E. M. Lifshitz, *Statistical Physics*, Oxford, 1958.
D. K. C. MacDonald, *Introductory Statistical Mechanics for Physicists*, New York, 1963.
J. E. and M. G. Mayer, *Statistical Mechanics*, New York, 1940.
A. Münster, *Statistische Thermodynamik*, Berlin, 1956.
G. S. Rushbrooke, *Introduction to Statistical Mechanics*, Oxford, 1949.
E. Schrödinger, *Statistical Thermodynamics*, Cambridge, 1948.
R. C. Tolman, *Statistical Mechanics*, Oxford, 1938.
G. E. Uhlenbeck and G. W. Ford, *Lectures in Statistical Mechanics*, Providence, R. I., 1963.

We further refer to the monographs and papers mentioned at the end of each chapter.

Sections printed in small type indicate arguments that can be omitted on first reading or that are slightly more advanced or complicated than the rest of the text. The decimal system is used to number the equations. The number before the decimal point refers to the chapter and the first decimal number (or in the case of Chapter 5, the first two decimal numbers) to the section. Figures are numbered consecutively within each chapter. The first occurrence of a symbol is given in the glossary at the back of the book.

ELEMENTS

OF

THERMOSTATISTICS

I

The Maxwell Distribution

1.1 THE MAXWELL DISTRIBUTION

Equilibrium statistical mechanics, or "thermostatistics," to use a term coined by Kramers,* is that branch of physics which attempts to derive the equilibrium or thermal properties of matter in bulk, and of radiation, from the properties of the constituent particles. Such an atomistic interpretation of the thermal properties of matter was first attempted in the kinetic theory of gases, developed in the nineteenth century by Clausius, Maxwell, and Boltzmann. In this theory it was shown how such phenomenological concepts as

* H. A. Kramers, *Nuovo cimento*, **6** Suppl., 158 (1949).

temperature and entropy could be interpreted in terms of the *average* properties of the particles that were the constituent parts of the systems under consideration. Kinetic theory could be applied as long as one could either completely neglect the interactions between the particles or could neglect them to a first approximation, taking them into account in second approximation. Such an approach leads, for instance, to the van der Waals law of an imperfect gas, as we shall see in Section 1.3 of this chapter. In most physical systems, however, this approach is much too simplified, and one needs to have recourse to more sophisticated methods of averaging.

This more sophisticated approach is statistical mechanics proper, a term coined by J. W. Gibbs in 1901 for that branch of rational mechanics which deals statistically with systems consisting of large numbers of constituents. In the first four chapters we shall consider systems of independent particles, a subject which really is part of kinetic theory, while in later chapters we shall consider statistical mechanics proper. The reason for using statistical methods to treat physical systems is partly because they are so complicated that they present us with "well-nigh unsurmountable mathematical difficulties"* if we try to solve their equations of motion exactly; and partly because, even if we could solve these equations of motion exactly, we have only incomplete experimental data from which we can obtain the boundary conditions for the equations of motion.

In the present chapter we shall confine ourselves to the simplest possible system: a monatomic gas, that is, a system of point particles, enclosed in a vessel of volume V. We shall assume that there are no external forces acting upon the gas apart from the forces which the walls of the vessel will exert on the gas and which in fact will keep the gas within the vessel. We introduce Cartesian coordinates x, y, and z to describe the system; the position of the ith particle is thus determined by its three position coordinates x_i, y_i, and z_i, or by the vector $\mathbf{r}_i$.† We denote by u, v, and w the x, y, and z components of a velocity $\mathbf{c}$, and u_i, v_i, and w_i are thus the components of the velocity $\mathbf{c}_i$, of the ith particle. Once the positions and velocities of all the atoms are given, the microscopic behavior of the system is completely determined, provided the interatomic forces and the forces exerted by the walls on the atoms are known. If N be the number of atoms in the system, we need $6N$ quantities, for instance, $x_i, y_i, z_i, u_i, v_i, w_i$ ($i = 1$ to N) to determine the microscopic behavior. However, usually we are interested only in a few combinations of these $6N$ quantities that will determine the macroscopic behavior of the system. We mentioned a moment ago that the exact knowledge of all $6N$ coordinates and velocities is outside the experimental possibilities, while the computation of their values from the equations of motion is outside our mathematical powers. As N is usually an extremely large number—of the order of 10^{19}

* H. A. Kramers, *ibid.*

† Vectors are denoted by boldface type.

for one cm^3 of a gas at NTP—we can use this fact and apply statistical methods in the safe knowledge that, because of the extremely large number of degrees of freedom, fluctuations will in general be small.

Let us for the moment neglect first of all the influence of the wall, and secondly possible fluctuations. In that case the number of atoms in a unit volume of the gas will be independent of the position of that unit volume in the gas. If we denote the number of atoms per unit volume by n, we have

$$n = \frac{N}{V}. \tag{1.101}$$

Let us denote by

$$f(u,v,w)\, du\, dv\, dw$$

the number of atoms per unit volume the velocity components of which lie in the specified intervals $(u,u+du)$, $(v,v+dv)$, and $(w,w+dw)$. The function $f(u,v,w)$ will be called the *distribution function.* It determines the fraction of atoms with velocities within given intervals. This fraction is obtained by dividing $f(u,v,w)\, du\, dv\, dw$ by n.

We shall call the Cartesian three-dimensional space in which we can plot the x, y, and z components of the velocities *velocity space*, and the point (u,v,w) in velocity space will be called the *representative point* of an atom with velocity components u, v, and w.

From the definition of $f(u,v,w)$ it follows that it satisfies the equation of *normalization*

$$\int_{-\infty}^{+\infty} du \int_{-\infty}^{+\infty} dv \int_{-\infty}^{+\infty} dw f(u,v,w) = n. \tag{1.102}$$

In the present chapter we shall assume that $f(u,v,w)$ is not only independent of x, y, and z, but also does not depend explicitly on the time t.

Let A be a quantity that is a function of the velocity components of an atom, but which does not explicitly depend on either x, y, and z or t. As an example we may give the kinetic energy of an atom. We can now ask for the *average value*, $\bar{A}$, of $A(u,v,w)$ where the average is taken over all the atoms of the gas and where the average value is equivalent to the arithmetic mean, that is, defined by the equation

$$\bar{A} = \frac{1}{n} \iiint_{-\infty}^{+\infty} du\, dv\, dw A(u,v,w) f(u,v,w). \tag{1.103}$$

Neither A nor f depends on x, y, z, or t, so that $\bar{A}$ will also be independent of x, y, z, and t.

In the following sections of this chapter it will be proved that the distribution function of a gas in equilibrium at an absolute temperature T will be given by the equation

$$f(u,v,w) = n\left(\frac{m}{2\pi kT}\right)^{3/2} e^{-m(u^2+v^2+w^2)/2kT}, \tag{1.104}$$

where m is the mass of one atom and k is Boltzmann's constant.* The distribution given by Eq. (1.104) is called the *Maxwell distribution* and was first introduced by Maxwell in 1859. It can easily be shown that the $f(u,v,w)$ given by Eq. (1.104) satisfies Eq. (1.102). Equations (1.103) and (1.104) can now be used to calculate average values.

Denoting by c the absolute value of the velocity of an atom,

$$c^2 = u^2 + v^2 + w^2, \tag{1.105}$$

and by $\mathscr{T}(=\frac{1}{2}mc^2)$ the kinetic energy of an atom, we have

$$\overline{c^2} = \frac{3kT}{m}, \tag{1.106}$$

$$\bar{c} = \sqrt{\frac{8kT}{\pi m}}, \tag{1.107}$$

$$\bar{u} = \bar{v} = \bar{w} = 0, \tag{1.108}$$

$$\overline{u^2} = \overline{v^2} = \overline{w^2} = \frac{1}{3}\overline{c^2} = \frac{kT}{m}, \tag{1.109}$$

$$\overline{\mathscr{T}} = \overline{\tfrac{1}{2}mc^2} = \tfrac{3}{2}kT. \tag{1.110}$$

There are three points in connection with Eqs. (1.106) to (1.110) worth noting.

(1) Compare Eqs. (1.106) and (1.107). We see that a difference exists between the mean absolute velocity $\bar{c}$ and the root-mean-square velocity $\sqrt{\overline{c^2}}$,

$$\sqrt{\overline{c^2}} = \sqrt{\frac{3\pi}{8}}\,\bar{c} = 1.085\bar{c}. \tag{1.111}$$

(2) Equation (1.110) shows that the mean kinetic energy is independent

* Boltzmann never deduced the value of k; this was first done by Planck in connection with his radiation law (1900). Smekal (*Enzyklopädie der mathematischen Wissenschaften*, Vol. V, Part 28, Leipzig-Berlin, 1926) and Meissner (*Science*, **113**, 78, 1951) suggest calling k the Boltzmann-Planck constant. It is interesting to note that Lorentz (see for example, *Lectures in Theoretical Physics*, Vol. II, p. 175, London, 1927) calls k Planck's constant! We refer to p. 303 for a list of constants and their values.

of the mass of the atoms. Thus we get the same result for any gas, provided the temperature is the same. We can therefore use Eq. (1.110) as a definition of the temperature of a gas.

(3) From Eq. (1.109) it follows that the average kinetic energy pertaining to the x, y, or z direction is the same and is equal to one third of the average total kinetic energy. This is an example of the *equipartition of kinetic energy*.

1.2 THE PERFECT GAS LAW

In this section, we shall assume that there is no interaction between the atoms of the gas which we are considering.

If we may assume that the distribution function is given by Eq. (1.104), we can calculate the pressure of the gas in terms of n and T. Two derivations of the formula for the pressure will be given. The first derivation is due to Clausius and the second one to Lorentz.

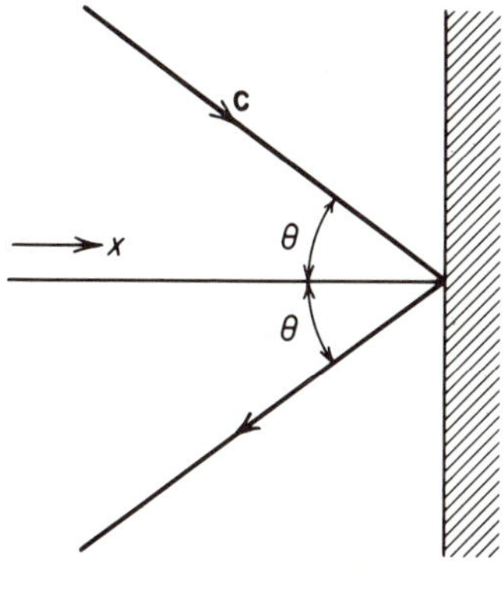

FIG. 1.1 The elastic reflection of a point particle by a wall. The normal to the wall is along the negative x axis, and a particle with velocity **c** *strikes the wall at an angle θ to the normal.*

In order to derive the formula in the way it was done by Clausius, we must remember that, according to Newton's second law, force is equivalent to the rate of change of momentum. The pressure of the gas is defined either as the force exerted by the gas on 1 cm^2 of the wall, or as the total transfer of momentum per sec from the gas to 1 cm^2 of the wall. The Cartesian axes will be chosen in such a way that the negative x axis falls along the normal to the wall (see Fig. 1.1). We shall again neglect any effect arising from a possible potential energy between the wall and the gas atoms, and we shall assume that the atoms will be perfectly reflected by the wall.

In the case of a perfect reflection, only the velocity component perpendicular to the wall will change during the collision of a gas atom with the wall. This component will change its sign, but the other components will remain unchanged. If an atom with velocity **c** strikes the wall at an angle to the normal, θ, the momentum transferred during the collision will be given by the expression

$$2mc \cos \theta = 2mu. \tag{1.201}$$

In order to calculate the total transfer of momentum, we must next determine the number of collisions per sec on 1 cm^2. For a given velocity **c** this number is zero if u is negative. If u is positive, the number of collisions is given by the total number of atoms contained in a cylinder with a base of 1 cm^2 and a slant height c, parallel to **c**. The volume of this cylinder is $c \cos \theta$. If the number of atoms per cm^3 with given velocity **c** is $n(\mathbf{c})$, the number of collisions N_{coll} per sec on 1 cm^2 will be given by the equation

$$N_{\text{coll}}(\mathbf{c}) = n(\mathbf{c})\, c \cos \theta = n(\mathbf{c})u. \tag{1.202}$$

The total momentum transferred to 1 cm^2 of the wall per sec by atoms of given velocity **c** is found by combining Eqs. (1.201) and (1.202) and is equal to

$$2m\, n(\mathbf{c})\, u^2. \tag{1.203}$$

The function $n(\mathbf{c})$ will be the distribution function $f(u,v,w)$, which is assumed to be given by Eq. (1.104). To obtain the pressure P we must integrate expression (1.203) over all velocities for which $u \geqslant 0$, so that we get

$$P = 2m \int_0^{+\infty} du \int_{-\infty}^{+\infty} dv \int_{-\infty}^{+\infty} dw\, f(u,v,w)u^2,$$

or,

$$P = nkT. \tag{1.204}$$

Equation (1.204) is called the *perfect gas law* or the Boyle-Gay-Lussac law.

Introducing the absolute gas constant R by the equation

$$R = kN_A, \tag{1.205}$$

where N_A is the Avogadro number, that is, the number of atoms in one mole, and introducing the molar volume V_m, we can write Eq. (1.204) in the form

$$PV_m = RT. \tag{1.206}$$

Clausius' derivation of the perfect gas law involves certain assumptions about the reflection of the atoms from the wall. These assumptions are unnecessary if the pressure is defined in a slightly different manner. Consider for this purpose a surface of unit area somewhere in the gas. The pressure can now be defined as the total momentum in the direction of the normal to the surface transported per sec through this surface in the direction of the positive normal.* Momentum transported in the opposite direction should be counted with a negative sign.

The above definition of the pressure is due to Lorentz, whose derivation of the perfect gas law proceeds as follows. Let the positive x axis be in the direction of the positive normal to the surface considered (see Fig. 1.2). An atom

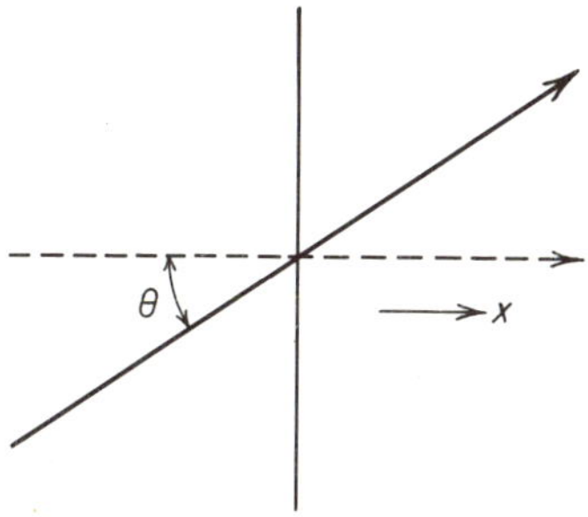

FIG. 1.2 The transfer of momentum through a unit area somewhere in the gas.

that crosses the chosen surface in the direction of the positive normal with a velocity **c** will transport in the $+x$ direction an amount of momentum in the $+x$ direction equal to the x component of its own momentum ($=mu$). The total number of atoms crossing the unit area per sec is again given by expression (1.202). In this way, the total momentum transported by atoms with a velocity **c**, the x component of which is positive, will be given by the expression

$$n(\mathbf{c})u\ mu. \qquad (u > 0). \tag{1.207}$$

The total momentum in $+x$ direction transported by atoms with velocities, the x component of which is negative, is also given by expression (1.207). The momentum is negative, but it is transported from the positive to the negative x axis and should thus be subtracted from the total momentum transported through the surface.

* Strictly speaking, one must still prove that the pressure defined in this way is identically equal to the force per cm^2 on the wall, since this force is the pressure that is measured. The proof follows if one considers a cylinder with cross section of 1 cm^2 and axis parallel to the normal to the wall, starting from the wall and extending into the gas. From the requirement that the gas inside this cylinder is in equilibrium, the equivalence of the two pressures follows.

The total pressure is thus obtained by integrating expression (1.207) over *all* possible **c**-values, or

$$P = \overline{mn(\mathbf{c})u^2} = m \iiint_{-\infty}^{+\infty} du\, dv\, dw\, f(u,v,w)u^2, \tag{1.208}$$

which leads again to Eq. (1.204).

1.3 THE VAN DER WAALS LAW

We mentioned at the beginning of this chapter that the simple approach of kinetic theory—with which we are concerned in this part of the book—entails that one either completely neglect all interactions between the atoms or at best take them into account as a small perturbation which leads to slight modifications of the results obtained when neglecting them. Along these lines we shall investigate in the present section qualitatively how the perfect gas law must be modified when the atoms in the gas exert forces upon one another. We shall assume that these forces are central forces and that they can be derived from a potential energy curve such as the one shown in Fig. 1.3.

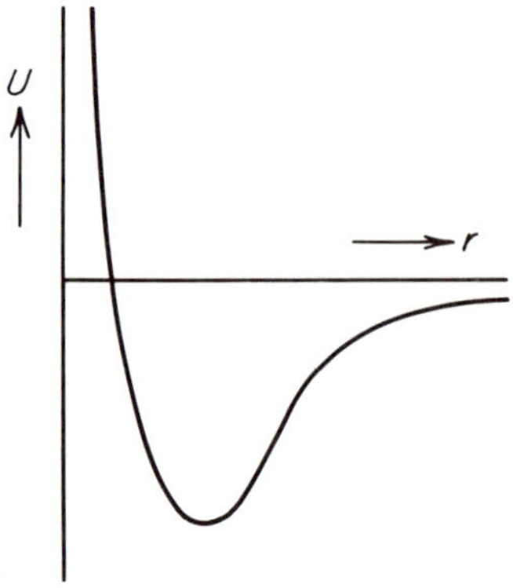

FIG. 1.3 The interatomic potential energy, U, as a function of the distance apart, r, of the atoms.

From this figure we see that at large distances apart two atoms will attract one another, while at small distances apart they repel one another. The attraction is due to the polarization of the electron cloud of the one atom by fluctuations in the charge distribution in the electron cloud of the other atom, while the repulsion occurs when the two electron clouds begin to overlap; the repulsion has the same effect as if the atoms had finite dimensions.

Various equations have from time to time been proposed for the potential energy $U(r)$. The most widely used are the Morse potential,*

$$U(r) = D[e^{-2a(r-r_1)} - 2e^{-a(r-r_1)}], \qquad (D > 0, a > 0) \qquad (1.301)$$

the Lennard-Jones potential†,

$$U(r) = \frac{A}{r^n} - \frac{B}{r^m}, \qquad (n > m, A > 0, B > 0) \qquad (1.302)$$

and the so-called "exp-six" potential‡,

$$U(r) = Ae^{-ar} - \frac{B}{r^6} \cdot (A > 0, B > 0, a > 0) \qquad (1.303)$$

The last potential approximates most closely the actual potential energies that can be derived from first principles, but the first two are more convenient to use for calculations.

As the repulsive forces between the atoms are equivalent to the atoms taking up a certain volume, this will lead to a smaller available volume for the gas. We must thus replace the factor V_m in Eq. (1.206) by a slightly smaller factor, say, $V_m - b$. If repulsive forces decrease the factor V_m, we should expect attractive forces to increase it; we must thus introduce a further correction and replace V_m by $V_m - b + a$, where we have now—completely artificially—separated the effects of the attractive and the repulsive forces. Instead of Eq. (1.206) we now have the following *equation of state*, as the relation between the pressure, the volume, and the temperature is called,

$$P(V_m - b + a) = RT. \qquad (1.304)$$

It is essential for our argument that b and a are small corrections to V_m a condition satisfied as long as the temperature or the density of the gas are sufficiently low. We can then rewrite Eq. (1.304) as follows:

$$\left(P + \frac{a'}{V_m^2}\right)(V_m - b) = RT, \qquad (1.305)$$

where the left-hand sides of Eqs. (1.304) and (1.305) differ only by terms of the second order in a/V_m or b/V_m, provided a' is given by the equation

$$a' = aRT.$$

* P. M. Morse, *Phys. Rev.*, **34**, 57 (1929).
† J. E. Jones, *Proc. Roy. Soc.* (*London*), **A106**, 463 (1924).
‡ R. A. Buckingham, *Proc. Roy. Soc.* (*London*), **A168**, 264 (1938).

The reason for rewriting Eq. (1.304) in the form (1.305) is purely a historic one. It is done in order to show that to a first approximation the changes introduced by considering the attractive and repulsive forces between the atoms in the gas will lead to the famous equation introduced by van der Waals in 1873 and hence called the *van der Waals law*. This equation was of the form (1.305). Its importance lies mainly in the very great number of successful empirical applications, but the reader may be reminded that neither Eq. (1.304) nor Eq. (1.305) gives us the correct equation of state.

It is instructive to write Eq. (1.304) in yet another way:

$$PV_m = RT\left(1 + \frac{b - a}{V_m} + \cdots\right). \tag{1.306}$$

Equation (1.306) is the form of the equation of state that is usually employed; this gives us a power series in descending powers of V_m, which can be written

$$PV_m = RT\left(1 + \frac{B}{V_m} + \frac{C}{V_m^{\,2}} + \cdots\right). \tag{1.307}$$

In Eq. (1.307) the quantities $B, C, \cdots$, which are functions of T, are called the *virial coefficients*, B being called the second virial coefficient, C the third virial coefficient, and so on. Comparing Eqs. (1.306) and (1.307), we see that the second virial coefficient is related to the b and a from the van der Waals equation by the equation

$$B = b - a. \tag{1.308}$$

In our present approximation we cannot say anything about the higher virial coefficients, and this must be left to a discussion where the interactions between the atoms are taken into account more accurately.

The term "virial" used in this connection arises from the function

$$\mathscr{V} = \sum_i (\mathbf{r}_i \cdot \mathbf{F}_i), \tag{1.309}$$

which was introduced into kinetic theory by Clausius and called by him the *virial*. In Eq. (1.309) $\mathbf{F}_i$ is the total force (hence the name "virial" from the Latin *vis* for force) acting upon the ith atom. The summation in Eq. (1.309) is over all atoms in the gas.

Clausius established an important theorem relating $\mathscr{V}$ to the pressure as follows. From Newton's second law we have

$$m_i \frac{d^2\mathbf{r}_i}{dt^2} = \mathbf{F}_i,$$

and hence

$$\mathscr{V} = \sum_i m_i\left(\mathbf{r}_i \cdot \frac{d^2\mathbf{r}_i}{dt^2}\right).$$

Taking the time average (denoted by a tilde) of the virial, we get

$$\begin{aligned}\tilde{\mathscr{V}} &= \frac{1}{\tau}\int_0^\tau \mathscr{V}\, dt = \frac{1}{\tau}\int_0^\tau \sum_i m_i\left(\mathbf{r}_i \cdot \frac{d^2\mathbf{r}_i}{dt^2}\right) dt \\ &= -\frac{1}{\tau}\int_0^\tau \sum_i m_i\left(\frac{d\mathbf{r}_i}{dt}\right)^2 dt + \frac{1}{\tau}\left[\sum_i m_i\left(\mathbf{r}_i \cdot \frac{d\mathbf{r}_i}{dt}\right)\right]_0^\tau. \end{aligned} \tag{1.310}$$

As the gas is confined to a finite volume, all $\mathbf{r}_i$ and $d\mathbf{r}_i/dt$ are always finite, and by letting τ tend to ∞ the integrated term can be made arbitrarily small compared to the first term on the right-hand side of Eq. (1.310). Thus we get

$$\tilde{\mathscr{V}} = -2N\bar{\bar{\mathscr{T}}}, \tag{1.311}$$

where N is again the number of atoms in the gas and $\bar{\bar{\mathscr{T}}}$ the average kinetic energy per atom; this last average is now an average both over time and over the system of all the atoms in the gas.

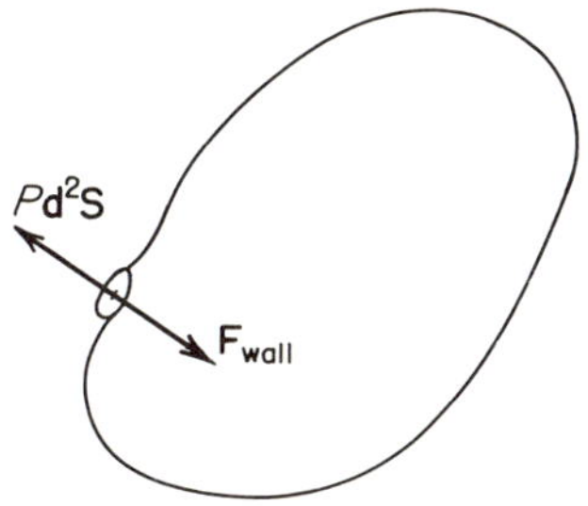

FIG. 1.4 The contribution from the pressure to the virial. The force on the element of the wall is **Pd²S** *and is directed along the normal to the wall, while the force on the gas,* $\mathbf{F}_{\text{wall}}$, *is equal to that force, but in the opposite direction.*

We can obtain another equation for $\tilde{\mathscr{V}}$ by dividing the virial into two parts, one part $\mathscr{W}$ due to the intermolecular forces and one part due to the pressure on the walls of the vessel (or the forces exerted by the wall on the gas):

$$\tilde{\mathscr{V}} = \tilde{\mathscr{W}} + \overline{\int_{\text{walls}} (\mathbf{r} \cdot \mathbf{F}_{\text{wall}})}, \tag{1.312}$$

where the integral is over all atoms in contact with the wall. As the time-averaged force exerted by the gas on an element of the wall is equal to $P\,\mathbf{d}^2\mathbf{S}$ (see Fig. 1.4), where $\mathbf{d}^2\mathbf{S}$ is a vector along the normal to the wall with a length equal to the area of the element, we have

$$\tilde{\mathscr{V}} = \tilde{\mathscr{W}} - \oint_{\text{walls}} P(\mathbf{r} \cdot \mathbf{d}^2\mathbf{S}), \tag{1.313}$$

or, after the usual transformation from a surface to a volume integral,

$$\tilde{\mathscr{V}} = \tilde{\mathscr{W}} - 3PV, \tag{1.314}$$

where V is the volume occupied by the gas.

Combining Eqs. (1.311) and (1.314) and using Eq. (1.110) for the average value of the kinetic energy per atom, we obtain the equation

$$PV = NkT + \tfrac{1}{3}\tilde{\mathscr{W}}. \tag{1.315}$$

If we compare Eqs. (1.307) and (1.315), it is clear why B, C, $\cdots$ are called virial coefficients.

1.4 COLLISIONS

In this section we shall consider the influence of collisions between the atoms on their velocity distribution. If we maintained the assumption of Section 1.2 that the atoms do not exert forces upon one another, there would be no

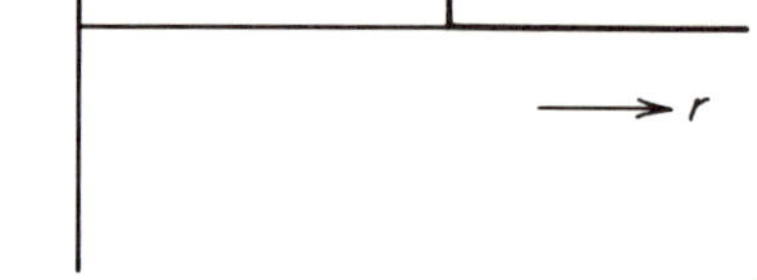

FIG. 1.5 The potential U as a function of the distance r of the centers of two elastic spheres of diameter D.

collisions. On the other hand, if we introduce interactions between the atoms, we are no longer dealing with a system of independent particles. We resolve this apparent contradiction by assuming that we may treat the interactions as being so weak that to a very good approximation the atoms are still independent, but at the same time there are a sufficient number of collisions to set up an equilibrium distribution as far as their velocities are concerned. To simplify our discussions we assume that the atoms in the gas may be treated as elastic spheres. The interatomic potential energy will then be of the form shown in Fig. 1.5.

If there are no external forces acting upon the system, it is plausible to assume that in any unit volume in the gas there will be $f(u,v,w)\,du\,dv\,dw$ atoms whose representative points in velocity space can be found within a given volume $du\,dv\,dw$ in velocity space; $f(u,v,w)$ is again the distribution function. If there are collisions in the gas, the velocities of the atoms will occasionally change and there will thus be atoms that leave the volume $du\,dv\,dw$* and atoms that enter the volume $du\,dv\,dw$. If the number of atoms per unit volume that are leaving $du\,dv\,dw$ per unit time is denoted by A and the number of atoms per unit volume that are entering $du\,dv\,dw$ per unit time by B, the following equation holds for the change of f per unit time:

$$\frac{\partial f}{\partial t}\,du\,dv\,dw = -A + B, \tag{1.401}$$

where the partial derivative indicates that we are at the moment interested only in the change of f due to collisions.

In order to calculate A and B we have to consider in detail collisions between two atoms (see Figs. 1.6 and 1.7). We denote by $\mathbf{c}_1$ and $\mathbf{c}_2$ the velocities of the two atoms before the collision and by $\mathbf{c}_1'$ and $\mathbf{c}_2'$ the velocities after the collision. The velocity of the center of mass will be denoted by $\mathbf{w}$ and is given by the equation

$$(m_1 + m_2)\mathbf{w} = m_1\mathbf{c}_1 + m_2\mathbf{c}_2. \tag{1.402}$$

The velocities $\mathbf{c}_1'$ and $\mathbf{c}_2'$ are not completely determined by $\mathbf{c}_1$ and $\mathbf{c}_2$, since there are only four equations from which the six components have to be determined. These four equations are

$$\tfrac{1}{2}m_1c_1^2 + \tfrac{1}{2}m_2c_2^2 = \tfrac{1}{2}m_1c_1'^2 + \tfrac{1}{2}m_2c_2'^2 \text{ (conservation of energy)} \tag{1.403}$$

and

$$m_1\mathbf{c}_1 + m_2\mathbf{c}_2 = m_1\mathbf{c}_1' + m_2\mathbf{c}_2'. \text{ (conservation of momentum)} \tag{1.404}$$

If we give the direction of the *line of centers*, that is, the line connecting the center of atom 1 with the center of atom 2, $\mathbf{c}_1'$ and $\mathbf{c}_2'$ are completely determined. Denoting by $\boldsymbol{\omega}$ a unit vector in the direction of the line of centers,

* In order to avoid cumbersome sentences, we shall use the expression "atoms leaving the volume *du dv dw*" when we want to say "atoms whose representative points are leaving the volume *du dv dw*."

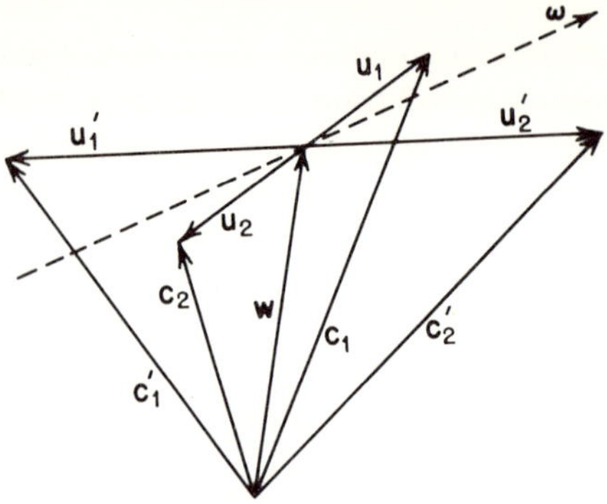

FIG. 1.6 The velocities of two colliding elastic spheres before and after their collision: $\mathbf{c}_1$ *and* $\mathbf{c}_2$ *are the velocities before, and* $\mathbf{c}_1'$ *and* $\mathbf{c}_2'$ *those after, the collision:* $\mathbf{w}$ *is the center of mass velocity;* $\mathbf{u}_1$ *and* $\mathbf{u}_2$ *are the velocities in the center of mass system before the collision, and* $\mathbf{u}_1'$ *and* $\mathbf{u}_2'$ *the velocities in the center of mass system after the collision; and* $\boldsymbol{\omega}$ *is a unit vector in the direction of the line of centers.*

we have (compare Figs. 1.6 and 1.7)

$$\boldsymbol{\omega} = \frac{\mathbf{c}_1 - \mathbf{c}_1'}{|\mathbf{c}_1 - \mathbf{c}_1'|}, \tag{1.405}$$

where $|\mathbf{c}|$ denotes the absolute magnitude of the vector $\mathbf{c}$, and where we have assumed that the two atoms collide like smooth spheres, so that there is no change in the components of the momentum which are perpendicular to the line of centers.

If we introduce the center of mass system, that is, the coordinate system in which the center of mass is at rest ($\mathbf{w} = 0$), the mathematical description of the collision is much simpler. Denoting the corresponding velocities by $\mathbf{u}_1$, $\mathbf{u}_2$, $\mathbf{u}_1'$, and $\mathbf{u}_2'$,* we have

$$\mathbf{u}_1 = \mathbf{c}_1 - \mathbf{w}, \qquad \mathbf{u}_2 = \mathbf{c}_2 - \mathbf{w}, \qquad \mathbf{u}_1' = \mathbf{c}_1' - \mathbf{w}, \qquad \mathbf{u}_2' = \mathbf{c}_2' - \mathbf{w}, \tag{1.406}$$

$$m_1\mathbf{u}_1 + m_2\mathbf{u}_2 = m_1\mathbf{u}_1' + m_2\mathbf{u}_2' = 0, \tag{1.407}$$

$$\tfrac{1}{2}m_1\mathbf{u}_1^2 + \tfrac{1}{2}m_2\mathbf{u}_2^2 = \tfrac{1}{2}m_1\mathbf{u}_1'^2 + \tfrac{1}{2}m_2\mathbf{u}_2'^2. \tag{1.408}$$

$$\boldsymbol{\omega} = \frac{\mathbf{u}_1 - \mathbf{u}_1'}{|\mathbf{u}_1 - \mathbf{u}_1'|}. \tag{1.409}$$

* The reader should note the difference between **u** (velocity in the center of gravity system) and u (x component of **c**) or between **w** and w.

Figures 1.6 and 1.7 show the velocities of the two atoms before and after the collision in the system where the center of mass has a velocity **w** and in the center of mass system.

The number of collisions per unit volume in a time interval dt between atoms with velocities between $\mathbf{c}_1$ and $\mathbf{c}_1 + \mathbf{dc}_1$* on the one hand and atoms

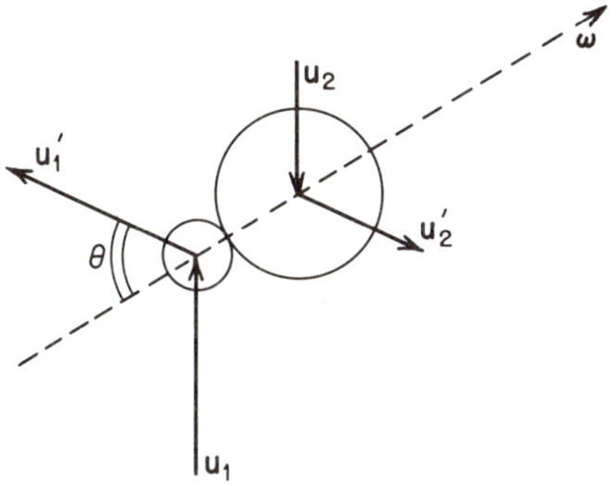

FIG. 1.7 The collision of two elastic spheres in the center of mass system.

with velocities between $\mathbf{c}_2$ and $\mathbf{c}_2 + \mathbf{dc}_2$ on the other hand, while the line of centers lies within a solid angle $d^2\boldsymbol{\omega}$,† is given by the expression

$$a_{12\to1'2'} f(u_1,v_1,w_1) f(u_2,v_2,w_2)\, du_1\, dv_1\, dw_1\, du_2\, dv_2\, dw_2\, d^2\boldsymbol{\omega}\, dt. \quad (1.410)$$

In the case of our spherical atoms $a_{12\to1'2'}$ can easily be calculated. It follows from the consideration that expression (1.410) is equal to the number per unit volume of atoms with velocities between $\mathbf{c}_1$ and $\mathbf{c}_1 + \mathbf{dc}_1$, which is $f(u_1,v_1,w_1)\, du_1\, dv_1\, dw_1$, multiplied by the number of atoms with velocities between $\mathbf{c}_2$ and $\mathbf{c}_2 + \mathbf{dc}_2$ that collide with one atom in an interval dt in such a way that the line of centers has the direction $\boldsymbol{\omega}$ within a margin $d^2\boldsymbol{\omega}$. If $\mathbf{c}_{\text{rel}}$ is the relative velocity of atom 2 with respect to atom 1($\mathbf{c}_{\text{rel}} = \mathbf{c}_1 - \mathbf{c}_2$) and if we denote its absolute value by c_{rel}, it can be seen from Fig. 1.8 that the number of atoms which collide with one atom is equal to the number of atoms which is contained in a cylinder with base $D^2\, d^2\boldsymbol{\omega}$ (D is the diameter of an atom; compare Fig. 1.5‡) and slant height $c_{\text{rel}}\, dt$. Since the number of atoms

* We shall use this shorthand notation to indicate velocities with an x component between u and $u + du$, a y component between v and $v + dv$, and a z component between w and $w + dw$.

† We shall denote an element of solid angle by $d^2\boldsymbol{\omega}$. If $\boldsymbol{\omega}$ is characterized by two angles ϑ and φ, ϑ being the angle between the $+z$ axis and $\boldsymbol{\omega}$, and φ being the angle between the $+x$ axis and the projection of $\boldsymbol{\omega}$ on the x,y plane, $d^2\boldsymbol{\omega}$ will be equal to $\sin\vartheta\, d\vartheta\, d\varphi$.

‡ If the radii of the two atoms are unequal, D will be equal to $r_1 + r_2$.

per unit volume with velocities between $\mathbf{c}_2$ and $\mathbf{c}_2 + \mathbf{dc}_2$ is equal to $f(u_2,v_2,w_2)$ $du_2\, dv_2\, dw_2$, we find for $a_{12\to1'2'}$

$$\left.\begin{aligned} a_{12\to1'2'} &= D^2 c_{\mathrm{rel}} \cos\theta, &\quad \cos\theta > 0,\\ a_{12\to1'2'} &= 0, &\quad \cos\theta \leqslant 0,\end{aligned}\right\} \tag{1.411}$$

where θ is the angle between $\boldsymbol{\omega}$ and $\mathbf{c}_{\mathrm{rel}}$.

If the atoms do not behave like rigid spheres, $a_{12\to1'2'}$ will be much more complicated, but will still only depend on θ and c_{rel}.

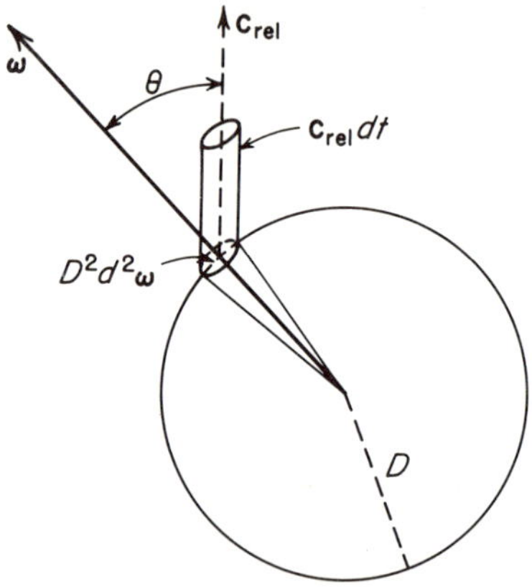

FIG. 1.8 Collisions between elastic sphere atoms. All atoms lying in a cylinder of base $D^2 d^2\boldsymbol{\omega}$ and slant height $c_{\mathrm{rel}}\, dt$ will undergo a collision with a given atom during the time interval dt.

The number of atoms A leaving the volume $du_1\, dv_1\, dw_1$ in velocity space per unit time will now be obtained by integrating expression (1.410) over all possible values of u_2,v_2,w_2 and over all allowed values of $\boldsymbol{\omega}$, and dividing by dt,

$$A = f(u_1,v_1,w_1)\, du_1\, dv_1\, dw_1 \int f(u_2,v_2,w_2)\, du_2\, dv_2\, dw_2 \int a_{12\to1'2'}\, d^2\boldsymbol{\omega}. \tag{1.412}$$

We must remark here that in deriving Eq. (1.412) we have tacitly assumed, as did Clausius in 1858, that there is no correlation whatever between velocities and positions of different atoms. More specifically, whereas $[f(u_2,v_2,w_2)$ $du_2\, dv_2\, dw_2] \cdot dV$ is the number of atoms with velocities between $\mathbf{c}_2$ and

$\mathbf{c}_2 + \mathbf{dc}_2$ that will be found in a volume element dV selected at random, it does not follow that we may use the same expression for the particular volume element (the volume of the cylinder which we just considered), since this is not selected at random. We must introduce as a basic assumption that we may still use the same expression. This assumption is called the *Stosszahlansatz.* It is also often called the assumption of *molecular chaos.* However, with Boltzmann and the Ehrenfests we shall distinguish between these two assumptions and use the German expression for the assumption introduced at this point, and use the English expression only in connection with the discussion of fluctuations in the number of collisions.

We must now calculate the number of collisions that will bring back atoms into the volume $du_1\, dv_1\, dw_1$. These are collisions where the velocity of one of the two colliding atoms lies between $\mathbf{c}_1$ and $\mathbf{c}_1 + \mathbf{dc}_1$ *after* the collision. We have just considered a collision where the velocities before the collision were $\mathbf{c}_1$ and $\mathbf{c}_2$, the line of centers was determined by $\boldsymbol{\omega}$, and the velocities after the collision were $\mathbf{c}_1'$ and $\mathbf{c}_2'$. Let us now consider a collision where the velocities before the collision were $\mathbf{c}_1'$ and $\mathbf{c}_2'$ and after the collision $\mathbf{c}_1$ and $\mathbf{c}_2$. From Eq. (1.405) it then follows that the new line of

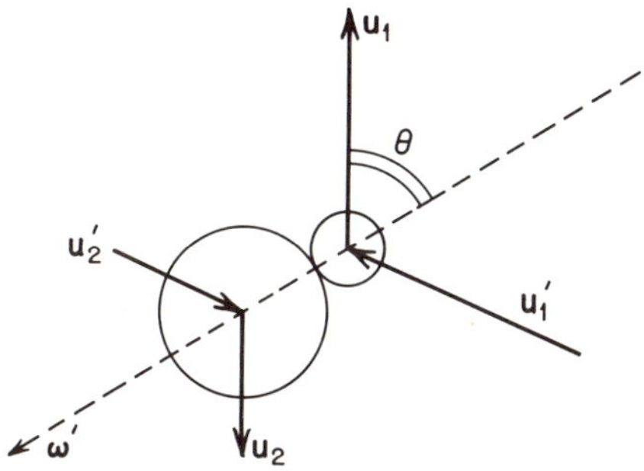

FIG. 1.9 The inverse collision in the center of mass system. This collision is the inverse of the one shown in Fig. 1.7 in that the velocities that are now before the collision were those after the collisions in Fig. 1.7, and vice versa.

centers ($\boldsymbol{\omega}'$) is determined by $-\boldsymbol{\omega}$ instead of $\boldsymbol{\omega}$ (see Fig. 1.9, where this inverse collision is pictured in the center of mass system). The second collision will be called the *inverse* collision corresponding to the original collision pictured in Fig. 1.7. It can easily be verified that there is a one-to-one correspondence between original and inverse collisions. This is a consequence of the fact that $\mathbf{c}_1$, $\mathbf{c}_2$, and $\boldsymbol{\omega}$ determine $\mathbf{c}_1'$, $\mathbf{c}_2'$, and $\boldsymbol{\omega}'(= -\boldsymbol{\omega})$.

The total number of inverse collisions per unit volume and unit time where the velocities before the collision lie between $\mathbf{c}_1'$ and $\mathbf{c}_1' + \mathbf{dc}_1'$, and $\mathbf{c}_2'$ and $\mathbf{c}_2' + \mathbf{dc}_2'$, while $\boldsymbol{\omega}'$ lies within a solid angle $d\boldsymbol{\omega}'$, is given by (compare expression (1.410))

$$a_{1'2'\to 12} f(u_1',v_1',w_1')\, du_1'\, dv_1'\, dw_1' f(u_2',v_2',w_2')\, du_2'\, dv_2'\, dw_2'\, d^2\boldsymbol{\omega}'. \tag{1.413}$$

In order to get B we must integrate expression (1.413) over all allowed values of $\mathbf{c}_1'$, $\mathbf{c}_2'$, and $\boldsymbol{\omega}'$. Only those values of $\mathbf{c}_1'$, $\mathbf{c}_2'$, and $\boldsymbol{\omega}'$ are allowed where one of the velocities after the collision lies between $\mathbf{c}_1$ and $\mathbf{c}_1 + \mathbf{dc}_1$. If we indicate this restriction by priming the integral sign, we have

$$B = \int' f(u_1',v_1',w_1')\, du_1'\, dv_1'\, dw_1' f(u_2',v_2',w_2')\, du_2'\, dv_2'\, dw_2' \times \int a_{1'2'\to 12}\, d^2\boldsymbol{\omega}'. \tag{1.414}$$

It is possible to simplify the expression for B, since $\mathbf{c}_1'$, $\mathbf{c}_2'$, and $\boldsymbol{\omega}'$ are completely determined by $\mathbf{c}_1$, $\mathbf{c}_2$, and $\boldsymbol{\omega}$ by Eqs. (1.403) to (1.405) and the equation $\boldsymbol{\omega}' = -\boldsymbol{\omega}$ which is a consequence of Eq. (1.405). If we change variables, using u_1,v_1,w_1,u_2,v_2,w_2, and $\boldsymbol{\omega}$ instead of the primed quantities, we can write Eq. (1.414) in the form

$$B = du_1\, dv_1\, dw_1 \int f(u_1',v_1',w_1') f(u_2',v_2',w_2')\, du_2\, dv_2\, dw_2 \int J a_{1'2'\to 12}\, d^2\boldsymbol{\omega}. \tag{1.415}$$

The arguments of the distribution functions in Eq. (1.415) are functions of the unprimed quantities through Eqs. (1.403) to (1.405), and J is the Jacobian of the transformation

$$J = \frac{\partial(u_1',v_1',w_1',u_2',v_2',w_2',\boldsymbol{\omega}')}{\partial(u_1,v_1,w_1,u_2,v_2,w_2,\boldsymbol{\omega})} = \begin{vmatrix} \dfrac{\partial u_1'}{\partial u_1} & \dfrac{\partial v_1'}{\partial u_1} & \dfrac{\partial w_1'}{\partial u_1} & \cdots & \dfrac{\partial w_2'}{\partial u_1} \\ \cdots & \cdots & \cdots & \cdots & \cdots \\ \cdots & \cdots & \cdots & \cdots & \cdots \\ \cdots & \cdots & \cdots & \cdots & \cdots \\ \dfrac{\partial u_1'}{\partial w_2} & \cdots & \cdots & \cdots & \dfrac{\partial w_2'}{\partial w_2} \end{vmatrix} \cdot \frac{\partial \boldsymbol{\omega}'}{\partial \boldsymbol{\omega}}. \tag{1.416}$$

It can be proved that J is always equal to one.*

We shall prove it in the case of an elastic collision between two particles with masses m_1 and m_2. We choose the $+x$ axis parallel to $\boldsymbol{\omega}$. Since the collision is

* This is a consequence of Liouville's theorem.

elastic, we have from Eqs. (1.403) to (1.405)

$$\left.\begin{array}{ll} v_1 = v_1', \quad w_1 = w_1'; & v_2 = v_2', \quad w_2 = w_2'; \\ u_1' = \dfrac{(m_1 - m_2)u_1 + 2m_2u_2}{m_1 + m_2}, & u_2' = \dfrac{(m_2 - m_1)u_2 + 2m_1u_1}{m_1 + m_2}. \end{array}\right\} \quad (1.417)$$

Equations (1.417) are most easily obtained by remembering that in the center of mass system, the y and z components of the velocities are unchanged while the x component only changes sign. The Jacobian now reduces to

$$J = \begin{vmatrix} \dfrac{\partial u_1'}{\partial u_1} & \dfrac{\partial u_2'}{\partial u_1} \\ \dfrac{\partial u_1'}{\partial u_2} & \dfrac{\partial u_2'}{\partial u_2} \end{vmatrix} \cdot \frac{\partial \boldsymbol{\omega}'}{\partial \boldsymbol{\omega}} = -\frac{\partial \boldsymbol{\omega}'}{\partial \boldsymbol{\omega}} = 1. \quad (1.418)$$

Since $a_{12\to1'2'}$ or $a_{1'2'\to12}$ depends only on c_{rel} and θ, neither of which changes by going over to the inverse collision, $a_{12\to1'2'} = a_{1'2'\to12}$, and we get from Eqs. (1.401), (1.412), and (1.415), dropping the subscripts of the a's, putting $J = 1$, and dividing by $du_1\, dv_1\, dw_1$,

$$\frac{\partial f_1}{\partial t} = -\int (f_1 f_2 - f_1' f_2') a \, du_2 \, dv_2 \, dw_2 \, d^2\boldsymbol{\omega}. \quad (1.419)$$

In Eq. (1.419) we used the abbreviated notation

$$f_i = f(u_i, v_i, w_i), \qquad f_i' = f(u_i', v_i', w_i'). \quad (1.420)$$

If the distribution function is given by Eq. (1.104), we have

$$f_1 f_2 = f_1' f_2', \quad (1.421)$$

due to the conservation of energy (Eq. (1.403)). Thus, for a Maxwell distribution, we find

$$\frac{\partial f}{\partial t} = 0. \quad (1.422)$$

This means that the Maxwell distribution is an equilibrium distribution: once it has been established, collisions will not alter it.

The distribution function can be generalized as follows:

$$f = Ke^{-(\mathbf{c} - \mathbf{c}_0)^2/2kT}, \quad (1.423)$$

where K is a normalizing constant and $\mathbf{c}_0$ a given constant vector. This distribution corresponds to a case where the system as a whole is moving with a constant velocity $\mathbf{c}_0$. This can be proved by performing a transformation to a coordinate system moving with velocity $\mathbf{c}_0$ with respect to our original system of reference; the distribution function f is then transformed into the distribution f', which is just the Maxwell distribution (1.104). Another proof is obtained by evaluating the average velocity $\bar{\mathbf{c}}$ of the atoms in the system; it is then found that

$$\bar{\mathbf{c}} = \mathbf{c}_0. \tag{1.424}$$

The distribution f is a stationary distribution, because by virtue of the conservation of energy (Eq. (1.403)) and the conservation of momentum (Eq. (1.404)) we have

$$f_1 f_2 = f_1' f_2' \quad \text{or} \quad \frac{\partial f_1}{\partial t} = 0. \tag{1.425}$$

In conclusion we draw attention to the fact that Eq. (1.421) means that there are as many inverse collisions per unit time as there are original ones; this is but one example of the far more general *principle of detailed balancing.*

1.5 THE *H*-THEOREM

In 1872 Boltzmann introduced his H-function defined by the equation

$$H = \overline{\ln f} = \int f \ln f \, du \, dv \, dw, \tag{1.501}$$

where f is a distribution function defined as in Section 1.1, and not necessarily the equilibrium distribution function given by Eq. (1.104) or Eq. (1.423). The integration in Eq. (1.501) is over the whole of velocity space and H will thus be a function of the time only. For its derivative with respect to the time we find*

$$\frac{dH}{dt} = \int \frac{\partial f}{\partial t} (\ln f + 1) \, d^3\mathbf{c}. \tag{1.502}$$

If we use Eq. (1.419) for $\partial f / \partial t$—which was derived without any restrictions upon f—we get from Eq. (1.502):

$$\frac{dH}{dt} = \int (f' f_1' - f f_1)(\ln f + 1) a \, d^3\mathbf{c} \, d^3\mathbf{c}_1 \, d^2\boldsymbol{\omega}. \tag{1.503a}$$

In Eq. (1.503a), $\mathbf{c}$ and $\mathbf{c}_1$ are integration variables, and both integrations are over the whole of velocity space; we can thus interchange them without affecting the result of the integration. As $a_{12 \to 1'2'}$ depends on c_{rel} and θ only,

* We use the abbreviation $d^3\mathbf{c}$ for $du \, dv \, dw$. It must be noted that $d^3\mathbf{c}$ is not a vector but a symbolical notation for a volume element in velocity space; it should not be confused with the vector $\mathbf{dc}$ which is an increment in $\mathbf{c}$.

it will not change when this interchange is performed; thus we get

$$\frac{dH}{dt} = \int (f'f_1' - ff_1)(\ln f_1 + 1) a \, d^3\mathbf{c} \, d^3\mathbf{c}_1 \, d^2\boldsymbol{\omega}. \tag{1.503b}$$

If in Eqs. (1.503a) and (1.503b) we denote the velocities before the collision by $\mathbf{c}'$ and $\mathbf{c}_1'$ and those after the collision by $\mathbf{c}$ and $\mathbf{c}_1$, and if we write $\boldsymbol{\omega}'$ for the unit vector along the line of centers and a' for $a_{1'2' \to 12}$, we have the equations

$$\frac{dH}{dt} = \int (ff_1 - f'f_1')(\ln f' + 1) a' \, d^3\mathbf{c}' \, d^3\mathbf{c}_1' \, d^2\boldsymbol{\omega}', \tag{1.503c}$$

$$\frac{dH}{dt} = \int (ff_1 - f'f_1')(\ln f_1' + 1) a' \, d^3\mathbf{c}' \, d^3\mathbf{c}_1' \, d^2\boldsymbol{\omega}'. \tag{1.503d}$$

We saw in the preceding section that the quantities $\mathbf{c}$, $\mathbf{c}_1$, and $\boldsymbol{\omega}(=(\mathbf{c}_1 - \mathbf{c}_1')/|\mathbf{c}_1 - \mathbf{c}_1'|)$ are completely determined by $\mathbf{c}'$, $\mathbf{c}_1'$, and $\boldsymbol{\omega}'(=(\mathbf{c}_1' - \mathbf{c}_1)/|\mathbf{c}_1' - \mathbf{c}_1| = -\boldsymbol{\omega})$ and that the Jacobian of the transformation from the former to the latter is equal to unity. If we perform the transformation from the primed to the unprimed variables in Eq. (1.503c) and (1.503d), we get

$$\frac{dH}{dt} = \int (ff_1 - f'f_1')(\ln f' + 1) a \, d^3\mathbf{c} \, d^3\mathbf{c}_1 \, d^2\boldsymbol{\omega}, \tag{1.503e}$$

$$\frac{dH}{dt} = \int (ff_1 - f'f_1')(\ln f_1' + 1) a \, d^3\mathbf{c} \, d^3\mathbf{c}_1 \, d^2\boldsymbol{\omega}. \tag{1.503f}$$

where the primed variables that still occur under the integral signs must be regarded as functions of the unprimed variables (see the remark after Eq. (1.415)).

Equations (1.503a), (1.503b), (1.503e), and (1.503f) are four equivalent equations for dH/dt. Adding these four equations and dividing by four, we get for dH/dt

$$\frac{dH}{dt} = -\frac{1}{4} \int (f'f_1' - ff_1) \ln \frac{f'f_1'}{ff_1} \, a \, d^3\mathbf{c} \, d^3\mathbf{c}_1 \, d^2\boldsymbol{\omega}. \tag{1.504}$$

Since a is a positive quantity and since $(p - q)\ln(p/q)$ is positive except when $p = q$, in which case it is zero, it follows from Eq. (1.504) that

$$\frac{dH}{dt} \geqslant 0. \tag{1.505}$$

The equality in Eq. (1.505) is valid only if for *every* two velocities $\mathbf{c}$ and $\mathbf{c}_1$ we have $ff_1 = f'f_1'$. Equation (1.505), which expresses that H can never increase, is known as Boltzmann's *H-theorem*.

One can prove that H is a bounded function, and Eq. (1.505) entails thus that H will decrease as long as the distribution function does not satisfy

Eq. (1.421). We see thus that Eq. (1.421) is not only a sufficient condition for an equilibrium distribution—as was shown in the preceding section—but also a necessary condition.

We can now prove that the generalized Maxwell distribution (1.423) is the only possible equilibrium distribution in the case of a gas which is not in an external field of force. We first of all note that we have just seen that for equilibrium the following relation must hold:

$$\ln f_1 + \ln f_2 = \text{the same before and after the collision.} \tag{1.506}$$

If $\psi(u,v,w)$ is a function of the velocity components of an atom such that $\psi(u_1, v_1, w_1) + \psi(u_2, v_2, w_2)$ remains unchanged by a collision, $\ln f = \psi$ will be a solution of Eq. (1.506). If there are several such functions, $\psi_1, \psi_2, \psi_3, \cdots$, the general solution of Eq. (1.506) will be

$$\ln f = c_0 + c_1\psi_1 + c_2\psi_2 + c_3\psi_3 + \cdots. \tag{1.507}$$

We know four ψ's having this property: the energy and the three components of the momentum. As the conservation laws for energy and momentum, together with the direction of the line of centers, completely determine the six velocity components after the collision, once the six velocity components before the collision are given, it follows that there can be no other ψ's but those four. From Eq. (1.507) it then follows that

$$\ln f = c_0 + c_1 \cdot \tfrac{1}{2}m(u^2 + v^2 + w^2) + c_2 \cdot mu + c_3 \cdot mv + c_4 \cdot mw,$$

or

$$f = Ke^{-a(\mathbf{c}-\mathbf{c}_0)^2} \tag{1.508}$$

where K is again a normalizing constant.

In order to prove that the distribution (1.508) is the same as (1.423) we must show that

$$a = \frac{m}{2kT}. \tag{1.509}$$

This can be done in different ways, depending on the way the temperature is introduced. If the temperature is introduced through Eq. (1.110) for the average value of the kinetic energy (or rather in the present case of a moving distribution, the average value of the random or *thermal* kinetic energy $\frac{1}{2}m(\mathbf{c} - \mathbf{c}_0)^2$) or through the equation of state for a perfect gas (1.204), Eq. (1.509) follows immediately, since Eq. (1.508) leads to

$$\overline{(\mathbf{c} - \mathbf{c}_0)^2} = \frac{3}{2a}. \tag{1.510}$$

If a distribution differs appreciably from the equilibrium distribution, the return to equilibrium is usually very rapid. The relaxation time, that is, the time necessary to return to equilibrium, is, for instance, in the case of a gas under normal conditions ($T = 300°$ K, $P = 1$ atm) of the order of 10^{-9} sec.

Another way to arrive at the generalized Maxwell distribution (1.423) is by showing that it is that distribution which makes H a minimum under the restricting conditions that the total number of atoms N in the gas, their total energy E, and their total momentum $\mathbf{P}$ are given. We thus take as the equilibrium condition the equation

$$H = \text{a minimum.} \tag{1.511}$$

If $f(u,v,w)$ is an equilibrium distribution, *any* small change δf (which itself will be a function of u, v, and w) will have to satisfy the following conditions:

$$\delta H = \int (\ln f + 1)\delta f \; d^3\mathbf{c} = 0 \quad \text{(equilibrium condition),} \tag{1.512a}$$

$$\delta E = \int (\tfrac{1}{2}m\mathbf{c}^2)\delta f \; d^3\mathbf{c} = 0 \quad \text{(total energy given),} \tag{1.512b}$$

$$\delta P = \int m\mathbf{c} \; \delta f \; d^3\mathbf{c} = 0 \quad \text{(total momentum given),} \tag{1.512c}$$

$$\delta N = V \int \delta f \, d^3\mathbf{c} = 0 \quad \text{(total number of atoms given),} \tag{1.512d}$$

where V is the volume of the gas.

Equations (1.512) express the fact that the equilibrium distribution f must satisfy Eq. (1.512a) for any small change in the distribution function $\delta f(u,v,w)$ which satisfies Eqs. (1.512b, 1.512c, and 1.512d). These restrictions on δf can be removed by the method of the *Lagrangian multipliers*. In using this method we multiply each of the four Eqs. (1.512) by a constant and add them. If we take

$$\delta H + \beta \delta E + (\boldsymbol{\lambda} \cdot m\mathbf{c}) - \frac{\alpha + 1}{V} \delta N,$$

which must be zero according to Eqs. (1.512), we get

$$\int \delta f[\ln f + \beta \cdot \tfrac{1}{2}mc^2 + (\boldsymbol{\lambda} \cdot m\mathbf{c}) - \alpha] \; d^3\mathbf{c} = 0, \tag{1.513}$$

and we can now treat δf as if it were an arbitrary change of f without any restrictions.

The justification of the method of the Lagrangian or undetermined multipliers is the following one.

Consider a function $f(x_1,x_2,\cdots,x_n)$ of n variables. Let us ask for which values of the x_i this function reaches an extremum under the following p restraining conditions:

$$g_j(x_1,x_2,\cdots,x_n) = 0; \qquad j = 1, 2, \cdots, p. \tag{1.514}$$

The extremum of f can be found by requiring that the variation of f will be zero for any possible variation of the x_i from their values corresponding to the extremum, or

$$\delta f = \sum_i \frac{\partial f}{\partial x_i} \delta x_i = 0. \tag{1.515}$$

If there were no restraining condition, the x_i could be varied independently and Eq. (1.515) would lead to the well-known conditions for an extremum:

$$\frac{\partial f}{\partial x_i} = 0, \qquad i = 1, \cdots, n. \tag{1.516}$$

However, if Eqs. (1.514) must be satisfied, the δx_i cannot be chosen arbitrarily, but must satisfy the p equations,

$$\delta g_j = 0 = \sum_i \frac{\partial g_j}{\partial x_i} \delta x_i, \qquad j = 1, \cdots, p. \tag{1.517}$$

From Eqs. (1.517) we can solve p of the δx_i in terms of the other $n - p$ variations and substitute for these p variations into Eq. (1.515). The remaining $n - p$ variations are then arbitrary and their coefficients must vanish identically in the x_i. We can also eliminate p of the δx_i, say δx_1, δx_2, $\cdots$, δx_p by multiplying the p equations (1.517) with constants λ_1, λ_2, $\cdots$, λ_p and adding them to Eq. (1.515). The result will be

$$\sum_i \left[\frac{\partial f}{\partial x_i} + \lambda_1 \frac{\partial g_1}{\partial x_i} + \lambda_2 \frac{\partial g_2}{\partial x_i} + \cdots + \lambda_p \frac{\partial g_p}{\partial x_i} \right] \delta x_i = 0. \tag{1.518}$$

We eliminate δx_1 to δx_p by determining the λ_j in such a way that

$$\frac{\partial f}{\partial x_i} + \lambda_1 \frac{\partial g_1}{\partial x_i} + \cdots + \lambda_p \frac{\partial g_p}{\partial x_i} = 0, \qquad i = 1, \cdots, p. \tag{1.519}$$

Once δx_1 to δx_p are eliminated, the remaining δx_i can be chosen arbitrarily, and Eq. (1.518) can only be satisfied for all choices if their coefficients vanish, or

$$\frac{\partial f}{\partial x_i} + \lambda_1 \frac{\partial g_1}{\partial x_i} + \cdots + \lambda_p \frac{\partial g_p}{\partial x_i} = 0, \qquad i = p + 1, \cdots, n. \tag{1.520}$$

Combining Eqs. (1.519) and (1.520), we see that we could have obtained the same result by assuming the δx_i in Eq. (1.518) to be independent. Equations (1.519) and (1.520) give us the values of the x_i for which f is an extremum; these values are functions of the undetermined multipliers λ_j which can be eliminated by substituting the x_i into Eqs. (1.514) and solving for the λ_j.

In the case just discussed, f depended on a set of continuous variables. If we group the values of these variables in finite intervals, we have the case for which we have just justified the method, and after that the transition to the continuous case can be made.

Since δf in Eq. (1.513) can be treated as an arbitrary change, this equation can be satisfied only if

$$\ln f = -\beta \cdot \tfrac{1}{2} mc^2 - (\boldsymbol{\lambda} \cdot m\mathbf{c}) + \alpha, \tag{1.521}$$

or

$$f = e^{\alpha} e^{-\beta mc2/2 - m(\boldsymbol{\lambda} \cdot \mathbf{c})}, \tag{1.522}$$

which is the same as expressions (1.423) or (1.508) with a slightly different notation. The Lagrangian multipliers α, β, and $\boldsymbol{\lambda}$ must be determined from the total number of particles, that is, the normalization condition (1.102), the total energy E of the system, and the total momentum $\mathbf{P}$ of the system. Comparing Eqs. (1.522) and (1.508) we see that $\boldsymbol{\lambda}$ will be proportional to the velocity $\mathbf{c}_0$ of the moving Maxwell distribution, and that

$$\beta = \frac{1}{kT}, \tag{1.523}$$

where we have used Eq. (1.509). For the case of a gas at rest we get for α the equation

$$e^{\alpha} = \left(\frac{\beta m}{2\pi}\right)^{3/2} n. \tag{1.524}$$

1.6 THE CONNECTION BETWEEN *H* AND ENTROPY

In this section, for the special case of a gaseous system of noninteracting particles, we shall discuss the connection between Boltzmann's *H*-function and the entropy.

From Eqs. (1.521) and (1.524) we get for the case of a gas at rest

$$\ln f = \ln n + \tfrac{3}{2} \ln \beta - \tfrac{1}{2}\beta mc^2 + \text{const}. \tag{1.601}$$

Introducing the volume v of 1 g of gas,

$$v = \frac{1}{nm}, \tag{1.602}$$

we can write Eq. (1.601) in the form

$$\ln f = -\ln v + \tfrac{3}{2} \ln \beta - \tfrac{1}{2}\beta mc^2 + \text{const}. \tag{1.603}$$

It follows now from the definition of H (Eq. (1.501)) that

$$\frac{H}{n} = -\ln v + \frac{3}{2} \ln \beta - \frac{1}{2}\beta m\overline{c^2} + \text{const}.$$

or, after evaluating $\overline{c^2}$,

$$\frac{H}{n} = -\ln v + \frac{3}{2}\ln \beta + \text{const.} \tag{1.604}$$

If we are dealing with a perfect gas, β will be equal to $1/kT$ (compare Eq. (1.523)) and we have from Eq. (1.604)

$$\frac{H}{n} = -\ln v - \frac{3}{2} \ln T + \text{const.} \tag{1.605}$$

If the entropy and the internal energy per unit mass are denoted by s and u, the second law of thermodynamics gives us the following relation between the changes in s, u, and the work done by the pressure,

$$ds = \frac{đq}{T} = \frac{du + P\,dv}{T}, \tag{1.606}$$

where $đq$ is the heat increase per unit mass.*

The perfect gas law can be written in the form

$$Pv = \frac{R}{M}T, \tag{1.607}$$

where M is the molecular weight of the gas. Since the internal energy of a perfect gas consists of the kinetic energy only, we have

$$du = c_v\,dT = \frac{3}{2}\frac{R}{M}\,dT, \tag{1.608}$$

where c_v is the specific heat (per unit mass) at constant volume.

From Eqs. (1.606), (1.607), and (1.608) it follows that

$$ds = c_v\frac{dT}{T} + \frac{R}{M}\frac{dv}{v}, \tag{1.609}$$

or

$$s = \frac{R}{M}\left(\frac{3}{2}\ln T + \ln v\right) + \text{const.} \tag{1.610}$$

* In this section, quantities referring to the whole system will be denoted by capital letters (S, V, U, Q, and so forth); quantities referring to unit mass, by small type (s, v, u, q, and so forth); and quantities referring to one mole by the subscript m (S_m, V_m, U_m, Q_m, and so forth). If M is the molecular weight of the gas, we have $S_m = Ms$, $V_m = Mv$, and so forth. Quantities referring to unit volume will be indicated by a subscript v (S_v, and so forth).

The symbols $đ$ or $\delta\!\!\!/$ will indicate that we are not dealing with a total differential or variation.

Using Eq. (1.206), we get for the entropy per unit volume

$$S_v = \frac{Mn}{N_A} s = nk\left(\frac{3}{2}\ln T + \ln v\right) + \text{const.} \tag{1.611}$$

If we compare Eqs. (1.605) and (1.611), we see that, apart possibly from an additive constant, there exists the following relation between H and the entropy per unit volume,

$$S_v = -kH. \tag{1.612}$$

The fact that H can never increase thus corresponds to the well-known statement of the second law of thermodynamics that the entropy will never decrease, if volume and internal energy are kept constant.*

It may be remarked here that, although in thermodynamics the entropy is, strictly speaking, defined only in the case of equilibrium, H is also defined for nonequilibrium situations and one can use Eq. (1.612) as a definition of the entropy in such nonequilibrium situations.

1.7 THE CONNECTION BETWEEN H AND PROBABILITY

To conclude this chapter, we shall discuss for the special case of a perfect gas the connection between Boltzmann's H and the probability that a given distribution is realized.

We must first of all define the probability of a distribution. To do this we divide velocity space into a number of cells Z_k, each cell being a volume element $du\,dv\,dw$, and we shall distribute the representative points of the N atoms in the system over velocity space. If this distribution results in N_1 points falling in Z_1, N_2 in Z_2, $\cdots$, we call this distribution an N_i-distribution. The probability $W(N_1,\cdots,N_k,\cdots)$ or $W(N_i)$ is defined as the fraction of all possible arrangements where this specially chosen distribution is realized. If we make the plausible assumption that the a priori probability for a representative point to fall into a cell Z_k will be proportional to its size Z_k, we find for $W(N_i)$ the equation

$$W(N_1,N_2,\cdots,N_i,\cdots) = CN!\prod_i \frac{Z_i^{N_i}}{N_i!}, \tag{1.701}$$

where C is a normalization constant.

In order to derive Eq. (1.701) we have to remember that the number of possible ways of choosing A objects out of B is given by the binomial coefficient

$$\binom{B}{A} = \frac{B!}{A!(B-A)!}. \tag{1.702}$$

* See, for example, D. ter Haar and H. Wergeland, *Elements of Thermodynamics*, Reading, Mass., 1966, Chap. 3.

In our case we have to choose first N_1 objects out of N, then N_2 objects out of $N - N_1$, N_3 out of $N - N_1 - N_2$, and so on; for the total number of possible ways we thus get

$$\binom{N}{N_1}\binom{N-N_1}{N_2}\binom{N-N_1-N_2}{N_3}\cdots$$

$$= \frac{N!}{N_1!(N-N_1)!} \cdot \frac{(N-N_1)!}{N_2!(N-N_1-N_2)!} \cdots = \frac{N!}{N_1!N_2!\cdots}. \qquad (1.703)$$

Multiplying this *multinomial* coefficient with the a priori probability, we get the expression (1.701).

The name "multinomial coefficient" arises from the fact that these coefficients appear in the multinomial expansion

$$(a_1 + a_2 + \cdots)^N = \sum \frac{N!}{N_1!N_2!\cdots} a_1^{N_1} a_2^{N_2} a_3^{N_3} \cdots \qquad (1.704)$$

where the summation extends over all nonnegative values of N_i in such a way that their sum is equal to N.

We now define the equilibrium distribution as that distribution which makes $W(N_i)$ a maximum under the restricting conditions that

$$\sum N_i = N \qquad (1.705)$$

and

$$\sum N_i \varepsilon_i = E, \qquad (1.706)$$

where ε_i is the energy of an atom in cell Z_i. Conditions (1.705) and (1.706) fix the total number of particles in the gas and the total energy of the system.

In order to find the distribution for which we are looking, we shall require $\ln W$ to be maximum instead of W. Now we have from Eq. (1.701)

$$\ln W = \ln N! + \sum_i [N_i \ln Z_i - \ln N_i!] + \text{const.} \qquad (1.707)$$

If we may assume that all of the N_i are large compared with unity, we can use for the factorials the Stirling formula,

$$x! \approx x^x e^{-x}. \qquad (1.708)$$

The Stirling formula in its form (1.708) can be derived as follows.*

Since $\ln x$ is a monotonically increasing function of x, we have

$$\sum_{k=1}^{N} \ln k < \int_0^N \ln x\, dx < \sum_{k=1}^{N} \ln (k+1),$$

* See P. Jordan, *Statistische Mechanik auf Quantentheoretischer Grundlage*, Braunschweig, 1933; J. Satterly, *Nature*, **111**, 220 (1923).

or

$$\int_0^{N-1} \ln x\,dx < \sum_{k=1}^{N} \ln k < \int_0^{N} \ln x\,dx,$$

and hence

$$(N-1)[\ln(N-1)-1] < \ln N! < N(\ln N - 1).$$

since $(N-1)[\ln(N-1)-1]$ and $N(\ln N - 1)$ differ relatively less the larger N is, we have asymptotically for large N

$$\ln N! \approx N \ln N - N,$$

which is the same as Eq. (1.708).

Using Eqs. (1.708) and (1.705), we get from Eq. (1.707):

$$\ln W = -\sum_i N_i \ln \frac{N_i}{NZ_i} + \text{const.} \tag{1.709}$$

Looking for the maximum of $\ln W$ under the conditions (1.705) and (1.706), we can again use Lagrange's method of undetermined multipliers. For a variation δN_i of the N_i, we have

$$\delta \ln W = 0 = -\sum_i \delta N_i \left[\ln \frac{N_i}{NZ_i} + 1\right], \tag{1.710a}$$

$$\delta N = 0 = \sum_i \delta N_i, \tag{1.710b}$$

$$\delta E = 0 = \sum_i \varepsilon_i \delta N_i. \tag{1.710c}$$

Next we multiply these equations by 1, $\alpha + 1$ and $-\beta$, respectively, add them, and obtain

$$\sum_i \delta N_i \left[-\ln \frac{N_i}{NZ_i} + \alpha - \beta\varepsilon_i\right] = 0. \tag{1.711}$$

Since in Eq. (1.711) the δN_i can be taken to be arbitrary, it follows that

$$\ln \frac{N_i}{NZ_i} = \alpha - \beta\varepsilon_i$$

or

$$N_i = NZ_i e^{\alpha - \beta\varepsilon_i}. \tag{1.712}$$

The quantities α and β can be determined from conditions (1.705) and (1.706).

We can now apply these formulas to the case of a perfect gas. The cells Z_i are then to be thought of as "elements of volume" in velocity space, or

$$Z_i = du\,dv\,dw. \tag{1.713}$$

As N is the total number of particles, N_i is the total number of particles with velocities between **c** and **c** + **dc**. Thus

$$\frac{N_i}{N} = \frac{1}{n} f(u,v,w)\,du\,dv\,dw, \tag{1.714}$$

while the energies ε_i of the atoms are given by

$$\varepsilon_i = \tfrac{1}{2}mc^2. \tag{1.715}$$

The transition to infinitesimal cells is trivial, and the distribution (1.712) goes over into the Maxwell distribution (1.104).

From Eqs. (1.709), (1.713), and (1.714) we then get

$$\ln W = -\int f \ln f\,du\,dv\,dw + \text{const.} \tag{1.716}$$

Comparing Eqs. (1.501) and (1.716), we see that, apart from an additive constant, $\ln W$ and $-H$ are the same. The tendency for a decrease of H (H theorem), or an increase of entropy, corresponds thus to a tendency for an increase in probability.

Combining Eqs. (1.612), (1.501), and (1.716) and omitting for the time being possible additive constants, we have

$$-kH = S_v = k \ln W, \tag{1.717}$$

which gives us the well-known relation between entropy and probability.

PROBLEMS*

1. Assume that the atoms in a gas satisfy the Maxwell distribution; evaluate the number of atoms that strike a unit area per unit time.
2. Use the result of the preceding problem to find
   (a) the velocity distribution of the atoms that escape through an (infinitesimal) hole of area A in the wall of a vessel, inside of which the velocity distribution is Maxwellian;
   (b) the root mean square speed of the escaping molecules;
   (c) the rate of loss of energy in the vessel;
   (d) the rate at which the temperature of the residual gas is changing.

* All problems, except the last one, are concerned with kinetic theory; they take the development of kinetic theory one stage further than was done in the text.

3. Prove Dalton's law for a mixture of noninteracting gases consisting of independent atoms.

4. For a gas of identical atoms in thermal equilibrium, evaluate the number of collisions per unit time and unit volume between atoms (taken to be perfectly elastic spheres) with relative velocities in the range c_{rel} to $c_{\text{rel}} + dc_{\text{rel}}$.
Determine the average length and duration of a free path.
Find an expression for the probability of an atom's describing a free path of a given length.

5. Find an expression for the mean free path of those atoms that cross a given plane with speed c, and explain why this differs from the mean free path of such atoms.

6. Explain the difference between the "Maxwell free path" (that is, the average free path per unit time) and the "Tait free path" (that is, the average free path of a given particle).

7. Consider a monatomic gas in a steady state in a vessel closed by a piston. Assume that the piston moves infinitesimally slowly, with speed U; then use kinetic theory to find a relation between the pressure P and the volume V during this quasistatic adiabatic process. Assume that through collisions the change in energy upon hitting the moving piston is distributed evenly over all degrees of freedom.

8. Consider the following simplified model of a gas, due to the Ehrenfests. In the plane of the paper there may be a large number of point particles, N per unit area, which we shall call the P-molecules. They do not interact with each other, but they collide elastically with another set of entities which we may call the Q-molecules. These Q-molecules are squares with edge length a, distributed at random over the plane. They are fixed in their positions in such a way that their diagonals are exactly parallel to the x and y axes. Their average surface density is n, and we shall let their mean distance apart be large compared to a.
Suppose that at a certain moment all the P-molecules have velocities which are of the same absolute magnitude, c, and limited in direction to (1) the positive x axis, (2) the positive y axis, (3) the negative x axis, and (4) the negative y axis (see Fig. 1.10). Due to the nature of our system and the interactions allowed in it, the situation at any other time will be similar to the one just described, the only possible difference being in the magnitude of the numbers of P-molecules moving in the four possible directions. Let f_1, f_2, f_3, and f_4 be the numbers of the P-molecules moving in those four directions. These numbers will be functions of t and together will take the place of the distribution function $f(u,v,w)$.
The equilibrium distribution will be given by the equation

$$f_1 = f_2 = f_3 = f_4 = \tfrac{1}{4}N.$$

Prove that if we assume that the number $N_{ij}\Delta t$ of P-molecules per unit area changing direction from i to j during a time interval Δt is given by the Stosszahlansatz expression

$$N_{ij}\Delta t = f_i S_{ij} n,$$

where S_{ij} is the area of a parallelogram of length $c\Delta t$ on that edge of one

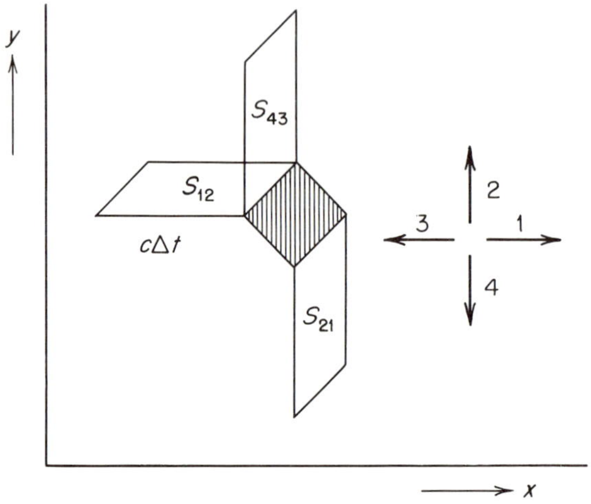

FIG. 1.10 The Ehrenfests' wind-wood model.

of the squares which is in the $-i, j$ quadrant (see Fig. 1.10), the f_i will approach their equilibrium values.

9. Consider the Lorentz model of a metal in which the electrons, which are supposed to be noninteracting, are scattered in such a way that the number per unit volume $N_{\omega,\omega'}\, dt\, d^2\omega\, d^2\omega'$ which during a time interval dt change their directions from within an element of solid angle $d^2\omega$ to within an element of solid angle $d^2\omega'$ is given by the equation

$$N_{\omega,\omega'}\, dt\, d^2\omega\, d^2\omega' = Af(\vartheta,\varphi)\frac{d^2\omega}{4\pi}\frac{d^2\omega'}{4\pi}\, dt,$$

where $f(\vartheta,\varphi)\sin\vartheta\, d\vartheta\, d\varphi/4\pi$ is the number of electrons per unit volume with velocities in a direction within the solid angle $d^2\omega$. It is assumed in this model that all electrons are moving with the same speed.
Find the equilibrium distribution of the electrons and show that the scattering mechanism will produce an exponential approach to this equilibrium distribution.

10. A vessel of volume V contains N atoms in thermal equilibrium. If n is the number of molecules in a subvolume v, and if the probability of finding any atom in v is equal to v/V, find the probabiltiy $f(n)$ for v containing exactly n atoms. Calculate the mean value $\bar{n}$ and also its dispersion, that is, the quantity $\overline{(n - \bar{n})^2}$, where the averages are all taken with $f(n)$ as the distribution function.
Prove that $f(n)$ is approximately Gaussian when both N and n are large. Use Stirling's formula in the form (1.708).
Show that in the limit as $v/V \to 0$ and $V \to \infty$, while N/V is kept constant, $f(n)$ tends to the Poisson distribution

$$f(n) = e^{-\bar{n}} \frac{\bar{n}^n}{n!}.$$

BIBLIOGRAPHICAL NOTES

General references dealing with the material of this chapter are

1. L. Boltzmann, *Vorlesungen über Gastheorie*, 2 vols., Leipzig, 1896–1898; *Lectures on Gas Theory*, Berkeley, California, 1964.
2. P. and T. Ehrenfest, *Enzyklopädie der mathematischen Wissenchaften*, Vol. IV, Part 32, Leipzig-Berlin, 1911; *The Conceptual Foundations of the Statistical Approach in Mechanics*, Ithaca, N. Y., 1959.
3. L. Boltzmann and J. Nabl, *Enzyklopädie der mathematischen Wissenschaften*, Vol. V, Part 8, Leipzig-Berlin, 1905.

For an up-to-date review of the historical development of kinetic theory leading to statistical mechanics, we refer to the work of Brush.

4. S. G. Brush, *Annals of Sc.* **13,** 188, 273, 1957; **14,** 185, 243, 1958; *Am. J. Phys.*, **29,** 593, 1961; *Am. Scientist* **49,** 202, 1961.
5. S. G. Brush, *Kinetic Theory*, 3 vols., Oxford, 1965–1966.

In these papers and books, Brush carefully traces the development of the subject. Many of the classical papers are reprinted in his books, for instance, References 6, 16, 17, 23, and 24 of the present bibliographical notes.

Section 1.1. The Maxwell distribution was introduced by Maxwell in 1859:

6. J. C. Maxwell, *Phil. Mag.*, **19,** 19, 1860. (This paper was read before the British Association in Aberdeen on September 21, 1859.)

In the paper, Maxwell derives his distribution from the following assumptions:

(a) The distribution function depends only on the absolute magnitude of the velocity.

(b) The probability that the x component of the velocity lies within a certain interval is independent of the value of the y or z component.

From these two assumptions the Maxwell distribution follows in a straightforward manner. The derivation which is given in Section 1.5, and which does not depend on the above assumptions, was given by Maxwell in 1867:

7. J. C. Maxwell, *Trans. Roy. Soc.* (*London*), **157,** 49, 1867 (= *Phil Mag.*, **35,** 129, 185, 1868).

See also

8. L. Boltzmann, *Wien. Ber.*,* **58,** 517, 1868.

Experimental verifications of the Maxwell distribution were obtained by Ornstein and van Wijk, who studied the shape of spectral lines, and by Stern and collaborators and by Kofsky and Levinstein, who used the molecular beam method:

9. L. S. Ornstein and W. R. v. Wijk, *Z. Physik*, **78,** 734, 1932.
10. I. Estermann, O. C. Simpson, and O. Stern, *Phys. Rev.*, **71,** 238, 1947.
11. I. L. Kofsky and H. Levinstein, *Phys. Rev.*, **74,** 500, 1948.

Equations (1.106) to (1.111) can be found in Maxwell's papers (References 6 and 7).

Section 1.2. The perfect gas law as a formula describing experimental data is due to the combined efforts of many scientists, such as Boyle, Mariotte, and Gay-Lussac. In Russian literature (L. Landau and E. Lifshitz, *Statistical Physics*, Oxford, 1938, p. 70; A. I. Khinchin, *Statistical Mechanics*, New York, 1949, p. 121) the perfect gas law is often called Clapeyron's law, probably because Clapeyron was the first to write the law in the form $pV = RT$:

12. E. Clapeyron, *J. d'école polytech.*, **14,** 153, 1834.

Clapeyron, however, explicitly mentions Mariotte and Gay-Lussac as the ones who discovered and stated the gas law.

The idea that the pressure of a gas is due to the movement of its constituent particles is very old. Both Maxwell (Reference 7) and Clausius (*Ann. Physik*, **115,** 2, 1862) give long lists of early papers discussing this idea. Herapath, for example, showed that the pressure was inversely proportional to the volume and independent of the form of the vessel:

13. J. Herapath, *Ann. Philos.*, **1,** 273, 340, 401, 1821.

Joule was the first to show that the pressure is proportional to the square of the velocity:

14. J. P. Joule, *Mem. and Proc. Manchester Lit. and Phil Soc.*, **9,** 107, 1851 (= *Phil. Mag.*, **14,** 211, 1857).

By assuming that all the atoms had the same speed c and that one sixth of all the atoms were flying in the $+x$ direction, one sixth in the $-x$ direction, and so on, Krönig derived the equation $pV = \frac{1}{3}nmc^2$, and from this equation deduced that mc^2 should be proportional to T:

15. A. Krönig, *Ann. Physik*, **99,** 315, 1856.

* Wien. Ber. = Sitzungsberichte der kaiserlichen Akademie der Wissenschaften in Wien, Klasse IIa.

Krönig can probably be considered to be the father of kinetic gas theory. In the introduction to his paper he writes, "··· die Bahn jedes Gasatoms muss deshalb eine so unregelmässige seyn, dass sie sich der Berechnung entzieht. Nach den Gesetzen der Wahrscheinlichkeitsrechnung wird man jedoch statt dieser vollkommenen Unregelmässigkeit eine vollkommene Regelmässigkeit annehmen dürfen."

Clausius derives the perfect gas law under slightly less stringent assumptions:

16. R. Clausius, *Ann. Physik*, **100**, 353, 1857 (= *Phil. Mag.*, **14**, 108, 1857).

In this paper, Clausius actually gives a definition of a perfect gas. There are three conditions that the gas must fulfill:

(a) Its own volume must be small compared to the total volume of the system.
(b) The period during which two atoms are actually colliding must be small compared to the period between two collisions (or the molecular dimensions must be small compared to the mean free path).
(c) There may only be a weak interaction between the atoms.

More precisely and in more modern language we can say: A perfect gas is a gas where the total potential energy of the system at any time is negligibly small compared to the total kinetic energy of the system.

In all the papers mentioned thus far, the perfect gas law is derived by considering the transfer of momentum from the gas to the wall. In describing pressure and tensions in liquids, Maxwell (Reference 7) uses the idea of transfer of momentum through an imaginary unit area. The use of this definition in deriving the gas law is usually attributed to Lorentz; see, for instance, his paper in *Collected Papers*, Vol. VI, The Hague, 1938, p. 143.

Section 1.3. The idea that the volume taken up by the atoms would change the gas law is very old. Bernoulli derived in 1738 that b should simply be equal to the volume ω of the atoms:

17. D. Bernoulli, *Hydrodynamica*, Argentorati, 1738, Chap. X.

Van der Waals showed, however, that as long as $b \ll V$, $b = 4\omega$:

18. J. D. v. d. Waals, *Over de Continuïteit van den Gas- en Vloeistoftoestand*, Thesis, Leiden, 1873.

For an account of van der Waals' thesis, see

19. J. C. Maxwell, *Nature*, **10**, 477, 1874.

For extensive bibliographies about the van der Waals law, see Reference 3 and

20. H. Kamerlingh Onnes and W. H. Keesom, *Enzyklopädie der mathematischen Wissenschaften*, Vol. V, Part 10, Leipzig-Berlin, 1912 (= Comm. Leiden Suppl. 23).
21. J. R. Partington, *Physical Chemistry*, Vol. I, London, 1949, Section VIIC.

The term "virial coefficient" was first introduced by Onnes:

22. H. Kamerlingh Onnes, *Proc. Kon. Ned. Akad. Wet.*, Amsterdam, **4**, 125, 1902 (= Comm. Leiden 71).

The use of the virial for calculating the pressure of a gas and the term "virial" are, however, due to Clausius:

23. R. Clausius, *Ann. Physik*, **141,** 124, 1870 (= *Phil Mag.*, **40,** 122, 1870).

Section 1.4. The derivation of the Maxwell distribution by considering collisions between elastic spheres was given by Maxwell (Reference 7) and Boltzmann (Reference 8). See also Reference 1.

The Stosszahlansatz was implied in many early calculations. For instance, Clausius used it in his calculations of the mean free path in the form: number of collisions per atom = (volume swept through by the atom) × (number of atoms per cm^3):

24. R. Clausius, *Ann. Physik*, **105,** 239, 1858 (= *Phil Mag.*, **17,** 81, 1859).

For a detailed discussion and bibliography we refer to

25. D. ter Haar, *Elements of Statistical Mechanics*, 1st ed., New York, 1954, Appendix I.

26. D. ter Haar, *Rev. Mod. Phys.*, **27,** 289, 1955.

Section 1.5. The function H and the H-theorem are both due to Boltzmann:

27. L. Boltzmann, *Wien. Ber.*, **66,** 275, 1872.

In this paper Boltzmann still uses the letter E (entropy) for H. See also References 25 and 26.

Sections 1.6 and 1.7. The considerations of these two sections can, for instance, be found in Boltzmann's papers (see also Reference 1).

2

The Maxwell-Boltzmann Distribution

2.1 THE BAROMETER FORMULA

In the preceding chapter we restricted our discussion to the case of a perfect gas which, moreover, was not under the influence of any external field of force. In the present chapter we want to lift some of these restrictions, and we shall consider the case where external fields of force are present. Also, we shall consider molecular gases. We shall, however, restrict ourselves to conservative systems where the force on each atom (or molecule) can be derived from a potential energy.

Before discussing the general case in detail we shall first consider in a rough, phenomenological manner the case of a perfect monatomic gas under the influence of a uniform gravitational field.

Consider a disk of cross section $\mathscr{A}$ and thickness dz (see Fig. 2.1). If this disk is in equilibrium, the following relation must hold:

$$(P + dP)\mathscr{A} = P\mathscr{A} - g\rho\mathscr{A}\, dz, \tag{2.101}$$

where g is the gravitational acceleration, ρ the density of the gas, and P and $P + dP$ the pressures in the gas at heights z and $z + dz$. If we assume that the

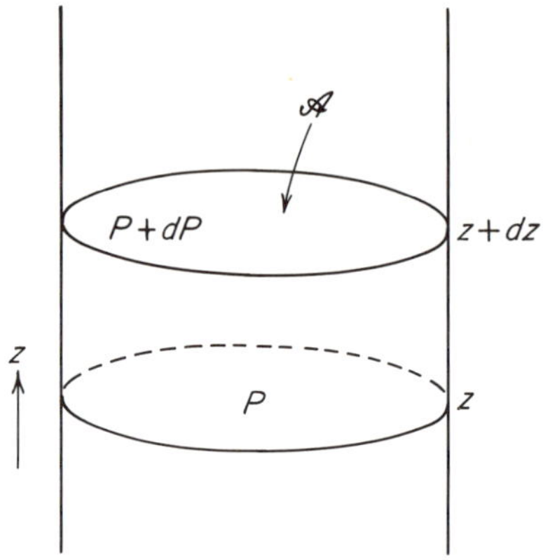

FIG. 2.1 Equilibrium of a gas in a uniform gravitational field. $\mathscr{A}$ is the cross-sectional area of the disk of gas between heights z and $z + dz$; P and $P + dP$ are the pressures at these heights.

pressure at any point in the gas is related to the density by the perfect gas law (1.204), which we can write in the form

$$P = \frac{\rho}{m} kT, \tag{2.102}$$

where m is the mass of one atom, we find from Eq. (2.101) that

$$\rho = \rho_0 e^{-mgz/kT}. \tag{2.103}$$

Equation (2.103) is known as the *barometer formula.*

We can interpret this formula in the following way. Let us introduce a generalized distribution function $f(x,y,z;u,v,w)$. This function is defined such that $f(x,y,z;u,v,w)\, dx\, dy\, dz\, du\, dv\, dw$ is the number of atoms with positions between $\mathbf{r}$ and $\mathbf{r} + \mathbf{dr}$* and velocities between $\mathbf{c}$ and $\mathbf{c} + \mathbf{dc}$.

* The vectors $\mathbf{r}$ and $\mathbf{r} + \mathbf{dr}$ have the components x, y, z and $x + dx$, $y + dy$, $z + dz$. A volume element $dxdydz$ will often symbolically be denoted by $d^3\mathbf{r}$ (not to be confused with $\mathbf{dr}$).

The density $\rho(x,y,z)$ can be obtained from $f(x,y,z;u,v,w)$ by integration over the whole of velocity space:

$$\rho(x,y,z) = m \iiint_{-\infty}^{+\infty} f(x,y,z;u,v,w)\, du\, dv\, dw. \qquad (2.104)$$

If we compare Eqs. (2.103) and (2.104) and make the reasonable assumptions that the velocity distribution is independent of the coordinate distribution and that the velocity distribution is again a Maxwell distribution, the function $f(x,y,z;u,v,w)$ will contain a factor

$$e^{-mc^2/2kT}; \qquad (2.105)$$

If we combine this result with Eq. (2.103), we find the distribution function

$$f(x,y,z;u,v,w) = Ce^{-(mc^2/2kT)-(mgz/kT)}, \qquad (2.106)$$

where C is a normalizing constant.

Expression (2.106) can be written in the form

$$f(x,y,z;u,v,w) = Ce^{-\varepsilon/kT}, \qquad (2.107)$$

where ε is the total energy of one atom in the gravitational field

$$\varepsilon = \mathcal{T} + U = \tfrac{1}{2}mc^2 + mgz, \qquad (2.108)$$

if $\mathcal{T}$ and U are the kinetic and potential energy of one atom.

2.2 THE μ-SPACE

We arrived at the distribution function (2.107) by a rather heuristic procedure. Before proving in the next section that the distribution (2.107)—the so-called *Maxwell-Boltzmann distribution*—is, indeed, the equilibrium distribution, we shall generalize it in such a way that ε is the energy of a particle in terms of the so-called generalized coordinates and momenta that are used in Hamiltonian mechanics.*

If the particles in our system have s degrees of freedom, the condition of one of the particles will completely be determined by the values of s suitably chosen parameters q_i ($i = 1,2,\cdots,s$) and the values of their time derivatives $\dot{q}_i$. The q_i are called *generalized coordinates* and their time derivatives *generalized velocities*.

Let us consider a few examples. One atom will have three degrees of freedom, if we may disregard its internal degrees of freedom, that is, if we may forget that an atom consists of a complicated nucleus with accompanying

* For an account of Hamiltonian mechanics see, for instance, D. ter Haar, *Elements of Hamiltonian Mechanics*, Amsterdam, 1961. Here we shall briefly summarize those results needed for our present discussion.

electrons.* In this case we can use for the q's the Cartesian coordinates of its center of mass x, y, and z and for the generalized velocities the linear velocities of the center of mass u, v, and w.

A more complicated case is the so-called dumbbell molecule, a diatomic molecule where the two atoms are a fixed distance, d, apart. The number of degrees of freedom here is five. It is possible to use the Cartesian coordinates of the centers of mass of the two atoms, x_1, y_1, z_1, x_2, y_2, and z_2, as the coordinates with the restriction, or constraint,

$$(x_1 - x_2)^2 + (y_1 - y_2)^2 + (z_1 - z_2)^2 = d^2. \tag{2.201}$$

However, it is much more convenient to introduce as q's the three coordinates of the center of mass X, Y, and Z and two angles θ and φ, defining the direction of the axis of the molecule. The generalized velocities are then $\dot{X}$, $\dot{Y}$, $\dot{Z}$, $\dot{\theta}$, and $\dot{\varphi}$. We may draw attention to the fact that, as one can see from this example, generalized coordinates have not necessarily the dimensions of a length.

The total energy of a particle consists of two parts: the kinetic energy $\mathcal{T}$ and the potential energy U. It can be shown that the kinetic energy $\mathcal{T}$ is a quadratic function of the $\dot{q}_i$, where the coefficients in the quadratic expression may still depend on the q_i:

$$\mathcal{T} = \mathcal{T}(q_i,\dot{q}_i) = \sum_{k,l=1}^{s} a_{kl}(q_i)\dot{q}_k\dot{q}_l. \tag{2.202}$$

We shall not at this point consider velocity-dependent potential energies; the potential energy U will then be a function of the q_i only:

$$U = U(q_i). \tag{2.203}$$

The total energy ε is given by the equation

$$\varepsilon = \varepsilon(q_i,\dot{q}_i) = \mathcal{T}(q_i,\dot{q}_i) + U(q_i). \tag{2.204}$$

Instead of the generalized velocities $\dot{q}_i$ it is more convenient to introduce a new set of variables, the *generalized momenta*, p_i, which are defined by the equations

$$p_i = \frac{\partial \mathcal{T}}{\partial \dot{q}_i} = \sum_j a_{ij}\dot{q}_j, \tag{2.205}$$

where we have used Eq. (2.202) for $\mathcal{T}$. If q_k is a Cartesian coordinate, p_k will be the linear momentum (=mass times velocity). Transforming from the

* The disregard of internal degrees of freedom, such as the degrees of freedom corresponding to the electron motion in an atom or the vibrational degree of freedom of a diatomic molecule, can only be justified by quantum statistics where one shows that these degrees of freedom are *frozen in*; for a discussion of the freezing-in of degrees of freedom leading to constraints in classical mechanics, see D. ter Haar, *Elements of Hamiltonian Mechanics*, Amsterdam, 1961, chap. II.

$\dot{q}_i$ to the p_i we have, instead of Eq. (2.204),

$$\mathcal{H}(p_i,q_i) = \mathcal{T}(p_i,q_i) + U(q_i), \tag{2.206}$$

where $\mathcal{T}$ is now a quadratic function of the p_i. In Eq. (2.206) the total energy is denoted by $\mathcal{H}$. If the energy is expressed in terms of the p_i and the q_i, it is called the *Hamiltonian* of the system, hence the notation $\mathcal{H}$. The $2s$ coordinates $p_1, p_2, \cdots, p_s, q_1, q_2, \cdots, q_s$ are called *canonical* coordinates. In theoretical mechanics it is shown that, expressed in the canonical coordinates, the equations of motion take on the canonical or Hamiltonian form, or

$$\dot{q}_i = \frac{\partial \mathcal{H}}{\partial p_i}, \qquad \dot{p}_i = -\frac{\partial \mathcal{H}}{\partial q_i}. \tag{2.207}$$

The s p_i and s q_i together form the *phase* of the molecule. Once they are given at one time, they follow at any other time from the equations of motion. The behavior of the molecule can be pictured by an orbit in the $2s$-dimensional space which is determined by the $2s$ coordinates $p_1, \cdots, p_s, q_1, \cdots, q_s$. This space is called the *phase space* of the molecule or, with the Ehrenfests, μ-space where μ stands for the first letter of *m*olecule.

A volume element in μ space will be denoted by $d\omega$,

$$d\omega = \prod_{i=1}^{s} dp_i\, dq_i. \tag{2.208}$$

If $f(p_i,q_i)d\omega$ is the number of molecules the p_i and q_i of which lie in the intervals $(p_i,p_i + dp_i)$ and $(q_i,q_i + dq_i)$, then in the case of equilibrium the distribution function $f(p_i,q_i)$ will be given by

$$f(p_i,q_i) = Ce^{-\mathcal{H}(p_i,q_i)/kT}, \tag{2.209}$$

where C is again a normalizing constant. This form of the distribution function was introduced by Boltzmann in 1871. In 1868 Boltzmann generalized the distribution function to include a potential energy, and in 1871 he extended it further to include the case of polyatomic molecules. The distribution of Eq. (2.209) is called the *Maxwell-Boltzmann distribution*.

The average value $\bar{G}$ of a function $G(p_i,q_i)$ of the p_i and q_i is now defined as

$$\bar{G} = \frac{\int G(p_i,q_i) f(p_i,q_i)\, d\omega}{\int f(p_i,q_i)\, d\omega}, \tag{2.210}$$

where the integrations extend over the whole of μ-space.

We can apply Eq. (2.210) to calculate the average value of the kinetic energy. Since $\mathcal{T}$ is a homogeneous, quadratic polynomial in the p_i, we have from Euler's theorem on homogeneous functions

$$\mathcal{T} = \frac{1}{2}\sum_i \frac{\partial \mathcal{T}}{\partial p_i} p_i. \tag{2.211}$$

For the average value of $\mathcal{T}$ we then find

$$\bar{\mathcal{T}} = s \cdot \tfrac{1}{2} kT. \tag{2.212}$$

The result of Eq. (2.212) is obtained as follows. If the expression (2.211) for $\mathcal{T}$ is substituted into Eq. (2.210), there will result s integrals of the type

$$\frac{1}{2}\int \frac{\partial \mathcal{T}}{\partial p_k} p_k e^{-\mathcal{H}/kT}\, dp_k\, d\omega', \tag{2.213}$$

where $d\omega'$ stands for the product of all the dp_i and dq_i except dp_k. Integrating by parts, we obtain

$$\tfrac{1}{2}kT \int f(p_i,q_i)\, d\omega, \tag{2.214}$$

since the integrated parts will be zero. Equation (2.212) then follows immediately.

Expression (2.212) means that for every degree of freedom there is a contribution $\frac{1}{2}kT$ to the average kinetic energy, which is called the *equipartition* of the kinetic energy. We met an example of this equipartition in Section 1.1 of the previous chapter when we considered a monatomic gas and found that the average kinetic energy pertaining to either the x, or the y, or the z direction was the same and equal to $\frac{1}{2}kT$. Another case would be the dumbbell molecule where the number of degreees of freedom is five and where we find for the average kinetic energy $\frac{5}{2}kT$.

2.3 THE *H*-THEOREM; *H* AND PROBABILITY

In order to prove that the distribution function given by Eq. (2.209) is the equilibrium distribution function for a system of independent particles, Boltzmann considered a function H

$$H = \overline{\ln f} = \int f \ln f\, d\omega, \tag{2.301}$$

where the integration extends over the whole of μ-space. As long as the distribution function f is not given by Eq. (2.209), dH/dt will be negative, provided a slightly modified Stosszahlansatz is valid. Furthermore, dH/dt will only be zero, if f is given by Eq. (2.209). This is a consequence of the fact that during any collision the total energy will be conserved. Although we shall not give a proof of these properties of H, it is along the same lines as that given in Section 1.5, albeit much more complicated. We may mention here that an essential point in the proof is that the integration in Eq. (2.301) is over μ-space and not over a $q_i,\dot{q}_i$-space.

The function H is again related to the entropy S_v per unit volume through Eq. (1.612)—if we forget about a possible additive constant. It is also possible to relate H again to the probability in phase space. The argument is the same as in Section 1.7, but now we replace the cells Z_i by volume elements in μ-space,

$$Z_i \to d\omega, \tag{2.302}$$

the N_i by the total number of particles with p_k and q_k in the intervals $(p_k, p_k + dp_k)$ and $(q_k, q_k + dq_k)$,

$$N_i \to f(p_k, q_k)\, d\omega, \tag{2.303}$$

and the ε_i by the total energy of a molecule,

$$\varepsilon_i \to \mathscr{H}(p_k, q_k). \tag{2.304}$$

2.4 APPLICATIONS OF THE MAXWELL-BOLTZMANN FORMULA

In many cases, the potential energy may contain a term that depends only on the position of the center of mass of the molecule. This will, for instance, be the case if we are considering a system in a gravitational field, or inside a vessel the walls of which exert forces which can be derived from a potential energy such as shown in Fig. 2.2. In such cases, we can take as three of the q_i (say q_1, q_2, and q_3) the x, y, and z coordinates of the center of mass. Integrating the distribution function over all the other q_i and over all the p_i, we get the density of the gas as a function of the position in space

$$\rho(x,y,z) = C \int dp_1 \cdots dp_s\, dq_4 \cdots dq_s e^{-\mathscr{H}/kT}, \tag{2.401}$$

or

$$\rho(x,y,z) = C' e^{-U(x,y,z)/kT}, \tag{2.402}$$

where we have put

$$\mathscr{H} = \mathscr{T}(p_i, q_4, \cdots, q_s) + U(q_1, q_2, q_3) + U'(q_4, \cdots, q_s). \tag{2.403}$$

In writing down Eq. (2.403) we have assumed that the potential energy can be split into two parts, one depending on the coordinates of the center of mass only and one not depending on those coordinates. It can be shown that the kinetic energy does not depend on the x, y, or z coordinate of the center of mass.

One instance of Eq. (2.402) is the barometer formula (2.103). In general we see that the density will be highest at places where the potential energy is lowest. In the case where the forces exerted by a wall are derived from a potential energy such as the one of Fig. 2.2, a certain amount of adsorption on the wall will take place.

Let us now consider the case where the potential energy is a quadratic function of the q_i,

$$U = \frac{1}{2} \sum_{k,l=1}^{s} b_{kl} q_k q_l, \tag{2.404}$$

and where the kinetic energy $\mathscr{T}$ does not contain the q_i. (For an example where the potential energy is of the form 2.404, but the kinetic energy contains

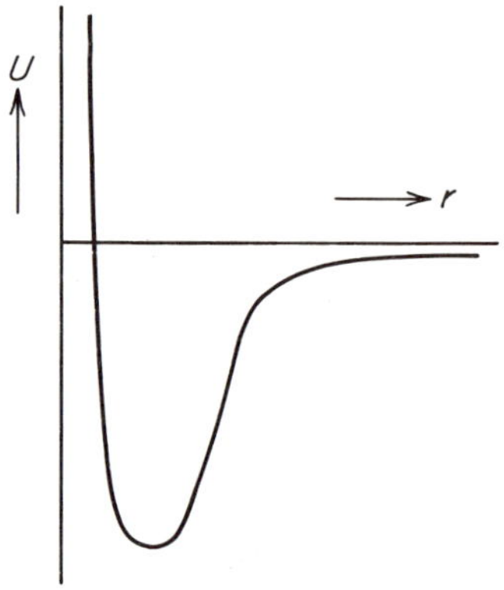

FIG. 2.2 The potential energy U of a particle near a wall as function of its distance r from the wall.

the q_i, see Problem 1 at the end of this chapter.) By a suitable choice of linear combinations of the q_i we can reduce U to a sum of t squares where $t \leqslant s$,

$$U = \frac{1}{2} \sum_{k=1}^{t} b_k q_k'^2, \tag{2.405}$$

where the q_i' are the linear combinations just mentioned, which we use as new coordinates.

One finds the average value of the potential energy in much the same way as was used in Section 2.2 to calculate the average value of the kinetic energy. We have therefore

$$\overline{U} = t \cdot \tfrac{1}{2} kT. \tag{2.406}$$

The average value of the total energy is now obtained by combining

Eqs. (2.212) and (2.406):

$$\bar{\varepsilon} = \bar{\mathscr{T}} + \bar{U} = (s + t) \cdot \tfrac{1}{2}kT. \tag{2.407}$$

Equation (2.407) expresses a more general equipartition of energy: each degree of freedom contributing to the kinetic energy and each degree of freedom contributing a quadratic term to the potential energy will give rise to a term $\frac{1}{2}kT$ in the average total energy, provided the kinetic energy does not contain the coordinates.

The specific heat per mole C_v will now be given by the equation

$$C_v = N_A \frac{\partial \bar{\varepsilon}}{\partial T} = \frac{1}{2}(s + t)R. \tag{2.408}$$

As an example we can calculate the average energy of an isotropic three-dimensional harmonic oscillator. The energy is given by the equation

$$\varepsilon = \frac{p_x{}^2 + p_y{}^2 + p_z{}^2}{2m} + \frac{1}{2}\alpha(x^2 + y^2 + z^2), \tag{2.409}$$

and its average value is given by the equation

$$\bar{\varepsilon} = 3kT, \tag{2.410}$$

leading to a specific heat of $3R$ per mole.

Let us now consider the case of a gas of dumbbell molecules, for example, HCl. If we treat the molecules as real dumbbells, that is, molecules with a fixed distance between the hydrogen atom and the chlorine atom, we have $s = 5$, $t = 0$, and the specific heat per mole will be $\frac{5}{2}R$, which is observed experimentally, at any rate, at not too high temperatures. If, however, we bear in mind that the idea of a completely fixed distance apart of the two atoms is an idealization and that there will always be the possibility of vibrations along the molecular axis, we use $s = 6$ and $t = 1$, leading to a specific heat per mole of $\frac{7}{2}R$, which is not found experimentally. The solution of this apparent contradiction—which upset Gibbs so much that he resolved to treat statistical mechanics purely as a branch of rational mechanics rather than as a branch of physics—can only be given by quantum mechanics and is related to the problem of the freezing-in of degrees of freedom, mentioned briefly at the beginning of Section 2.2.

A similar problem arises in the case of nonlinear polyatomic molecules. If all the distances in the molecule are completely fixed, it will behave as a solid body, and, since it then would have three rotational and three translational degrees of freedom, we should expect a specific heat per mole equal to $3R(s = 6, t = 0)$, which has been found experimentally. However, if there are N atoms in the molecule, and if we take into account all possible vibrations, we find $3N$ degrees of freedom of which $3N - 6$ contribute to the potential energy: only the translational and rotational degrees of freedom do not make a contribution to the potential energy. If all $3N - 6$ terms in the potential energy are quadratic, we should expect a specific heat per mole of $3(N - 1)R$, which is not found experimentally. The situation becomes even more complicated and leads to even larger specific heats* if we take the degrees of freedom belonging to the electronic and nuclear movements into account.

Let us now consider the case of a dumbbell molecule that possesses a magnetic (or electric) dipole moment, and let us calculate its average moment in a homogeneous field. For the sake of simplicity we shall choose as our q_i the three coordinates of the center of mass X, Y, Z, the angle θ between the magnetic moment $\boldsymbol{\mu}$, and the magnetic field $\mathbf{B}$ and the angle φ between the projection of $\boldsymbol{\mu}$ on a plane perpendicular to $\mathbf{B}$ and a fixed axis in this plane. The corresponding momenta will be denoted by p_X, p_Y, p_Z, p_θ, and p_φ. If the absolute values of the permanent magnetic moment and of the homogeneous magnetic field are denoted by μ and B, the potential energy will be given by the expression

$$U = -\mu B \cos\theta. \tag{2.411}$$

For the average magnetic moment in the direction of the field we get

$$\frac{\bar{\mu}}{\mu} = \frac{\int \cos\theta e^{-(\mathscr{T}+U)/kT}\, d\omega}{\int e^{-(\mathscr{T}+U)/kT}\, d\omega} = \mathscr{L}\left(\frac{\mu B}{kT}\right), \tag{2.412}$$

where $\mathscr{L}(x)$ is the Langevin function:

$$\mathscr{L}(x) = \coth x - \frac{1}{x}. \tag{2.413}$$

The behavior of $\mathscr{L}(x)$ is illustrated in Fig. 2.3.

In order to derive Eq. (2.412) we have to remember that the kinetic energy $\mathscr{T}$ expressed in p_X, p_Y, p_Z, p_θ, and p_φ is

$$\mathscr{T} = \frac{p_X{}^2 + p_Y{}^2 + p_Z{}^2}{2M} + \frac{p_\theta{}^2}{2A} + \frac{p_\varphi{}^2}{2A\sin^2\theta}, \tag{2.414}$$

where M is the total mass of the molecule and where A is the moment of inertia of the molecule with respect to an axis through the center of mass, perpendicular to the axis of the molecule.

Integration over p_φ produces, apart from constants that cancel in numerator

* We must warn our readers here that, in the discussion of the paradox of the specific heats, we have presented the situation as being simpler than it really is. The reader can, however, easily verify the real state of affairs, once the discussion of the next chapter is understood. It is then also possible to see for what temperatures our present considerations will hold.

and denominator, a factor $\sin\theta$. In this way, we have finally to evaluate the expression

$$\int_0^\pi \cos\theta e^{\mu B\cos\theta/kT}\sin\theta\,d\theta\Big/\int_0^\pi e^{\mu B\cos\theta/kT}\sin\theta\,d\theta, \tag{2.415}$$

which leads to Eq. (2.412).*

Equation (2.412) was derived by Langevin in 1905, and in 1912 Debye derived a similar formula for a system of electric dipoles. From Fig. 2.3 we

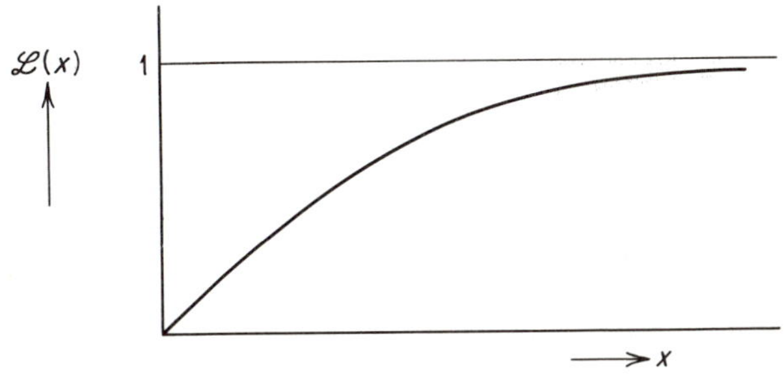

FIG. 2.3 The Langevin function $\mathscr{L}(x) = \coth x - (1/x)$.

see that $\bar{\mu}$ approaches μ for low temperatures (or large fields), that is, saturation sets in at low temperatures, while at higher temperatures thermal agitation produces a smaller mean dipole moment.

2.5 THE BOLTZMANN TRANSPORT EQUATION

In Section 1.5 we saw that collisions change the distribution function in such a way as to decrease the value of H, and we mentioned in Section 2.3 of the present chapter that this remains true for the case where the distribution function depends on the coordinates as well as on the velocities. In that discussion we were not interested in the actual change in the distribution function itself; nevertheless, it is sometimes important to evaluate this change in f, for instance, if we wish to know how soon equilibrium will be reached.

In order to find the general equation that must be satisfied by the distribution function, we must consider not only the change of this distribution

* We note that Eq. (2.412) will also hold for the more general case where the particle is a symmetric top with its dipole moment $\boldsymbol{\mu}$ along the axis of symmetry; see Problem 8 at the end of this chapter.

function by collisions, but also the change due to the continous motion of the particles. For the sake of simplicity we shall consider only the case of systems of point particles; the distribution function will then be a function of the coordinates x, y, and z, and of the velocity components u, v, and w.

We shall consider those atoms that have coordinates between $\mathbf{r}$ and $\mathbf{r} + \mathbf{dr}$ and velocities between $\mathbf{c}$ and $\mathbf{c} + \mathbf{dc}$. At time t there will be*

$$f(\mathbf{r},\mathbf{c},t)\, d^3\mathbf{r}\, d^3\mathbf{c} \tag{2.501}$$

such atoms, and at time $t + dt$ there will be

$$f(\mathbf{r},\mathbf{c},t + dt)\, d^3\mathbf{r}\, d^3\mathbf{c} \tag{2.502}$$

such atoms. The difference between these two numbers (2.501) and (2.502), which is equal to

$$\frac{\partial f}{\partial t}\, dt\, d^3\mathbf{r}\, d^3\mathbf{c}, \tag{2.503}$$

is due to two causes: (1) there may not be a balance between atoms entering the element $d^3\mathbf{r}\, d^3\mathbf{c}$ during the time interval t, $t + dt$ because of collisions and those atoms leaving this element during the given time interval as the result of collisions (compare Eq. (1.401)), and (2) the number of atoms that have acquired positions and velocities during t, $t + dt$ such that their representative points falls at the end of the time interval inside $d^3\mathbf{r}\, d^3\mathbf{c}$ may not be equal to the number of atoms that had positions and velocities at the beginning of the time interval within the element $d^3\mathbf{r}\, d^3\mathbf{c}$ but the representative points of which have left $d^3\mathbf{r}\, d^3\mathbf{c}$ during the time interval. We can express this fact by writing

$$\frac{\partial f}{\partial t} = \left(\frac{\partial f}{\partial t}\right)_{\text{drift}} + \left(\frac{\partial f}{\partial t}\right)_{\text{coll}}. \tag{2.504}$$

We can find for $(\partial f/\partial t)_{\text{drift}}$ the following equation:

$$-\left(\frac{\partial f}{\partial t}\right)_{\text{drift}} = \frac{\partial f}{\partial x}u + \frac{\partial f}{\partial y}v + \frac{\partial f}{\partial z}w + \frac{\partial f}{\partial u}\frac{F_x}{m} + \frac{\partial f}{\partial v}\frac{F_y}{m} + \frac{\partial f}{\partial w}\frac{F_z}{m}, \tag{2.505}$$

where $\mathbf{F}$ is the external force on the particles and m their mass.

Equation (2.503) can be proved by considering the flow of representative points out of and into the element $d^3\mathbf{r}\, d^3\mathbf{c}$ in the six-dimensional space of which x, y, z, u, v, and w are the coordinates. The total number of representative points that cross the "face" which is "orthogonal" to the x axis during a time interval dt is equal to the product of the area of the face, $dy\, dz\, du\, dv\, dw$, the distance traveled in the x direction by the representative point during dt,

* We use the abbreviated expression $f(\mathbf{r},\mathbf{c},t)$ for $f(x,y,z;u,v,w;t)$.

$u\,dt$, and the number of representative points per unit volume in the **r**, **c**-space, f. The net decrease of representative points in $d^3\mathbf{r}\,d^3\mathbf{c}$ during that time interval is the difference between the number crossing the face at $x + dx$ and the number crossing the "opposite" face at x; that is, it is equal to $[f(x + dx,y,z;u,v,w;t) - f(x,y,z;u,v,w;t)]\,u\,dy\,dz\,du\,dv\,dw\,dt$, or, $(\partial f/\partial x)\;u\,dx\,dy\,dz\,du\,dv\,dw\,dt$. From the flow through the faces orthogonal to the y, the z, the u, the v, and the w axes, we get similar contributions to the net decrease in the number of representative points in $d^3\mathbf{r}d^3\mathbf{c}$, that is, to $(\partial f/\partial t)_{\mathrm{drift}}\;d^3\mathbf{r}d^3\mathbf{c}\,dt$; these are equal to $(\partial f/\partial y)\;v\;d^3\mathbf{r}\;d^3\mathbf{c}\;dt, \cdots, (\partial f/\partial w)\;a_z\;d^3\mathbf{r}\;d^3\mathbf{c}\;dt$, where a_z is the z component of the acceleration which by Newton's second law is equal to F_z/m.

If we use shorthand notation and introduce the symbolical vectors $\mathbf{\nabla}$ and $\mathbf{\nabla}_c$ with components $\partial/\partial x$, $\partial/\partial y$, $\partial/\partial z$ and $\partial/\partial u$, $\partial/\partial v$, $\partial/\partial w$, respectively ($\mathbf{\nabla}$ is the gradient vector operator), we have from Eq. (2.505)

$$-\left(\frac{\partial f}{\partial t}\right)_{\mathrm{drift}} = (\mathbf{c}\cdot\mathbf{\nabla})f + \frac{1}{m}(\mathbf{F}\cdot\mathbf{\nabla}_c)f. \tag{2.506}$$

For $(\partial f/\partial t)_{\mathrm{coll}}$ we can use Eq. (1.419), since the collision mechanism is independent of where the collisions take place. The only difference between the present situation and the one leading to Eq. (1.419) is that the distribution function is now a function of x, y, and z as well as of u, v, and w, which means that f_1', for instance stands for $f(x,y,z;u_1',v_1',w_1';t)$, and so on. Combining Eqs. (2.504), (2.506), and (1.419) we get the following integro-differential equation for the distribution function:

$$\frac{\partial f}{\partial t} + (\mathbf{c}\cdot\mathbf{\nabla})f + \frac{1}{m}(\mathbf{F}\cdot\mathbf{\nabla}_c)f + \int (ff_1 - f'f_1')a\;d^3\mathbf{c}_1\;d^2\boldsymbol{\omega} = 0. \tag{2.507}$$

Equation (2.507) is known as *Boltzmann's transport equation* or, more simply, as *Boltzmann's equation.*

If there are no external forces, the Maxwell distribution will be an (equilibrium) solution of Eq. (2.507). Since f does not depend on x, y, or z, $\mathbf{\nabla}f = 0$, also $\mathbf{F} = 0$ and $(\partial f/\partial t)_{\mathrm{coll}} = 0$, as we saw in Section 1.4, and thus $\partial f/\partial t = 0$.

If the external forces can be derived from a potential energy function $U(x,y,z)$, the Maxwell-Boltzmann distribution (2.107) will be an equilibrium solution. We have in that case:

$$\mathbf{F} = -\mathbf{\nabla}U, \tag{2.508}$$

$$\mathbf{\nabla}f = -\frac{1}{kT}f\mathbf{\nabla}U, \tag{2.509}$$

$$\mathbf{\nabla}_c f = -\frac{1}{kT}f\mathbf{\nabla}_c\mathcal{T} = -\frac{m\mathbf{c}}{kT}f, \tag{2.510}$$

and

$$ff_1 = f'f_1', \tag{2.511}$$

because of the conservation of energy. From Eqs. (2.507) to (2.511) it then follows again that $\partial f/\partial t = 0$.

Equation (2.507) can, for instance, be used to estimate the time necessary to re-establish equilibrium. Its most important applications have, however, been in the discussion of steady nonequilibrium states by means of which Chapman and Enskog, in the years 1916 and 1917 working independently and using powerful approximation methods, derived theoretical expressions for the viscosity coefficient, thermal conductivity, and other kinetic properties of gases.

2.6 EXTERNAL PARAMETERS

In Section 2.4 we met a case where the energy of a molecule depended not only on the p_i and q_i, but also on the magnitude of the magnetic (or electric) field. In general, it is possible for the energy of a molecule to depend on a number of *external parameters*, a_i. Examples of such parameters are the volume of a gas, the electric field strength, and so on. We have then

$$\varepsilon = \varepsilon(p_1,\cdots,p_s,q_1,\cdots,q_s;a_1,a_2,\cdots). \tag{2.601}$$

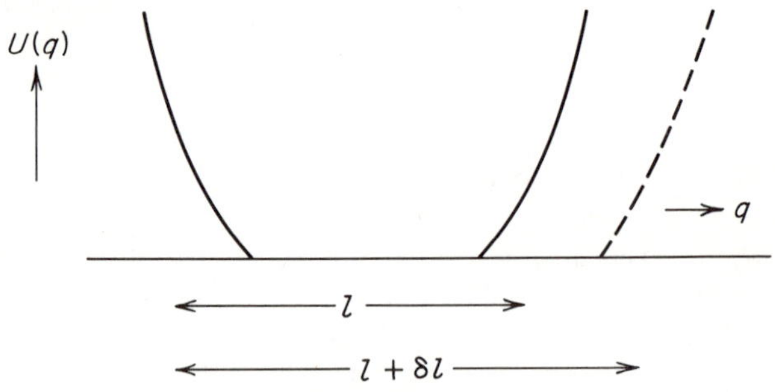

FIG. 2.4 External parameters. The potential energy U(q) of a particle in a one-dimensional "box" of "volume" l. The dotted curve indicates the potential energy when the "volume" is increased to $l + \delta l$.

Let us first consider the case where a is the "volume" of a one-dimensional gas (see Fig. 2.4). If $U(q)$ is the potential energy of the molecule, then $-\partial U/\partial q$ will be the force exerted by the wall on the molecule. Apart from

q, U will also depend on the "volume" l of the one-dimensional vessel: $U = U(q,l)$. Let us now displace the wall a distance δl. If we displace the molecule at the same time, a distance δq equal to δl, there will be no change in potential energy, or

$$\delta U = 0 = \frac{\partial U}{\partial q}\,\delta q + \frac{\partial U}{\partial l}\,\delta l, \qquad \text{if } \delta q = \delta l, \tag{2.602}$$

or

$$\frac{\partial U}{\partial l} = -\frac{\partial U}{\partial q}. \tag{2.603}$$

Since $-\partial U/\partial q$ is the force exerted by the "volume" on the molecule, $-\delta U/\delta l$ will be the force exerted by the molecule "on the volume." In general we can introduce quantities A_i defined by the equation

$$A_i = -\frac{\partial \varepsilon}{\partial a_i}. \tag{2.604}$$

These A_i can be called the *generalized forces* (per molecule) exerted by the system tending to increase the parameters. If N is the total number of particles in the system, the total force exerted by the system tending to increase the parameter a_i will be equal to $N\bar{A}_i$.

We see that the total work done by the molecule "on the parameters," which is equal to minus the change in the energy ε of the molecule, when the external parameters change from a_i to $a_i + \delta a_i$ is given by the equation

$$-\delta\varepsilon = \sum_i A_i \delta a_i. \tag{2.605}$$

In the case considered in Section 2.4 the parameter was the magnetic field and the generalized force was there $-\partial\varepsilon/\partial B = \mu \cos\theta$, that is, the magnetic moment in the direction of the magnetic field. In that case $-\delta\varepsilon$ was equal to $\mu \cos\theta\, \delta B$.

2.7 THE PHASE INTEGRAL; CONNECTION WITH THERMODYNAMICS

Let us consider the following quantity:

$$Z_\mu = e^{-\beta F} = K \int e^{-\beta\varepsilon}\, d\omega, \tag{2.701}$$

where the suffix μ indicates that the integration is over the whole of μ-space, and where now and henceforth β is related to the absolute temperature T

through the equation

$$\beta = \frac{1}{kT}. \tag{2.702}$$

The quantity Z_μ is called the *phase integral* or *partition function* of one molecule. Both Z_μ and F, the physical meaning of which we shall see in a moment, are functions of the temperature (or of β) and of the external parameters. The const K, which we shall leave undetermined for the moment, enters in order that Z_μ may be made a dimensionless quantity.

We can write Eq. (2.701) in the form

$$\frac{1}{K} = \int e^{\beta(F-\varepsilon)}\, d\omega. \tag{2.703}$$

If we vary the temperature and the external parameters, we get from Eq. (2.703)

$$\delta\frac{1}{K} = 0 = \int \delta(\beta F) e^{\beta(F-\varepsilon)}\, d\omega - \int \varepsilon\delta\beta e^{\beta(F-\varepsilon)}\, d\omega - \int \beta\delta\varepsilon e^{\beta(F-\varepsilon)}\, d\omega. \tag{2.704}$$

Since

$$\delta\varepsilon = \sum_i \frac{\partial\varepsilon}{\partial a_i}\delta a_i = -\sum_i A_i \delta a_i, \tag{2.705}$$

Eq. (2.704) leads to

$$\delta(\beta F) = \bar{\varepsilon}\delta\beta - \beta\sum_i \bar{A}_i\delta a_i, \tag{2.706}$$

where we have used Eqs. (2.701), (2.703), and the definition (2.210) of the average value of a physical quantity.

To find the physical meaning of F we shall express the right-hand side of Eq. (2.706) in terms of the internal energy U and the entropy S per molecule.* This can be done by considering the second law of thermodynamics in the form

$$\delta S = \frac{\not{\delta} Q}{T} = \frac{\delta U + \not{\delta} A}{T}, \tag{2.707}$$

where A is the work done by the system.

For one molecule we have the relations

$$U = \bar{\varepsilon} \qquad \text{and} \qquad \not{\delta} A = \sum_i \bar{A}_i\delta a_i, \tag{2.708}$$

where the stroke through the δ indicates that $\not{\delta} A$ $(= -\overline{\delta\varepsilon})$ is not necessarily a total variation.

* In Section 1.6, S and U referred to the entropy and internal energy of the whole system, but here they refer to one particle only.

From Eqs. (2.707) and (2.708) we find that the right-hand side of Eq. (2.706) can be written in the form

$$\bar{\varepsilon}\delta\beta - \beta \sum_i \bar{A}_i \delta a_i = \delta\left(\frac{U}{kT} - \frac{S}{k}\right), \tag{2.709}$$

and it follows that F/T, apart from a possible additive constant, satisfies the equation

$$\frac{F}{T} = \frac{U}{T} - S, \qquad \text{or,} \qquad F = U - TS. \tag{2.710}$$

Thus F corresponds to the free energy per molecule (apart possibly from an additive multiple of the temperature).

In the case where the volume V is the only external parameter, Eqs. (2.706) and (2.707) can be written in the form (compare Eq. (1.606))

$$\delta \frac{S}{k} = \beta\delta\bar{\varepsilon} + \frac{P}{N}\beta\delta V \tag{2.711}$$

and

$$\delta(\beta F) = \bar{\varepsilon}\delta\beta - \frac{P}{N}\beta\delta V, \tag{2.712}$$

where all quantities refer to one molecule except the pressure (hence the factor N).

The temperature (or β) and the external parameters determine the *thermodynamic state* of the system. Once they are given, one can calculate the phase integral Z_μ and thus the free energy F. All thermodynamic quantities can then be derived from Z_μ or F. Since $F = -kT \ln Z_\mu$, we shall give here only the equations involving F. From Eq. (2.706) it follows that

$$\bar{\varepsilon} = \frac{\partial(\beta F)}{\partial\beta} = -T^2 \frac{\partial(F/T)}{\partial T}, \tag{2.713}$$

$$\bar{A}_i = -\frac{1}{\beta}\frac{\partial(\beta F)}{\partial a_i} = -T\frac{\partial(F/T)}{\partial a_i}, \tag{2.714}$$

while we get from Eq. (2.713) for the specific heat per molecule, c_v,

$$c_v = \frac{\partial\bar{\varepsilon}}{\partial T} = -\beta^2\frac{\partial\bar{\varepsilon}}{\partial\beta} = -\beta^2\frac{\partial^2(\beta F)}{\partial\beta^2} = -2T\frac{\partial(F/T)}{\partial T} - T^2\frac{\partial^2(F/T)}{\partial T^2}. \tag{2.715}$$

If one of the a_i is the volume, we get for the pressure P the equation

$$P = -\frac{N}{\beta}\frac{\partial(\beta F)}{\partial V}. \tag{2.716}$$

Equation (2.716) can be used to obtain the equation of state. The equation of state can also be obtained from the equation

$$P = NT \frac{\partial S}{\partial V}, \tag{2.717}$$

once the entropy is known as a function of the volume; Eq. (2.717) follows from Eq. (2.711).

We note that the integrand in Eq. (2.701) is proportional to the distribution function f given by Eq. (2.107). As it follows from the definition of f that $\int f\,d\omega$, integrated over the whole of μ-space must be equal to N, we find from Eqs. (2.701) and (2.107) that

$$f = KNe^{\beta(F-\varepsilon)}, \tag{2.718}$$

and for Boltzmann's H-function we get

$$H = \int f \ln f\,d\omega = \beta(F - \bar{\varepsilon}) + \text{const} = -\frac{S}{k} + \text{const}. \tag{2.719}$$

This shows once again that kH, apart from a possible additive const, is equal to $-S$.

In conclusion let us apply our formulas to the case of a monatomic perfect gas. The energy of one atom is given by the equation

$$\varepsilon = \frac{p_x^{\,2} + p_y^{\,2} + p_z^{\,2}}{2m} + U(x,y,z), \tag{2.720}$$

where the potential energy is such that the gas stays inside the volume V. This can be realized, if U is given by the equations

$$\left.\begin{aligned} U(x,y,z) &= 0 \qquad \text{inside } V; \\ U(x,y,z) &= \infty \qquad \text{outside } V. \end{aligned}\right\} \tag{2.721}$$

Introducing expression (2.720) into Eq. (2.701) for Z_μ we get after integrating over x, y, z, p_x, p_y, and p_z

$$Z_\mu = KV(2\pi mkT)^{3/2}. \tag{2.722}$$

From Eqs. (2.722), (2.713), and (2.715) we get for the energy and specific heat:

$$\bar{\varepsilon} = \tfrac{3}{2}kT, \tag{2.723}$$

$$c_v = \tfrac{3}{2}k. \tag{2.724}$$

Finally, we get from Eq. (2.716) for the equation of state the perfect gas law (1.204).

PROBLEMS

1. Consider a two-dimensional harmonic oscillator with energy

$$\varepsilon = \frac{p_x^2 + p_y^2}{2m} + \frac{1}{2}\alpha(x^2 + y^2).$$

According to Eq. (2.407), the average energy will be equal to $2kT$. Introducing polar coordinates, $x = r\cos\varphi$, $y = r\sin\varphi$, the energy can be expressed in the form

$$\varepsilon = \frac{1}{2m}\left[p_r^2 + \frac{p_\varphi^2}{r^2}\right] + \frac{1}{2}\alpha r^2$$

There are now two kinetic energy terms and one, quadratic, potential energy term. Verify by direct calculation, using polar coordinates, that just the same the average energy equals $2kT$, and comment on the result.

2. From the example of the preceding problem it is clear that the equipartition theorem must be formulated with caution. Prove the following general statement, due to Tolman,* stating clearly the conditions under which it is valid:

$$\overline{q_i \frac{\partial \mathscr{H}}{\partial q_j}} = kT\delta_{ij}, \qquad \overline{p_i \frac{\partial \mathscr{H}}{\partial p_j}} = kT\delta_{ij},$$

where δ_{ij} is the Kronecker symbol, which is equal to 1 if $i = j$ and equal to 0 if $i \neq j$.

3. Show that the following distribution function satisfies the transport equation (2.507) for the case of point particles moving either in force-free space or in a central field of force:

$$f(x,y,z;u,v,w;t) = A\exp[-\alpha(u^2 + v^2 + w^2) + \gamma(uy - vx)].$$

Discuss the physical meaning of this distribution.

4. Consider the Lorentz model† of a metal in which the transport properties are ascribed to classical, charged particles ("electrons") that do not interact with one another, but only with the ions in the lattice. The electron-lattice collisions are supposed to be elastic and isotropic. This means that if $\Theta(\mathbf{c},\mathbf{c}')\,d^3\mathbf{c}'$ denotes the probability that per unit time an electron will change its velocity from $\mathbf{c}$ to a velocity in the range $\mathbf{c}'$, $\mathbf{c}' + \mathbf{dc}'$, we assume that $\Theta(\mathbf{c},\mathbf{c}')$ can be written in the form

$$\Theta(\mathbf{c},\mathbf{c}') = \delta(c - c')\eta/c^2,$$

* R. C. Tolman, *Phys. Rev.*, **11**, 261, 1918.
† H. A. Lorentz, *The Theory of Electrons*, Leipzig, 1909, pp. 47–50.

where $\delta(c - c')$ is Dirac's δ-function* and where we assume η to be constant.

Consider the transport equation (2.504) with the first term on the right-hand side given by Eq. (2.505) and the second term by Eq. (1.401). Let the system be subject to an external electromagnetic field so that the force **F** acting on the electrons is the Lorentz force:

$$\mathbf{F} = -e\{\mathbf{E} + [\mathbf{c} \wedge \mathbf{B}]\},$$

where $-e$ is the charge of an electron. Find a stationary solution of the transport equation, assuming (1) that the electric field **E** is uniform and lies in the xy plane, (2) that the magnetic field **B** is uniform and along the z axis, and (3) that we may write the distribution function f in the form

$$f = f_0 + u\chi_1(c) + v\chi_2(c),$$

where f_0 is the Maxwell distribution and where we assume the terms with the χ_i to be small compared to similar terms involving f_0 so that we may neglect them. Using the relation that exists between η and the mean free path λ, express the solution in terms of f_0, λ, E_x, E_y, and the ratio s of λ and the radius of curvature of an electron moving with speed c in the magnetic field B.

5. Use the solution obtained in the previous problem to evaluate the electrical conductivity of a (classical) metal. In this case we put $E_y = B = 0$, and find the conductivity σ from Ohm's law, $j_x = \sigma E_x$, where

$$j_x = -e \int uf\, d^3\mathbf{c}.$$

* Dirac's δ-function is defined by the equations (see, for example, P. A. M. Dirac, *Quantum Mechanics*, 2d ed., Oxford, 1935, Sections 20, 21)

$$\delta(x) = 0, \qquad x \neq 0; \qquad \int_{-\infty}^{+\infty} \delta(x)\, dx = 1. \tag{A}$$

From Eqs. (A) it follows that for any function $f(x)$ we have

$$\left.\begin{aligned} &\int_{-\infty}^{+\infty} f(x)\delta(x)\, dx = f(0); \qquad \int_{-\infty}^{+\infty} f(x)\delta(x-a)\, dx = f(a); \\ &\int_a^b f(x)\delta(x-c)\, dx = f(c), \qquad \text{if } a < c < b; \\ &\int_a^b f(x)\delta(x-c)\, dx = 0, \qquad \text{if either } b < c \text{ or } c < a. \end{aligned}\right\} \tag{B}$$

For a discussion of the mathematical aspects of the δ-function see, for instance, L. Schwartz, *Théorie des Distributions*, Paris, 1950–1951; I. Halperin, *Introduction to the Theory of Distributions*, Toronto, 1952; or J. P. Marchand, *Distributions*, Amsterdam, 1962.

6. Use the solution obtained in Problem 4 to evaluate the isothermal Hall coefficient R, defined by the equation

$$E_y = Rj_xB_z,$$

for the case where a current j_x is flowing in the x direction, where there is a magnetic field B_z in the z direction, and where E_y is the electromotive force in the y direction needed to prevent current from flowing in that direction.

7. Consider the transport equation for the electrons in a metal for the case where $B = 0$, $E_y = 0$, but where there is a temperature gradient dT/dx in the x direction so that the equilibrium distribution f_0 does now depend on x. Solve the transport equation under the same conditions as those in Problem 4. Find the coefficient of thermal conductivity κ, defined by the equation

$$w_x = -\kappa \frac{dT}{dx},$$

where

$$w_x = \int (\tfrac{1}{2}mc^2) f \, d^3\mathbf{c}$$

is the thermal current in the metal. Bear in mind that we have to find w_x under the condition that there is no electrical current flowing in the metal ($j_x = 0$).
Having found κ, combine the result with that of Problem 5 to find the *Wiedemann-Franz ratio** $\kappa/\sigma T$.

8. Evaluate the phase integral for the following systems:
   (a) a perfect gas of particles for which the energy ε is proportional to the absolute magnitude of their momentum ($\varepsilon = sp$);
   (b) a perfect gas of point particles in a spherical container of volume V, using spherical polars (compare your result with Eq. (2.722));
   (c) a perfect gas of magnetic dipoles in an external magnetic field $\mathbf{B}$;
   (d) a perfect gas of symmetric top molecules, the kinetic energy of which is

$$\mathscr{T} = \frac{1}{2m}(p_x^2 + p_y^2 + p_z^2) + \frac{1}{2}\left[\frac{p_\vartheta^2}{A} + \frac{(p_\varphi - p_\psi \cos\vartheta)^2}{A \sin^2\vartheta} + \frac{p_\psi^2}{C}\right],$$

   where m is the mass of the top; A, A, and C are the principal moments of inertia about the principal axes through the center of mass, the Cartesian coordinates of which are x, y, and z, while ϑ, φ, and ψ are Euler angles;

* G. Wiedemann and R. Franz, *Ann. Physik*, **89**, 497, 1853.

(e) a perfect gas of asymmetric top molecules, the kinetic energy of which is

$$\mathcal{T} = \frac{1}{2m}(p_x{}^2 + p_y{}^2 + p_z{}^2)$$
$$+ \frac{1}{2}\left\{\frac{[(p_\varphi - p_\psi \cos \vartheta)\cos\psi - p_\vartheta \sin\vartheta \sin\psi]^2}{A \sin^2 \vartheta}\right.$$
$$\left.+ \frac{[(p_\varphi - p_\psi \cos\vartheta)\sin\psi + p_\vartheta \sin\vartheta\cos\psi]^2}{B \sin^2\vartheta} + \frac{p_\psi{}^2}{C}\right\},$$

where A, B, and C are the principal moments of inertia, while the remainder of the notation is the same as under (d);

(f) a perfect gas of point particles moving in a one-dimensional potential energy field given by the equation

$$U(x) = \tfrac{1}{2}ax^2 - bx^3 - cx^4,$$

assuming that the phase integral can be written as a power series in the supposedly small quantities $b^2/\beta a^3$ and $c/\beta a^2$ and that only the first nontrivial term needs to be retained;

(g) a perfect gas of point particles in an infinitely high cylindrical vessel in a uniform gravitational field along the axis of the cylinder;

(h) a perfect gas of symmetric top molecules, each of which carries along its axis of symmetry a permanent magnetic dipole moment $\boldsymbol{\mu}$, in a uniform external magnetic field $\mathbf{B}$.

9. Use the results of the preceding problem
   (a) to calculate the specific heat for Cases (a), (f), and (g);
   (b) to calculate the magnetization in Case (h). Find also the asymptotic high-temperature expression for the magnetization.

10. Prove the Bohr-van Leeuwen theorem, which states that the magnetic susceptibility of a system of charged-point particles obeying classical mechanics and classical statistics vanishes rigorously. Introduce the magnetic field by means of the vector potential, so that only the kinetic energy contains the magnetic field in the form*

$$\mathcal{T} = \frac{(\mathbf{p} + e\mathbf{A})^2}{2m}.$$

11. Consider a system consisting of two rigid point-dipoles at a fixed distance apart. Assuming these dipoles to be in thermal equilibrium as far as their orientation is concerned, find an expression (in the form of integrals) for

* See, for instance, D. ter Haar, *Elements of Hamiltonian Mechanics*, Amsterdam, 1961, p. 113.

the average force acting between them and evaluate the high-temperature limit of that expression. The force between two dipoles $\boldsymbol{\mu}$ and $\boldsymbol{\mu}'$ at a distance R follows from the potential energy

$$U = -\frac{\mu\mu'}{R^3}[2\cos\vartheta\cos\vartheta' - \sin\vartheta\sin\vartheta'\cos(\varphi - \varphi')],$$

where ϑ, φ (ϑ',φ') are the angles defining the direction of $\boldsymbol{\mu}$ ($\boldsymbol{\mu}'$).

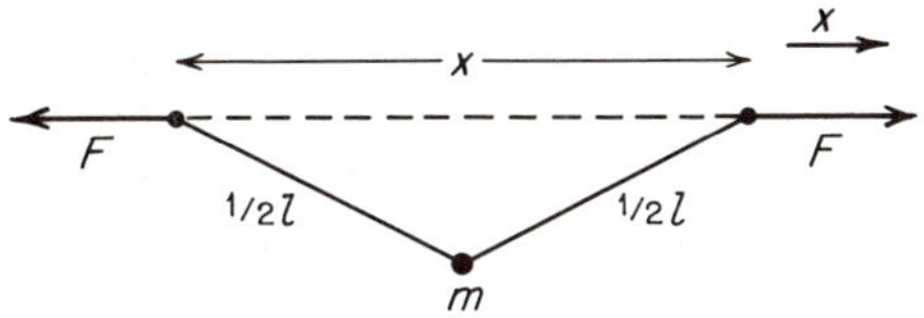

FIG. 2.5 The mass m is fixed at the middle of the string and is rotated around the x axis. The length of the string is l and the force needed to keep the ends at a distance x is F.

12. A point particle of mass m is fixed to the middle of a thin, inextensible string of length l (Fig. 2.5). The mass is rotated about the x axis connecting the two end points of the string, and the system is in contact with its surroundings at a temperature T. Calculate the force needed to keep the end points at a distance x.

BIBLIOGRAPHICAL NOTES

General references for this chapter are

1. L. Boltzmann *Vorlesungen über Gastheorie* Leipzig, 1896–1898; *Lectures on Gas Theory*, University of California Press, 1964.
2. S. Chapman and T. G. Cowling, *The Mathematical Theory of Non-Uniform Gases*, Cambridge, 1939.
3. J. W. Gibbs, *Elementary Principles in Statistical Mechanics* (Vol. II of his Collected Works), New Haven, 1948.
4. S. G. Brush, *Kinetic Theory*, 3 Vols., Oxford, 1965–1966.

Section 2.1. A rather crude attempt to derive the barometer formula can be found in Krönig's paper on kinetic theory:

5. A. Krönig, *Ann. Physik*, **99**, 315, 1856.

The exact barometer formula was first given by Boltzmann:

6. L. Boltzmann, *Wien. Ber.*, **78,** 7, 1879.

Section 2.2. The term "μ-space" was introduced by the Ehrenfests:

7. P. and T. Ehrenfest, *Enzyklopädie der mathematischen Wissenschaften*, Vol. IV, Part 32, Leipzig-Berlin, 1911, p. 36.

The Maxwell-Boltzmann distribution was derived by Boltzmann in 1871:

8. L. Boltzmann, *Wien. Ber.*, **58,** 517, 1868.
9. L. Boltzmann, *Wien. Ber.*, **63,** 397, 1871.

In Reference 8 Boltzmann gives the distribution in the case of an external field of force, while in Reference 9 he also takes into account the internal degrees of freedom of a molecule. The name "Maxwell-Boltzmann distribution" was suggested by G. H. Bryan in a paper read before the British Association at Oxford in 1894.

The equipartition of kinetic energy over the different degrees of freedom was proved by Boltzmann:

10. L. Boltzmann, *Wien. Ber.*, **63,** 712, 1871.

Compare also:

11. R. C. Tolman, *Phys. Rev.*, **11,** 261, 1918.

Section 2.3. The generalized H-theorem is also due to Boltzmann:

12. L. Boltzmann, *Wien. Ber.*, **72,** 427, 1875.

Section 2.4. Equation (2.412) was first derived by Langevin for the case of molecules possessing magnetic dipole moments:

13. P. Langevin, *J. phys.*, **4,** 678, 1905.
14. P. Langevin, *Ann. chim. et phys.*, **5,** 70, 1905.

The case of electric dipoles was studied by Debye, see, for example,

15. P. Debye, *Physik. Z*, **13,** 97, 1912.

Section 2.5. The transport equation is due to Boltzmann:

16. L. Boltzmann, *Wien, Ber.*, **74,** 503, 1876.

Probably the most important papers on the solution of the equation are those by Chapman and Enskog*:

17. S. Chapman, *Trans. Roy. Soc.* (*London*), **A216,** 279, 1916.
18. S. Chapman, *Trans. Roy. Soc.* (*London*), **A217,** 115, 1917.
19. D. Enskog, *Dissertation*, Uppsala, 1917.

For a thorough discussion of the whole subject, we refer to Reference 2.

Section 2.6. The introduction of external parameters into the statistical discussion was made by Gibbs, Reference 3, p. 4.

* Born (*Nuovo cimento*, **6,** Suppl. 296, 1949) has drawn attention to the fact that Hilbert had developed methods similar to those of Chapman and Enskog for solving the Boltzmann equation. His results were given in a series of lectures at Göttingen and only a short account of them was published in 1912 (*Math. Ann.*, **72,** 562, 1912).

Section 2.7. The phase integral and its relation to the free energy were introduced and used by Gibbs, Reference 3, Chaps. IV *et seq.*

Darwin and Fowler first introduced the term "partition function":

20. C. G. Darwin and R. H. Fowler, *Phil. Mag.*, **44,** 450, 1922.

Tolman uses in his textbook the term "sum over states" (which is certainly *not* meant as a translation of *Zustandssumme*, as Partington suggests (*Physical Chemistry*, London 1949, p. 305)).

3

The Partition Function

3.1 THE PARTITION FUNCTION

So far we have used classical mechanics to describe the systems in which we were interested. We know, however, that in many instances classical mechanics is inadequate to describe the behavior of physical systems, and that only a quantum-mechanical treatment can give correct results. We should thus build statistical mechanics upon a quantum-mechanical basis and afterwards show that, indeed, the classical formulas that we derived in the preceding chapters are limiting cases of the quantum-mechanical formulas. We shall see in the next chapter that the statistical methods used are the same for all cases and that the term "quantum statistics" is thus a misnomer.

We remind ourselves that there are two aspects in which quantum mechanics differs from classical mechanics: "diffraction" effects and "symmetry" effects. The first arise because of the wave nature of matter; they lead to the Schrödinger wave equation and thus to the existence of energy levels for the systems considered. The second occur because the wave function of a system of identical particles must be totally symmetric or totally antisymmetric. In the present chapter we shall be concerned with the diffraction effects only, postponing until the next chapter the discussion of the symmetry effects.

Let us consider a system of particles, each of which has a discrete energy spectrum. Let the energy levels be ε_n, all of which we shall for the moment assume to be nondegenerate.* We can now ask the question: "How many particles will be in each of the various levels, if we know that the total energy of the system consisting of N particles is E?" This problem is similar to the one treated in Section 1.7. The cells Z_i introduced there correspond now to the discrete energy levels. Assuming that the a priori probability for a particle to be in a particular energy level is the same for all levels, we may put all Z_i equal to one another, and we find for the probability W that there are N_1 particles in the energy level ε_1, N_2 in the level ε_2, and so on (compare Eq. (1.701)),

$$W = C \frac{N!}{\prod_i N_i!}. \tag{3.101}$$

The assumption of equal a priori probabilities for the occupation of different nondegenerate energy levels is a basic assumption in quantum-statistical mechanics. To a large extent its justification lies in the fact that the results of quantum-statistical mechanics are in accordance with experimental facts. We shall discuss this basic assumption in some detail in Section 5.13. At this point we may draw attention to the fact that its classical counterpart lies in the assumption that the a priori probability for a representative point to fall into a cell Z_i will be proportional to its size. In Section 3.4 we shall, indeed, see the connection between these two assumptions.

Having obtained an expression for W, we can use the same arguments as in Section 1.7: we find for the equilibrium value of N_i the expression

$$N_i = e^{\alpha - \beta\varepsilon_i}, \tag{3.102}$$

where we have again identified the most probable distribution with the equilibrium condition, and where we have imposed the constraints of constant total energy and a constant total number of particles,

$$\sum_i N_i = N, \qquad \sum_i N_i\varepsilon_i = E. \tag{3.103}$$

* We shall not consider here the complications of degeneracy or a continuous spectrum; we leave the generalization to the case where there is a continuous spectrum to the reader. The case of degenerate levels will be considered later on in this chapter. We refer to the quantum-mechanics textbooks mentioned in the Introduction (for example, A. S. Davydov, *Quantum Mechanics*, Oxford, 1965) for details about how one obtains the energy spectrum of a particle moving in a given potential energy.

The quantities α and β can be evaluated as functions of N and E by substituting Eq. (3.102) into Eqs. (3.103).

The average energy $\bar{\varepsilon}$ per particle can be obtained from the equation

$$\bar{\varepsilon} = \frac{\sum_i N_i \varepsilon_i}{\sum_i N_i} = \frac{\sum_i \varepsilon_i e^{\alpha - \beta\varepsilon_i}}{\sum_i e^{\alpha - \beta\varepsilon_i}} = -\frac{\partial \ln Z_\mu^{(\mathrm{qu})}}{\partial \beta}, \tag{3.104}$$

where the *partition function* $Z_\mu^{(\mathrm{qu})}$ is defined by the equation

$$Z_\mu^{(\mathrm{qu})} = K' \sum_i e^{\alpha - \beta\varepsilon_i}, \tag{3.105}$$

where K' is an as yet unspecified constant.

If we introduce a function F by the relation

$$e^{-\beta F} = Z_\mu^{(\mathrm{qu})}, \tag{3.106}$$

we can write Eq. (3.104) in the form (compare Eq. (2.713)):

$$\bar{\varepsilon} = \frac{\partial \beta F}{\partial \beta}, \tag{3.107}$$

and we see that, provided F is the free energy and β satisfies the equation

$$\beta = \frac{1}{kT}, \tag{3.108}$$

this equation is just the Gibbs-Helmholtz equation of thermodynamics. At this point we shall use this correspondence as a sufficient justification for the identification of F with the Helmholtz free energy and of β with $1/kT$; we shall discuss this point in more detail in the next chapter.

Since the Helmholtz free energy as a function of the temperature (or β) and of the external parameters, which enter into F through the dependence of the ε_i on them, completely determines the thermodynamic state of the system, we should be able to derive all thermodynamic quantities from F. Indeed, Eqs. (2.713) to (2.716) are again valid, and it follows that, once we have evaluated the partition function $Z_\mu^{(\mathrm{qu})}$, we can find the average energy, the specific heat, the generalized forces such as the magnetization or pressure, and so on.

3.2 THE HARMONIC OSCILLATOR

It is of historical interest to consider for a moment the old quantum theory or model theory. A discussion of the connection between the old quantum theory and statistical mechanics is particularly appropriate because it was considerations of a statistical nature that led Planck to introduce his quantum of action h, thus starting quantum theory.

We shall first of all consider a one-dimensional oscillator. It is defined by a quadratic potential energy function $U(q)$,

$$U(q) = \tfrac{1}{2}\alpha q^2, \tag{3.201}$$

where α is a constant and q the coordinate describing the position of the oscillator. If the mass of the oscillator is m, its frequency ω will be related to α by the equation

$$\omega = \sqrt{\frac{\alpha}{m}}. \tag{3.202}$$

Its total energy can then be written in the form

$$\varepsilon(p,q) = \frac{p^2}{2m} + \frac{1}{2}\, m\omega^2 q^2. \tag{3.203}$$

In this case, μ-space, which is now the p,q plane, has only two dimensions. We can draw curves of constant energy in this phase plane. These curves will be ellipses, a few of which are drawn in Fig. 3.1. As long as the oscillator is not disturbed from outside, it will move in such a way that its representative point will stay on one ellipse.

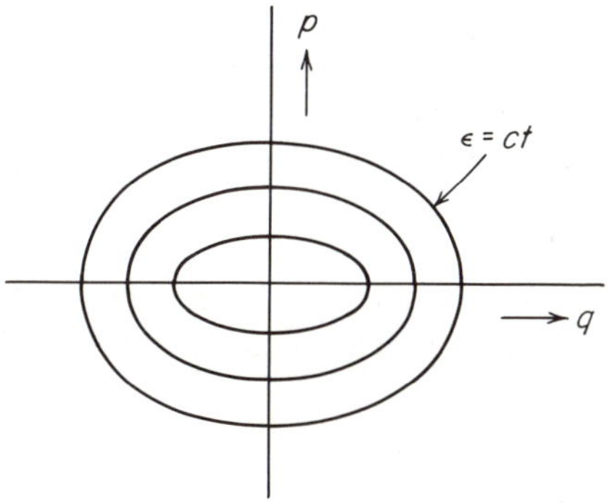

FIG. 3.1 Constant energy curves in the μ-space (the pq-plane) of a one-dimensional harmonic oscillator.

The area enclosed by such an ellipse is given by the following equation:*

$$I = \oint p\, dq, \tag{3.204}$$

where the integral extends over one period of the oscillator.

* This integral is sometimes called the *phase integral*. Since we have already used this term in a different connection, we shall not use it for I to avoid confusion.

Since p and q satisfy the equations

$$q = q_0 \sin \omega t, \qquad p = m\dot{q} = m\omega q_0 \cos \omega t, \tag{3.205}$$

we find for I

$$I = \int_0^{2\pi/\omega} p\dot{q}\, dt = \pi m \omega q_0^{\,2} = \frac{\varepsilon}{\nu}, \qquad \nu = \frac{\omega}{2\pi}. \tag{3.206}$$

Although in classical mechanics any ellipse in the phase plane could represent a state of the oscillator, this will no longer be true in quantum theory. In order to find out how to select the possible energy levels we shall first recall briefly how Planck was led to the introduction of his quantum of action.

As Kirchhoff had shown that the energy density in the radiation field in thermal equilibrium was independent of the material of the surrounding walls, Planck considered a system of one-dimensional harmonic oscillators in equilibrium with the radiation field. To find the thermodynamic properties of the radiation field, Planck considered its entropy and evaluated it using Boltzmann's formula (1.717); to do this he needed to find the probability function W. He then divided the phase plane into sections of area h, allotting to each cell a representative energy (compare the discussion at the end of Section 2.3). He originally intended to take the limit $h \to 0$, but found—rather to his surprise—that the radiation density calculated in that way corresponded to the experimentally observed one, provided h was not allowed to tend to zero, but was put equal to $6.6 \cdot 10^{-27}$ erg sec.

A few years later, in 1915, Sommerfeld and Wilson suggested that this division of the phase plane into cells of area h was the correct way to derive the possible values which the energy of a one-dimensional periodic system could assume. In fact, Sommerfeld's quantization rule is

$$I = nh, \tag{3.207}$$

where n is a nonnegative integer.

If one uses the quantization rule (3.207), the energy corresponding to the nth cell is determined by Eq. (3.106) and given by

$$\varepsilon_n = n\hbar\omega, \qquad n = 0, 1, 2, \cdots, \tag{3.208}$$

where $\hbar$ $(= h/2\pi)$ is Dirac's constant.

In this way we have allotted to each energy level a cell of equal area (equal to h) in μ-space. The energy corresponding to a cell is that corresponding to the inner boundary (ellipse) of the cell, if we call ε_n the $n + 1$-st level, n indicating not the actual number of the level, but the value of the corresponding *quantum number*.

In order to get a better agreement with the experimental evidence regarding the three-dimensional rotator (compare Eq. (3.504)), Sommerfeld later introduced instead of Eq. (3.207) the so-called *halbzahlige* quantization rule,

$$I = (n + \tfrac{1}{2})h. \tag{3.209}$$

For the energy levels we then get the same values as those following from the Schrödinger equation,

$$\varepsilon_n = (n + \tfrac{1}{2})\hbar\omega, \tag{3.210}$$

corresponding to a value inside the original cells. We see that the lowest energy available to a one-dimensional oscillator is not zero, but $\frac{1}{2}\hbar\omega$, the so-called *zero-point energy*.

That such a zero-point energy is not just a special way of choosing the energy scale, but a physical reality, can, for example, be seen in the case of liquid helium where the zero-point energy is sufficiently great to prevent the liquid from solidifying under its own vapor pressure.

Using Eq. (3.105), we can evaluate the partition function; the result is

$$Z_\mu^{(\mathrm{qu})} = K' \sum_{n=0}^{\infty} e^{-\beta\hbar\omega}[e^{-\beta\hbar\omega/2}], \tag{3.211}$$

where the factor between the square brackets occurs if we use Eq. (3.210), and does not occur if we use Eq. (3.108).

From Eq. (3.104) we get for the average energy per oscillator

$$\bar{\varepsilon} = \frac{\hbar\omega}{e^{\beta\hbar\omega} - 1}\left[+\frac{1}{2}\hbar\omega\right]. \tag{3.212}$$

Equation (3.212) can be written as a series expansion, both at high temperatures ($\beta\hbar\omega \ll 1$) and at low temperatures ($\beta\hbar\omega \gg 1$), as follows,

$$\hbar\omega \ll kT : \bar{\varepsilon} = kT - \frac{1}{2}\hbar\omega\left[+\frac{1}{2}\hbar\omega\right] + \text{series in } \frac{\hbar\omega}{kT}; \tag{3.213}$$

$$\hbar\omega \gg kT : \bar{\varepsilon} = [\tfrac{1}{2}\hbar\omega] + \hbar\omega \cdot e^{-\beta\hbar\omega} + \cdots. \tag{3.214}$$

Figure 3.2 shows the behavior of $\bar{\varepsilon}$ as a function of temperature. From the considerations of the previous chapter we might have expected $\bar{\varepsilon} = kT$ (Eq. (2.407) with $s = t = 1$). We see that this behavior is realized at high temperatures. This is not surprising, as we should expect classical behavior in the limit as $h \to 0$, and this means $\beta\hbar\omega \ll 1$, a condition also realized at high temperatures for the given, fixed value of h. We may remark here that it also gives us an example of the *correspondence principle*, enunciated by Bohr in 1918. This principle states that in the limit of large quantum numbers, the results obtained from quantum mechanics will be the same as those obtained from classical mechanics. That the requirement of large quantum numbers will lead in the present case to the same result as taking the limit $h \to 0$ can be seen from Eq. (3.208) where the condition $n \to \infty$ would entail $h \to 0$ for a given energy.

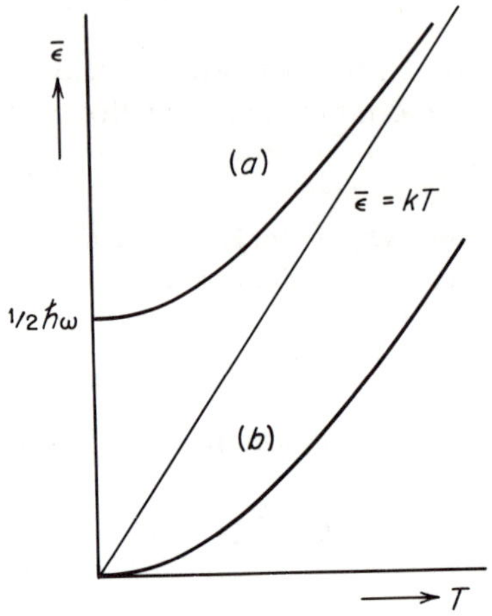

FIG. 3.2 The average energy of a one-dimensional harmonic oscillator: (a) with zero-point energy; (b) without zero-point energy.

We shall now consider the case of the two-dimensional oscillator. Its Hamiltonian is given by the equation

$$\varepsilon = \frac{1}{2m}(p_1^2 + p_2^2) + \frac{1}{2} m\omega_1^2 q_1^2 + \frac{1}{2} m\omega_2^2 q_2^2, \tag{3.215}$$

and the energy levels of the stationary states, which follow from the corresponding Schrödinger equation, are given by

$$\varepsilon_{n_1 n_2} = (n_1 + \tfrac{1}{2})\hbar\omega_1 + (n_2 + \tfrac{1}{2})\hbar\omega_2. \tag{3.216}$$

For the partition function we have from Eq. (3.105)

$$Z_\mu^{(\mathrm{qu})} = K' \sum_{n_1, n_2} \exp[-\beta\varepsilon_{n_1 n_2}], \tag{3.217}$$

which leads to the expression

$$Z_\mu^{(\mathrm{qu})} = K' e^{-\beta\hbar(\omega_1 + \omega_2)/2} [1 - e^{-\beta\hbar\omega_1}]^{-1} \cdot [1 - e^{-\beta\hbar\omega_2}]^{-1}. \tag{3.218}$$

For a three-dimensional oscillator the corresponding equations will be

$$\varepsilon = \frac{1}{2m}(p_1{}^2 + p_2{}^2 + p_3{}^2) + \frac{1}{2}m[\omega_1{}^2 q_1{}^2 + \omega_2{}^2 q_2{}^2 + \omega_3{}^2 q_3{}^2], \quad (3.219)$$

$$\varepsilon_{n_1 n_2 n_3} = (n_1 + \tfrac{1}{2})\hbar\omega_1 + (n_2 + \tfrac{1}{2})\hbar\omega_2 + (n_3 + \tfrac{1}{2})\hbar\omega_3, \quad (3.220)$$

$$Z_\mu{}^{(\mathrm{qu})} = K' \sum_{n_1 n_2 n_3} \exp[-\beta\varepsilon_{n_1 n_2 n_3}], \quad (3.221)$$

or

$$Z_\mu{}^{(\mathrm{qu})} = K' e^{-\beta\hbar(\omega_1+\omega_2+\omega_3)/2}[1 - e^{-\beta\hbar\omega_1}]^{-1}[1 - e^{-\beta\hbar\omega_2}]^{-1}[1 - e^{-\beta\hbar\omega_3}]^{-1}. \quad (3.222)$$

These formulas become interesting only if we consider isotropic oscillators. For the two-dimensional case, Eqs. (3.216) to (3.218) then become

$$\varepsilon_n = (n + 1)\hbar\omega, \qquad (\omega_1 = \omega_2 = \omega), \quad (3.223)$$

$$Z_\mu{}^{(\mathrm{qu})} = K' \sum_n (n + 1)e^{-\beta\varepsilon_n}, \quad (3.224)$$

or

$$Z_\mu{}^{(\mathrm{qu})} = K' e^{-\beta\hbar\omega}[1 - e^{-\beta\hbar\omega}]^{-2}. \quad (3.225)$$

Equation (3.224) follows from the fact that there are $n + 1$ different ways of writing n as the sum of two nonnegative integers n_1 and n_2. Equation (3.225) follows either from Eq. (3.218) by putting $\omega_1 = \omega_2 = \omega$ or directly from Eq. (3.224) by observing that

$$\sum_n n e^{-nx} = -\frac{d}{dx}\left[\sum_n e^{-nx}\right]. \quad (3.226)$$

In the three-dimensional case, we have

$$\varepsilon_n = (n + \tfrac{3}{2})\hbar\omega, \qquad (\omega_1 = \omega_2 = \omega_3 = \omega), \quad (3.227)$$

$$Z_\mu{}^{(\mathrm{qu})} - K' \sum_n \tfrac{1}{2}(n + 1)(n + 2)e^{-\beta\varepsilon_n}, \quad (3.228)$$

or

$$Z_\mu{}^{(\mathrm{qu})} = K' e^{-3\beta\hbar\omega/2}[1 - e^{-\beta\hbar\omega}]^{-3}. \quad (3.229)$$

In Eq. (3.228) we used the fact that the number of different ways of writing n as the sum of three nonnegative integers is $\frac{1}{2}(n + 1)(n + 2)$ (compare the derivation of Eq. (3.224)). Equation (3.229) then follows either from Eq. (3.222) or from Eq. (3.328) by using an equation similar to Eq. (3.226).

The easiest way to compute the number of ways of writing n as a sum of three nonnegative integers is by thinking of it as a division of n dots by 2 strokes (Fig. 3.3). Since there are $(n+2)!$ different ways of arranging these n dots and 2 strokes, and since a rearrangement of the dots or of the strokes does not interest us, the total number of different divisions will be equal to $(n+2)!/n!2! = \frac{1}{2}(n+1)(n+2)$.

As long as the oscillators are anisotropic, the situation is practically the same as in the case of the one-dimensional oscillator. For instance in the three-dimensional case, μ-space has six dimensions and each energy level corresponds to one cell with a (six-dimensional) volume h^3. We could think of obtaining these cells (and their corresponding energy levels) by applying the Sommerfeld rules for each of the three independent oscillations.

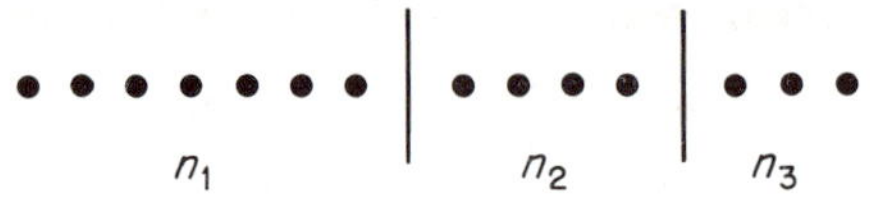

FIG. 3.3 Partition of an integer n into three nonnegative integers n_1, n_2, and n_3.

However, as soon as we turn to the isotropic case, the situation becomes different. In the three-dimensional case an energy level ε_n does not correspond to one cell but to $\frac{1}{2}(n+1)(n+2)$ cells. The Schrödinger equation in this case allows $\frac{1}{2}(n+1)(n+2)$ linearly independent solutions for the same energy level, and the energy level is said to be g-fold *degenerate* with $g = \frac{1}{2}(n+1)(n+2)$.

3.3 PLANCK'S RADIATION LAW

In this section we shall derive an expression for the energy density distribution in a radiation field. We shall assume with Planck and Rayleigh that the radiation field can be considered as consisting of one-dimensional oscillators with frequencies $\omega_1, \omega_2, \cdots$. If the first oscillator is in the state corresponding to quantum number n_1, the second one in that corresponding to n_2, and so on, the total energy of the system will be given by the equation

$$E = \sum_i \varepsilon_i = \sum_i n_i \hbar\omega_i \tag{3.301}$$

in which it will be noticed that we have omitted the zero-point energies*. The reason why this leads to a correct result is that one really should discuss

* For a discussion of the infinite zero-point self-energy of the electromagnetic field which would occur, if we took the zero-point energy into account see, for instance, H. A. Kramers, *Quantum Mechanics*, Amsterdam, 1957, Section 87.

the radiation field in terms of light quanta (compare the discussion in Section 4.1). From Eq. (3.212) we have for the average energy of the ith oscillator

$$\bar{\varepsilon}_i = \frac{\hbar\omega_i}{e^{\beta\hbar\omega_i} - 1}. \tag{3.302}$$

We must draw attention to one important difference between Eqs. (3.212) and (3.302). Equation (3.212) gives us the average energy when the average is taken over a large number of identical oscillators, that is, we get a system average. In the present case, however, we have only one oscillator with a frequency ω_i and expression (3.302) must be considered to give an average over a large period rather than over a large number, that is, we have a time average. That we may use the time average and the system average interchangeably follows from the general equivalence of these two kinds of average. This equivalence will be discussed in Section 5.7 and 5.14.

If the frequencies are lying very densely, the sum in Eq. (3.301) is virtually an integral. In order to evaluate this sum (or integral), we must know the number of frequencies $D(\omega)d\omega$ between ω and $\omega + d\omega$. This can be done in the following way.

In order to simplify the considerations, we assume that the volume V is a cube of edge length l. It can be proved that this simplification does not affect the final result for $D(\omega)$. Since we are dealing with standing waves,

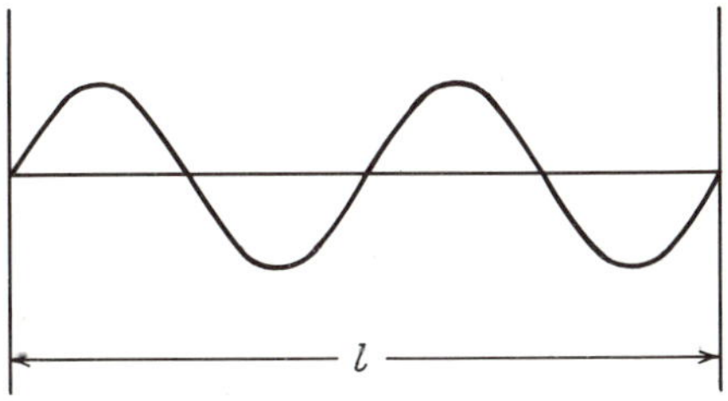

FIG. 3.4 The fitting of a standing wave in a length l ($n = 5$).

the wavelength λ of the waves in the cube must be such that the amplitude at the walls is zero. In the one-dimensional case (Fig. 3.4) this leads to the equation

$$l = \frac{1}{2} n\lambda = \frac{\pi n c}{\omega}, \tag{3.303}$$

where we have used for the velocity of light c the relation $2\pi c = \omega\lambda$.

The three-dimensional case is much more complicated. Essentially, one must fit standing waves into the cube so that they fit in the x, the y, and the z direction. This leads to the following requirement for the frequency ω:

$$\frac{\omega^2 l^2}{\pi^2 c^2} = n_1^{\,2} + n_2^{\,2} + n_3^{\,2}, \tag{3.304}$$

where n_1, n_2, and n_3 are positive integers.

To prove Eq. (3.304) we shall consider the solutions of the Maxwell equations for the case where the walls of the volume containing the radiation are metallic, so that the normal component of the electrical field must vanish at the boundaries. As the electrical field strength **E** must also satisfy the condition $(\nabla \cdot \mathbf{E}) = 0$ in empty space, we find for the components of **E** the expressions

$$\begin{aligned} E_x &= A_x \cos(n_1 \pi x/l) \sin(n_2 \pi y/l) \sin(n_3 \pi z/l) \sin(\omega t - \delta), \\ E_y &= A_y \sin(n_1 \pi x/l) \cos(n_2 \pi y/l) \sin(n_3 \pi z/l) \sin(\omega t - \delta), \\ E_z &= A_z \sin(n_1 \pi x/l) \sin(n_2 \pi y/l) \cos(n_3 \pi z/l) \sin(\omega t - \delta). \end{aligned} \tag{3.305}$$

Substituting expressions (3.305) into the wave equation, we find

$$\nabla^2 \mathbf{E} = \frac{1}{c^2} \frac{\partial^2 \mathbf{E}}{\partial t^2}, \tag{3.306}$$

Eq. (3.304) follows.

If we wish to know how many frequencies there are between ω and $\omega + d\omega$, we must calculate the number of lattice points with positive integer coordinates in a spherical shell of radius $\omega l/\pi c$ and thickness $l\, d\omega/\pi c$. We get then for $D(\omega)$ the equation

$$D(\omega)\, d\omega = 2 \cdot \frac{1}{8} \cdot \frac{4\pi \cdot \omega^2 l^2 \cdot l d\omega}{\pi^3 c^3} = \frac{\omega^2 V}{\pi^2 c^3}\, d\omega. \tag{3.307}$$

The factor 2 in the second member of Eq. (3.307) arises from the fact that the transverse light waves have two polarization degrees of freedom. The factor $\frac{1}{8}$ arises, since we are dealing with only that part of the spherical shell for which n_1, n_2, and n_3 are all positive.

The energy density per unit volume due to oscillators with frequencies between ω and $\omega + d\omega$ can now be obtained from Eqs. (3.301), (3.302), and (3.307), thus,

$$\rho(\omega)\, d\omega = \frac{\hbar}{\pi^2 c^3} \frac{\omega^3\, d\omega}{e^{\beta \hbar \omega} - 1}. \tag{3.308}$$

Equation (3.308) was derived by Planck in 1900 and is called *Planck's radiation law*. For high frequencies ($\beta\hbar\omega \gg 1$) it goes over into *Wien's law*,

$$\rho(\omega) = \frac{\hbar}{\pi^2 c^3}\,\omega^3 e^{-\beta\hbar\omega}, \tag{3.309}$$

while for low frequencies ($\beta\hbar\omega \ll 1$) it becomes *Rayleigh's law*,

$$\rho(\omega) = \frac{kT}{\pi^2 c^3}\,\omega^2, \tag{3.310}$$

the same result that Planck achieved by letting $h \to 0$. Since by letting h go to zero he could obviously not obtain Wien's law for high frequencies, Planck was obliged to let h remain finite.

For the total energy density per cm³ we get

$$\rho = \int_0^\infty \rho(\omega)\,d\omega = \frac{\hbar}{\pi^2 c^3}\int_0^\infty \frac{\omega^3\,d\omega}{e^{\beta\hbar\omega} - 1},$$

or

$$\rho = aT^4, \tag{3.311}$$

where a is given by the equation

$$a = \frac{\pi^2 k^4}{15\hbar^3 c^3} = 7.56 \cdot 10^{-15} \text{ erg degree}^{-4}\text{ cm}^{-3}. \tag{3.312}$$

In the calculations leading to Eq. (3.311) we must evaluate the integral

$$\int_0^\infty \frac{x^3\,dx}{e^x - 1} = \int_0^\infty e^{-x}x^3(1 + e^{-x} + e^{-2x} + \cdots)\,dx$$
$$= \sum_{n=1}^\infty \int_0^\infty e^{-nx}x^3\,dx = 6\zeta(4) = \frac{\pi^4}{15},$$

where

$$\zeta(s) = \sum_{n=1}^\infty n^{-s} = \text{Riemann's } \zeta\text{-function}.$$

Equation (3.311) is called the *Stefan-Boltzmann law*, which states that the total energy density in a radiation field in temperature equilibrium is proportional to the fourth power of the temperature.

Let us now consider briefly the case where we have an equilibrium between atoms and a radiation field. From the known quantum-mechanical transition probabilities we can then again derive the radiation law (3.308). We assume that there is only one kind of atom present that can be in stationary states with energies ε_k. According to Bohr's postulate (1913) the transition from a state k to a state l will be accompanied by the absorption or emission of a light quantum of frequency ω_{kl},

$$\omega_{kl} = \left| \frac{\varepsilon_k - \varepsilon_l}{\hbar} \right|, \tag{3.313}$$

where emission takes place when $\varepsilon_k > \varepsilon_1$, and absorption when $\varepsilon_k > \varepsilon_l$. If N_k denotes the number of atoms in the kth state, the number of atoms $N_{k \to l}$ making a transition from k to l per sec will be given by the equation

$$N_{k \to l} = N_k A_{kl} + N_k B_{kl} \rho(\omega_{kl}), \tag{3.314}$$

where we have assumed that $\varepsilon_k > \varepsilon_l$. The first term on the right-hand side of Eq. (3.213) corresponds to spontaneous transitions and will be proportional to N_k, the factor of N_k being denoted by A_{kl}. The second term corresponds to the induced emissions (or negative absorptions) and will be proportional both to N_k and to the density of the radiation field at the frequency concerned, the factor here being denoted by B_{kl}.

The number of atoms $N_{l \to k}$ making the transition $l \to k$ per sec is given by the equation

$$N_{l \to k} = N_l B_{lk} \rho(\omega_{kl}), \tag{3.315}$$

since in this case no spontaneous transition can take place.

The coefficients A_{kl} and B_{kl} were introduced by Einstein in 1917 and are called the *Einstein transition probabilities.* In the quantum theory of radiation it is proved that*

$$B_{lk} = B_{kl}, \tag{3.316}$$

and

$$A_{kl} = \frac{\hbar \omega^3}{\pi^2 c^3} B_{kl}. \tag{3.317}$$

We should perhaps mention here that the A_{kl} and B_{kl} are atomic quantities that do *not* depend on the temperature.

In equilibrium there will be as many transitions from k to l as form l to k (principle of detailed balancing; see Section 6.5), so that we have the equation

$$N_k[A_{kl} + B_{kl} \rho(\omega_{kl})] = N_l B_{lk} \rho(\omega_{kl}). \tag{3.318}$$

As N_k and N_l will satisfy Eq. (3.102), we have

$$N_k : N_l = e^{-\beta \varepsilon_k} : e^{-\beta \varepsilon_l}, \tag{3.319}$$

and, combining Eqs. (3.316) to (3.319), we can calculate the equilibrium density in the radiation field for which we again obtain Eq. (3.308).

* See, for example, H. A. Kramers, *Quantum Mechanics*, Amsterdam, 1957, Section 82; or A. S. Davydov, *Quantum Mechanics*, Oxford, 1965, Section 78.

3.4 THE TRANSITION TO CLASSICAL STATISTICS

The Schrödinger equation of a particle will give us the energy levels of the stationary states ε_n and their degree of degeneracy g_n where n stands for all the quantum numbers that determine the energy level. We can introduce the partition function $Z_\mu^{(\text{qu})}$ by the equation

$$Z_\mu^{(\text{qu})} = K' \sum_n g_n e^{-\beta\varepsilon_n}, \tag{3.401}$$

which comprises all expressions for $Z_\mu^{(\text{qu})}$ that we have met in this chapter up to now: Eqs. (3.105), (3.211), (3.217), (3.221), (3.224), and (3.228).

Each level corresponds to a volume $g_n h^s$ in μ-space, where s is the number of degrees of freedom of the particle. If we accept, as we did in Sections 1.7 and 2.3, that the a priori probability of finding the representative point of the particle inside a specified volume of μ-space is proportional to that volume, we see that nondegenerate energy levels all have equal a priori probabilities of being realized, but that the a priori probability of a degenerate level is weighted with a factor g_n. For this reason the degree of degeneracy g_n is often called the (*statistical*) *weight* of the level ε_n.

Let us now return to the partition function and let us, as in Section 3.1, introduce a function F by the equation

$$e^{-\beta F} = Z_\mu^{(\text{qu})}. \tag{3.402}$$

From Eqs. (3.102), (3.401), and (3.402) it follows that

$$\bar{\varepsilon} = \frac{\sum_n g_n e^{-\beta\varepsilon_n}\varepsilon_n}{\sum_n g_n e^{-\beta\varepsilon_n}} = -\frac{1}{Z_\mu^{(\text{qu})}}\frac{\partial Z_\mu^{(\text{qu})}}{\partial\beta} = \frac{\partial\beta F}{\partial\beta}, \tag{3.403}$$

and recognizing this as the Gibbs-Helmholtz equation, we see again that F is the (Helmholtz) free energy per particle. In the same way as was done in Section 2.7 we can now derive all thermodynamic quantities from F or $Z_\mu^{(\text{qu})}$.

In Section 2.4 we saw that often some degrees of freedom did not contribute to the average energy or the specific heat. In order to see how this can happen, let us consider the case of the anisotropic two-dimensional harmonic oscillator, and let us assume that ω_1 is much smaller than ω_2.

The partition function in this case is given by Eq. (3.218), and for the average energy we get from Eqs. (3.218) and (3.403)

$$\bar{\varepsilon} = \frac{1}{2}\hbar\omega_1 + \frac{1}{2}\hbar\omega_2 + \frac{\hbar\omega_1}{e^{\beta\hbar\omega_1}-1} + \frac{\hbar\omega_2}{e^{\beta\hbar\omega_2}-1}. \tag{3.404}$$

Let us now consider three temperature regions around T_1, T_2, and T_3 where

$$kT_1 \ll \hbar\omega_1 \ll kT_2 \ll \hbar\omega_2 \ll kT_3. \tag{3.405}$$

For $\bar{\varepsilon}$ we get, in these three regions,

$$T \sim T_1 : \bar{\varepsilon} = \tfrac{1}{2}\hbar\omega_1 + \tfrac{1}{2}\hbar\omega_2 + \hbar\omega_1 e^{-\beta\hbar\omega_1} + \cdots, \tag{3.406}$$

$$T \sim T_2 : \bar{\varepsilon} = \frac{1}{2}\hbar\omega_2 + kT + \hbar\omega_2 e^{-\beta\hbar\omega_2} + \frac{(\hbar\omega_1)^2}{12kT} + \cdots, \tag{3.407}$$

$$T \sim T_3 : \bar{\varepsilon} = 2kT + \frac{\hbar^2\omega_1{}^2 + \hbar^2\omega_2{}^2}{12kT} + \cdots. \tag{3.408}$$

The specific heat follows from the equation

$$c_v = \frac{\partial\bar{\varepsilon}}{\partial T}, \tag{3.409}$$

or

$$T \sim T_1 : c_v = k[(\beta\hbar\omega_1)^2 e^{-\beta\hbar\omega_1} + \cdots], \tag{3.410}$$

$$T \sim T_2 : c_v = k[1 - \tfrac{1}{12}(\beta\hbar\omega_1)^2 + (\beta\hbar\omega_2)^2 e^{-\beta\hbar\omega_2} + \cdots], \tag{3.411}$$

$$T \sim T_3 : c_v = k[2 - \tfrac{1}{12}(\beta\hbar\omega_2)^2 + \cdots]. \tag{3.412}$$

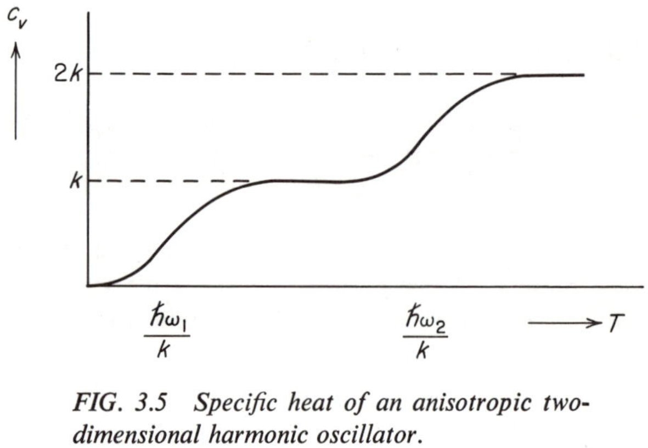

FIG. 3.5 Specific heat of an anisotropic two-dimensional harmonic oscillator.

Figure 3.5 shows the behavior of c_v as a function of the temperature. From this figure we see the following:

(a) If the temperature is sufficiently high, the specific heat will reach its classical value $2k$ (compare Eq. (2.408) with $s = t = 2$).

(b) At temperatures that are large compared to $\hbar\omega_1/k$, but small compared to $\hbar\omega_2/k$, the specific heat is only k. It is as if one degree of freedom were frozen-in.

(c) Finally, at very low temperatures the specific heat is practically equal to zero.

That the classical behavior follows at sufficiently high temperatures is again an instance of the correspondence principle. Indeed, from Eqs. (3.408), (3.407), and (3.216) it follows that if $kT \gg \hbar\omega_2$, then both $\bar{n}_1$ and $\bar{n}_2$ are large compared to unity; while if $\hbar\omega_2 \gg kT \gg \hbar\omega_1$, then, although $\bar{n}_1$ is large as compared to unity, $\bar{n}_2$ is small and only the first degree of freedom will give the classical contribution k.

The problem with which we are thus left is to explain why at low temperatures the specific heat is so much smaller than its classical value. This can best be done by considering the transition from the quantum-mechanical formulas to the classical ones.

We can start with the partition function

$$Z_\mu^{(\text{qu})} = K' \sum_n g_n e^{-\beta\varepsilon_n}. \tag{3.413}$$

The sum in Eq. (3.413) must be replaced by an integral over μ-space if we wish to go over to the classical case. Since each nondegenerate energy level corresponds to a volume h^s in μ-space, we should expect

$$\sum_n g_n \to \int \frac{d\omega}{h^s}. \tag{3.414}$$

We do not have to worry about degeneracy. A degenerate level corresponds to a volume $g \cdot h^s$ so that g disappears in the final result. We see that by making the transition in this way we get, indeed, the classical formula (2.718) for the partition function, and we have now a definite relation between K in (2.718) and K' in Eq. (3.413).

According to the old quantum statistics one can justify the transition (3.414) rather easily as long as one is dealing with multiple periodic systems. One introduces the phase integrals

$$J_k = \oint p_k \, dq_k,$$

where the p_k and q_k are the generalized momenta and coordinates and where the integration is extended over one complete period. It is possible to perform a canonical transformation from the p_k and q_k to the J_k and their conjugate coordinates w_k.* The w_k measure the phase during the periodical motion and

* See J. M. Burgers, *Het Atoommodel van Rutherford-Bohr*, Thesis, Leiden, 1918.

change by 1 over one period. Because the transformation is canonical, we have

$$\prod_k dp_k \, dq_k = \prod_k dJ_k \, dw_k.$$

The volume Ω of phase space corresponding to J_k values between $J_{k'}$ and $J_{k'}$ and w_k between $w_{k'}$ and $w_{k'} + 1$ is given by the equation

$$\Omega = \prod_k (J_{k''} - J_{k'}).$$

If we are in the region of high quantum numbers, we have $J_{k''} = n_{k''}h$, $J_{k'} = n_{k'}h$ according to the Sommerfeld rules. The volume is thus equal to

$$\Omega = h^s \prod_k (n_{k''} - n_{k'}).$$

The number of stationary states in the volume under consideration is clearly

$$\prod_k (n_{k''} - n_{k'}).$$

We see thus that each stationary state corresponds to a volume h^s in μ-space. We do not wish to discuss here the importance of the theorem of the invariance of the statistical weights, but refer the reader to the literature.*

The situation becomes more complicated when we consider modern quantum theory. In many textbooks on statistical mechanics one finds the statement that the transition (3.414) is connected with the Heisenberg relations, which are then written in the form

$$\Delta p \cdot \Delta q \geqslant h.$$

It can readily be seen that this statement can reveal only part of the true state of affairs, since the Heisenberg relations in their most rigorous form read

$$\Delta p \cdot \Delta q \geqslant \tfrac{1}{2}\hbar.$$

One is thus led to ask why the conversion factor should be h^s and not, for instance, $(\frac{1}{2}\hbar)^s$.

It seems to us, however, that one can justify the factor h^s in the following way. In classical mechanics the phase of the particle is given by the representative point in μ-space. One can consider this point in μ-space to be the combination of a point in coordinate- or q-space and a point in momentum- or p-space. In quantum mechanics, however, one must consider a probability density both in q-and in p-space. These two probability densities are not independent. The density in q-space is determined by the wave function ψ, and is in fact $|\psi|^2$. If one transforms the wave function from a coordinate representation to a momentum representation,† one obtains the probability amplitude in p-space, A. The probability density in p-space is then given by $|A|^2$.

* See, for instance, P. Ehrenfest, *Collected Scientific Papers*, Amsterdam, 1959.

† For a more thorough discussion of the details of this transformation and proofs of the subsequent statements, we refer to W. Heisenberg, *The Physical Principles of Quantum Theory*, Chicago, 1932.

From the transformation theory it follows that A and ψ are Fourier transforms of one another, if we consider q and p/h as the variables. (The h enters here through the commutation expressions of which the Heisenberg relations are also a consequence.) Since ψ must be normalized to represent the actual state of a particle,

$$\int |\psi|^2 dq = 1,$$

it follows from the theory of Fourier transforms* that

$$\int |A|^2 dp = h^s.$$

The total volume in phase space corresponding to one stationary state is now obtained by multiplying the probability density in q-space with the probability density in p-space and integrating over the whole of μ-space. The result is then h^s†.

The next question is, "Under what conditions is the approximation of the sum in Eq. (3.413) by an integral a good one?" The condition for this is that the successive terms in the sum do not differ appreciably or, in other words, that

$$\beta(\varepsilon_n - \varepsilon_{n-1}) \ll 1. \tag{3.415}$$

Condition (3.415) will certainly be realized in the limit where $h \to 0$, since ε_n and ε_{n-1} are the representative energies of neighboring cells in μ-space and will become the energies of neighboring representative points after $h \to 0$.

We also see that the higher the temperature the greater the chance that condition (3.415) will be satisfied. In general, then, we may expect that the higher the temperature the closer the approach to the classical formulas.

If, however, condition (3.415) is not fulfilled, we are not allowed to substitute an integral for the sum. Let us assume for the sake of simplicity that $\varepsilon_0 = 0$ (no zero-point energy) and that all levels are nondegenerate. The partition function can then be written in the form

$$Z_\mu^{(\mathrm{qu})} = K'[1 + e^{-\beta\varepsilon_1} + e^{-\beta\varepsilon_2} + \cdots]. \tag{3.416}$$

Since the terms in Eq. (3.416) are decreasing, we get to a first approximation and, assuming that $\beta\varepsilon_1 \gg 1$, $\beta(\varepsilon_2 - \varepsilon_1) \gg 1, \cdots$,

$$Z_\mu^{(\mathrm{qu})} \approx K'[1 + e^{-\beta\varepsilon_1}], \tag{3.417}$$

and for the average energy and specific heat we get from Eqs. (3.403), (3.409),

* The functions ψ and A are functions of $q_1, \cdots, q_s$ and $p_1, \cdots, p_s$, respectively. We write abbreviatedly only one p or q. Also dq and dp are abbreviated notations for volume elements in q- and p-space.

† Similar considerations are to be found in P. A. M. Dirac, *The Principles of Quantum Mechanics*, Oxford, 1935, Section 37. See also J. E. Moyal, *Proc. Cambridge Phil. Soc.* **45**, 99, 1949.

and (3.417)

$$\bar{\varepsilon} \approx \varepsilon_1 e^{-\beta\varepsilon_1}, \tag{3.418}$$

and

$$c_v \approx k(\beta\varepsilon_1)^2 e^{-\beta\varepsilon_1}. \tag{3.419}$$

Since according to our assumptions $\beta\varepsilon_1 \gg 1$, both $\bar{\varepsilon}$ and c_v will be very small. We may therefore make the general statement that *a degree of freedom for which* $\beta(\varepsilon_1 - \varepsilon_0) \gg 1$ *will not make any appreciable contribution to the specific heat.** This explains why the specific heat of diatomic gases is $\frac{5}{2}k$ per molecule. The energies corresponding to vibrations along the axis of the molecule are of the order of magnitude of electron volts, that is, they correspond to temperatures of the order 10,000° K. The degree of freedom that such vibrations represent is thus completely frozen-in and will not contribute a term k to the specific heat.

3.5 THE RIGID ROTATOR: THE HYDROGEN MOLECULE

Let us consider a dumbbell molecule. At room temperatures the electronic degrees of freedom and the degree of freedom corresponding to a vibration along the axis of the molecule are frozen-in and the molecule may be regarded as a *rigid rotator*, that is, a particle that has two rotational degrees of freedom around its center of gravity. The three degrees of freedom relating to the movement of the center of mass may be treated classically and will contribute $\frac{3}{2}kT$ to the average energy and $\frac{3}{2}k$ to the specific heat.

The energy levels of a point particle in a cubical box of edge length a are given by the equation (compare Eq. (4.601))

$$\varepsilon = \varepsilon_0(k^2 + l^2 + m^2) \quad \text{where} \quad \varepsilon_0 = \frac{h^2}{8ma^2} \tag{3.501}$$

Since $\varepsilon_0/k \approx 10^{-15}$ °K, if $m = 10^{-23}$ g, and $a = 1$ cm, we see that we are always in the classical region.

If we consider for the moment only the rotational part of the energy, we have (compare Eq. (2.414)),

$$\varepsilon = \frac{p_\theta{}^2}{2A} + \frac{p_\varphi{}^2}{2A\sin^2\theta}, \tag{3.502}$$

* The statement in this form presupposes that the total energy of the particle is built up *additively* from the energies corresponding to the different degrees of freedom. That this situation is often realized follows from a consideration of the Schrödinger equation in the case where the classical energy is built up in this way.

where A is the moment of inertia of the molecule with respect to an axis through the center of mass, perpendicular to the axis of the molecule, and where θ and φ are two angles determining the direction of the axis of the molecule. For θ we can take the angle between the axis of the molecule and the z axis and for φ the angle between the x axis and the projection of the molecular axis on the x,y-plane. If m_1 and m_2 are the masses of the two atoms out of which the molecule is built up and if r_0 is their equilibrium distance apart, we have for A

$$A = \frac{m_1 m_2}{m_1 + m_2} r_0^2. \tag{3.503}$$

From the Schrödinger equation it follows that the energy levels of the rotator and their weights are given by the equations

$$\varepsilon_j = \frac{\hbar^2}{2A} j(j+1), \qquad g_j = 2j+1, \qquad j = 0, 1, 2, \ldots, \tag{3.504}$$

where j is the rotational quantum number.

In order to determine the temperature above which classical behavior may be expected, we introduce the "rotational" temperature $\Theta = \hbar^2/2Ak$. Expressing both m_1 and m_2 in units of proton mass and r_0 in Ångströms, we find for this rotational temperature:

$$\Theta = 24.2 \frac{m_1 + m_2}{m_1 m_2 r_0^2} \,{}^\circ\mathrm{K}. \tag{3.505}$$

From Eqs. (3.504) and (3.415) it follows that classical behavior may be expected when $T \gg \Theta$. From Eq. (3.505) it follows that above room temperatures the classical formulas will probably be satisfactory, but at lower temperatures the quantum-mechanical formulas must be used.

The partition function is in this case

$$Z_\mu^{(\mathrm{qu})} = K' \sum_{j=0}^{\infty} (2j+1) e^{-j(j+1)\beta k \Theta}. \tag{3.506}$$

At high temperatures ($T \gg \Theta$) the sum can be replaced by an integral,

$$Z_\mu^{(\mathrm{qu})} \approx K' \int_0^\infty (2x+1) e^{-\beta k \Theta x(x+1)}\, dx = K' \frac{T}{\Theta} \tag{3.507}$$

and at low temperatures ($T \ll \Theta$) we get a power series,

$$Z_\mu^{(\mathrm{qu})} = K'[1 + 3e^{-2\beta k\Theta} + 5e^{-6\beta k\Theta} + \cdots]. \tag{3.508}$$

For the average rotational energy we get from Eqs. (3.507) and (3.508)

$$\text{if } T \gg \Theta, \qquad \bar{\varepsilon} \approx kT, \tag{3.509}$$

$$\text{and if } T \ll \Theta, \qquad \bar{\varepsilon} \approx k\Theta[6e^{-2\beta k\Theta} - 18e^{-4\beta k\Theta} + \cdots]. \tag{3.510}$$

Finally, for the rotational specific heat we get

$$\text{if } T \gg \Theta, \qquad c_v \approx k; \tag{3.511}$$

$$\text{and if } T \ll \Theta, \qquad c_v = k(\beta k\Theta)^2[12e^{-2\beta k\Theta} - 72e^{-4\beta k\Theta} + \cdots]. \tag{3.512}$$

Equations (3.509) and (3.511) correspond to the classical expressions, since there is no potential energy and there are two degrees of freedom.

We shall now consider the case of the hydrogen molecule. At room temperatures and below, the vibrational degrees of freedom are frozen-in, and if we forget about the translational degrees of freedom, which contribute the usual $\frac{3}{2}k$ term to the specific heat, we can regard this molecule as a rigid rotator.

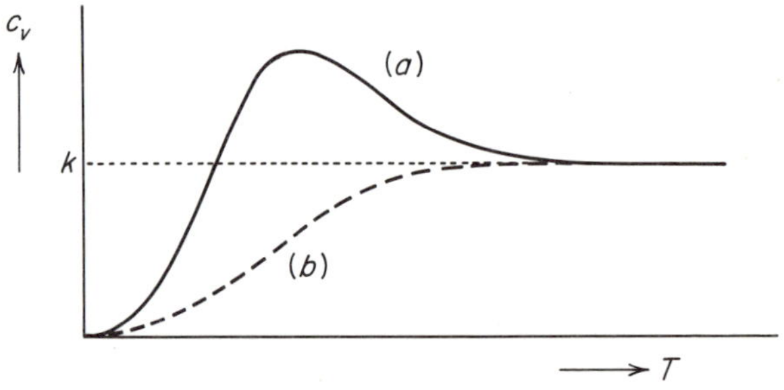

FIG. 3.6 The specific heat (a) of a rigid rotator according to Eq. (3.511) and (3.512), and (b) of hydrogen (experimentally).

On curve b of Fig. 3.6 we have drawn the specific heat of molecular hydrogen as determined experimentally by Eucken, and we see that there is a large discrepancy between this experimental curve and the theoretical curve a calculated from Eqs. (3.511) and (3.512). The difference between the calculated and observed curves cannot be due to the fact that we have neglected the vibrational and electronic energies. First, these energies are far too large to have an effect at temperatures where the difference occurs and, secondly, if they were important, the observed curve should lie higher than the calculated one and not lower.

However, we have neglected two other factors. First of all, we have neglected the fact that both hydrogen nuclei have spin $\frac{1}{2}$.* Secondly, we have neglected the fact that the hydrogen molecule is built up out of two identical atoms. Since both the hydrogen nuclei have spin $\frac{1}{2}$, the resultant nuclear spin S of the molecule is 0 (antiparallel spins) or 1 (parallel spins).

* The value of the spin is always expressed in units of $\hbar$.

The weight of a state with spin S is $2S+1$ corresponding to the $2S+1$ possibilities of orientation of the total spin.* The state with $S=0$, which is called the *para*-state, has thus a weight $g_S{}^p = 1$, while the *ortho*-state with $S=1$ has a weight $g_S{}^o = 3$. To a very good approximation the energy of the molecule will be independent of the value of S and the spin weight factors therefore have no effect on the average energy if the two atoms are not identical. If they are not identical, the partition function is multiplied by a constant factor $(2s_1+1)(2s_2+1)(=4=g_S{}^p+g_S{}^o$ in our case), where s_1 and s_2 are the spins of the two atoms in the molecule. However, when the two atoms are identical, the Pauli principle intervenes.

This principle states that the final total wave function of the hydrogen molecule should be antisymmetrical in the two atoms. The vibrational and electronic parts of the wave function are symmetrical in the ground state, which is the only state of the electronic and vibrational parts that has to be considered The spin function corresponding to para-states is antisymmetrical, while the spin functions corresponding to ortho-states are symmetrical. The rotational wave functions are symmetrical or antisymmetrical according as j is even or odd. Since the total wave function has to be antisymmetrical, we see that we have the following combinations:

$$\left.\begin{array}{lll} j=0,2,4,\cdots, & S=0, & g_S{}^p=1, \\ j=1,3,5,\ldots, & S=1, & g_S{}^o=3, \end{array}\right\} \tag{3.513}$$

and the partition function is not given by Eq. (3.506) but by the equation

$$Z_\mu{}^{(\text{qu})} = Z_p + 3Z_o, \tag{3.514}$$

where

$$Z_p = K' \sum_{j=\text{even}} (2j+1)e^{-\beta k\Theta j(j+1)} \tag{3.515}$$

and

$$Z_o = K' \sum_{j=\text{odd}} (2j+1)e^{-\beta k\Theta j(j+1)}. \tag{3.516}$$

Using expression (3.514) for the partition function, one can again calculate the specific heat; however, once more the calculated curve does not agree with the experimental one.

The final solution to this problem was given by Dennison in 1927. In order to understand it, we must first calculate the ratio, r, of the total number of ortho- to the total number of parahydrogen molecules at room temperature. To do this we use Eq. (3.102) for the number of particles in a given energy

* More precisely, there are $2S+1$ linearly independent spin functions corresponding to the same total spin S.

3.102 $N_i = e^{\alpha-\beta\varepsilon_i}$ particles in ε_i

$\sum N_i = N$

level, and, remembering that that expression has to be multiplied by the degree of degeneracy, we get

$$r = \frac{N_{\text{ortho}}}{N_{\text{para}}} = \frac{\sum_{j=\text{odd}} 3(2j+1)e^{-\beta k\Theta j(j+1)}}{\sum_{j=\text{even}} (2j+1)e^{-\beta k\Theta j(j+1)}}. \tag{3.517}$$

From Eq. (3.517) we see, first of all, that as $T \to 0$, r tends to zero and, secondly, that as soon as T is well above Θ, so that the sums can be replaced by integrals, r becomes equal to 3, as the two sums lead to the same integrals, apart from the spin degeneracy factor. At room temperature, then, as this is well above Θ (compare Eq. (3.505)), there will be three times as many orthohydrogen as parahydrogen molecules present in the gas. Now if the temperature is lowered, one should expect the ratio to decrease in the appropriate manner according to Eq. (3.517). However, one runs into the following difficulty. Since the transition from ortho- to parahydrogen involves the turning-over of the spin of one of the atoms, the transition probability will be small and one cannot expect equilibrium with regard to the ortho-para ratio to have been attained in a short period. Indeed, the periods concerned are of the order of years, even at relatively high temperatures such as room temperature. If one therefore performs a measurement of the specific heat of hydrogen at low temperatures, one is dealing with a mixture of ortho- and parahydrogen in the ratio of 3 to 1, and one should expect for the specific heat

$$c_v = 3c_{vo} + c_{vp}, \tag{3.518}$$

where c_{vo} and c_{vp} are the specific heats calculated from the ortho- and para-partition functions (3.516) and (3.515). The specific heat curve calculated in this way agrees well with the observed specific heat curve.

Further evidence that Dennison's explanation is the correct one can be obtained by performing experiments with ortho-para mixtures where r has a value different from 3. One can speed up the ortho-para conversion by passing the hydrogen over activated charcoal. By doing this at various temperatures and afterward removing the catalyst, one can fix r at any value between 0 and 3. The specific heat will then follow a curve obtained by mixing c_{vo} and c_{vp} with appropriate weights. If one measures c_v in such a way that r has at every temperature the value corresponding to Eq. (3.517), it will follow the curve calculated from the expression for the partition function given in Eq. (3.514).

Analogous considerations apply to the case of heavy hydrogen D_2. In the case of HD there are no complications, since the two constituent atoms are different and the specific heat can be calculated from expression (3.506) for the partition function.

PROBLEMS

1. Derive Eq. (3.229) directly from Eq. (3.228).
2. Prove Eq. (3.305).
3. Use Eqs. (3.319) and (3.308) to prove the relations (3.316) and (3.317) between the Einstein transition probabilities.
4. Prove the following relation between the entropy S and the energy E of the radiation field:

$$S = \frac{4E}{3T}.$$

5. Consider radiation in a dispersive medium, in which the refractive index $n(\omega)$ depends on the frequency ω. Derive an expression for the radiation energy density in such a medium.
6. Consider a Debye solid. The Debye solid is a model in which the eigenvibrations of the N atoms in a solid are replaced by the elastic vibrations of an isotropic continuum. Moreover, it is assumed that we need only consider the $3N$ lowest eigenfrequencies. Of these $3N$ eigenvibrations, $2N$ will be transverse and N longitudinal vibrations. If, as is the case in an isotropic elastic continuum, the velocity of the waves is independent of the frequency, prove that the number of vibrational modes $g(\omega)\,d\omega$ with frequencies between ω and $\omega + d\omega$ is given by (compare Eq. (3.307))

$$g(\omega) = \frac{3V}{2\pi^2}\frac{\omega^2}{s^3} = \frac{9N}{{\omega_m}^3}\,\omega^2, \qquad \omega < \omega_m; \tag{A}$$

$$= 0 \qquad\qquad\qquad , \qquad \omega > \omega_m; \tag{B}$$

here s is the velocity of the waves and ω_m is the cut-off frequency, defined by the relation

$$\int_0^{\omega_m} g(\omega)\, d\omega = 3N. \tag{C}$$

Discuss the change to be made in Eq. (A) when the velocities s_{trans} and s_{long} of the transverse and longitudinal waves are different, but Eqs. (B) and (C) are retained.
Use the above expression for $g(\omega)$ to evaluate the (Debye) specific heat of a solid, assuming that the average energy of an eigenvibration with frequency ω is again given by Eq. (3.212).
Discuss the temperature dependence of the specific heat at low temperatures ($T \ll \Theta$) and high temperatures ($T \gg \Theta$), where the *Debye tempera-*

ture Θ is defined by the relation

$$k\Theta = \hbar\omega_m. \tag{D}$$

7. Discuss the specific heats of two-dimensional and one-dimensional Debye solids.
8. Figure 3.7 shows the specific heat curve of a real solid. The horizontal line represents the classical limit of $6N$ (Dulong-Petit law; see Problem 6).

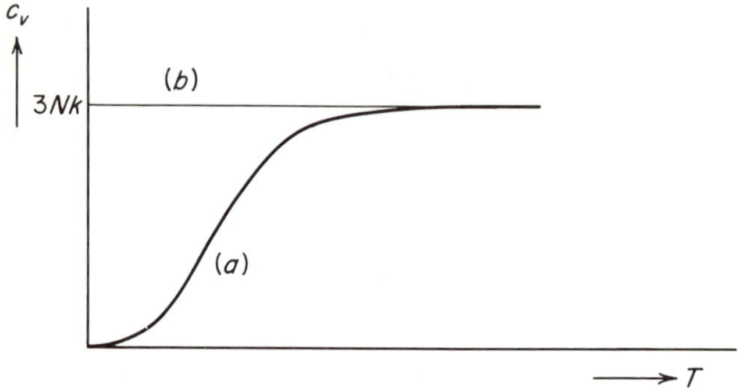

FIG. 3.7 Specific heat of a solid: (a) actual curve; (b) curve corresponding to the classical Dulong-Petit law.

Prove that the area between the specific heat curve and the horizontal line equals the zero-point energy of the solid, by expressing the specific heat of the solid as the sum of the specific heats of the $3N$ normal modes, each of which has an energy given by Eq. (3.212). Note that no special assumptions need be made about the shape of $g(\omega)$.

9. Assume that instead of a dispersion relation (relation between frequency ω and wave vector $\mathbf{q}$) $\omega = sq$ ($q = |\mathbf{q}|$) which is valid for the Debye solid, we have a dispersion relation

$$\omega = Cq^n, \qquad C = \text{const.}$$

Find the low-temperature behavior of the specific heat of a solid with such a dispersion relation.

10. Sometimes the energy levels of a quantum-mechanical system will be such that the first excited level lies close to the lowest level, while all other levels are far away. In that case, we can at most temperatures neglect all levels except the lowest and the first excited one. Find an expression for the specific heat of a system with two energy levels only. Find the value of the temperature for which the specific heat has a maximum, and discuss

the physical reasons for the shape of the specific heat curve, which represents the so-called *Schottky anomaly*.

11. Consider a system with three energy levels 0, ε, and 2ε, of which the middle one is doubly degenerate, while the other two are nondegenerate. Derive an expression for the specific heat of this system.

12. Consider a system of independent spin-$\frac{1}{2}$ point particles that have a magnetic moment $\boldsymbol{\mu}$. Calculate the entropy and magnetic susceptibility of this system at temperature T_1 in a magnetic field $\mathbf{B}_1$. What is the final temperature of the system if the magnetic field is now brought to zero adiabatically?

13. An atom with angular momentum J (in units $\hbar$) has an energy in a magnetic field $\mathbf{B}$ equal to $g\mu_B MB$, where M is the magnetic quantum number which can be equal to $-J$, $-J+1, \cdots, J-1, J$ and μ_B is the Bohr magneton. Prove that the magnetic moment $\mathcal{M}$ of a system of N such atoms, assumed to be noninteracting with one another, is equal to $\mathcal{M} = Ng\mu_B J \cdot B_J(x)$, where $B_J(x)$ is the so-called *Brillouin function* and $x = \beta g\mu_B JB$. The Brillouin function is given by the equation

$$B_J(x) = \frac{2J+1}{2J}\coth\left(\frac{2J+1}{2J}x\right) - \frac{1}{2J}\coth\frac{x}{2J}.$$

Figure 3.8 shows $B_J(x)$ as the function of x for $J = \frac{1}{2}, 1, \frac{5}{2}$, and ∞. In the last case, one also puts in the expression for $\mathcal{M}$: $\mu_B \to 0$, while $g\mu_B J \to \mu_0$. Consider the high-temperature ($\beta g\mu_B JB \ll 1$) limit of the magnetic susceptibility.

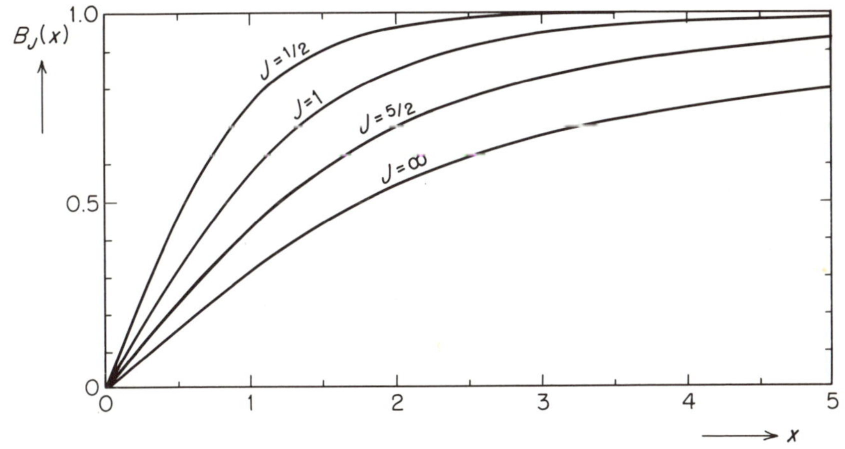

FIG. 3.8 The Brillouin function $B_J(x)$ for $J = \frac{1}{2}, 1, \frac{5}{2}$, and ∞.

14. A crystal contains ions in a state $J = 1$. An internal electric field partly lifts the degeneracy of this state, leaving a doublet lowest and a singlet higher by an energy D. When a small magnetic field $\mathbf{B}$ is applied perpendicularly to the crystal axis, the energies of these states become: $0, -(g\mu_B B)^2/D$, and $D + (g\mu_B B)^2/D$, where we assume that $g\mu_B B \ll D$. Find the partition function for such an ion. Find the magnetic moment $\mathcal{M}$ of a system of N such ions, assuming them to be noninteracting. Find both the high temperature ($\beta D \ll 1$) and the low temperature ($\beta D \gg 1$) behavior of the susceptibility.

15. Discuss the specific heat of a gas of noninteracting heteronuclear diatomic molecules, stating clearly what approximations are made.
Consider the case of nitric oxide NO. In this case the rotational temperature given by Eq. (3.505) is equal to 2.5° K, the vibrational temperature defined by $\hbar\omega/k$, were ω is the frequency of the vibrations in the direction of the molecular axis that are assumed to be harmonic, is equal to 2700° K. At the same time, the electronic energy spectrum is such that apart from high-lying levels, which we can neglect, there is one excited level at a distance δ from the ground state level, and $\delta/k = 45°$ K. Assume that the rotational and vibrational properties of the molecules are unaffected by the electronic state; give a rough sketch of the specific heat curve in its dependence on temperature.

16. Find an expansion in powers of Θ/T (Θ: the rotational temperature) of the rotational partition function (3.507) and hence estimate how accurately Eq. (3.511) gives the specific heat at $T/\Theta = 0.1$. To find the expansion, use the Euler-MacLaurin summation formula:*

$$\sum_{n=0}^{\infty} f(n) = \int_0^{\infty} f(x)\,dx + \frac{1}{2}f(0) - \frac{1}{12}f'(0) + \frac{1}{720}f'''(0) - \frac{1}{30240}f'''''(0) + \cdots$$

17. Discuss the specific heat of a gas of noninteracting HD and D_2 molecules.

18. To a very high degree of approximation the energy levels of the Morse potential (1.301), which is a reasonable approximation for the vibrational potential energy, are given by the equation

$$\varepsilon_n = \hbar\omega(n + \tfrac{1}{2})[1 - \gamma(n + \tfrac{1}{2})] - D,$$

where $\gamma = a^2\hbar/2\mu\omega$ where μ is the reduced mass.
For most diatomic molecules γ is a small quantity. Calculate the effect of the term with γ in the expression for the energy levels, that is, the effect of the anharmonicity of the vibrations upon the vibrational specific heat, retaining only the first-order term in γ.

* See, for example, J. E. and M. G. Mayer, *Statistical Mechanics*, *New York*, 1940, p. 431.

BIBLIOGRAPHICAL NOTES

General references for this chapter are

1. P. Jordan, *Statistische Mechanik auf Quantentheoretischer Grundlage*, Braunschweig, 1933.
2. M. Planck, *Theorie der Wärme*, Leipzig, 1930; *Theory of Heat*, London, 1932.

Section 3.1. The story of the introduction of h is beautifully told by Planck himself in

3. M. Planck, *Naturwiss.*, **31,** 153, 1943.

For an account of the old quantum theory we may refer to

4. A. Sommerfeld, *Atombau und Spektrallinien*, Vol. I, Braunschweig, 1919; *Atomic Structure and Spectral Lines*, London, 1924.
5. D. ter Haar, *The Old Quantum Theory*, Oxford, 1966.

In Reference 5 some of the papers (References 16 and 22) mentioned in these Bibliographical Notes are reprinted.

The Sommerfeld quantization rules were independently proposed by Sommerfeld and Wilson:

6. A. Sommerfeld, *Münchener Ber.*, **1915,** 425, 459; **1916,** 131.
7. W. Wilson, *Phil. Mag.*, **29,** 795, 1915.

The "halbzahlige" quantization was introduced in 1920:

8. A. Sommerfeld, *Ann. Physik*, **63,** 221, 1920.

The correspondence principle was first clearly stated by Bohr in 1918:

9. N. Bohr, *Trans. Dan. Acad. Sci.*, **(8)4,** nr 1, 1918.

A discussion of the reasons why and when the Sommerfeld rules are exactly correct is usually given by the W-K-B or semiclassical method:

10. G. Wentzel, *Z. Physik*, **38,** 518, 1926.
11. H. A. Kramers, *Z. Physik*, **39,** 828, 1926.
12. L. Brillouin, *Compt. rend.*, **183,** 24, 1926.

See also

13. A. Zwaan, *Intensitäten im Ca-Funkenspektrum*, Thesis, Utrecht, 1929.

For a general account of this method see, for instance,

14. A. S. Davydov, *Quantum Mechanics*, Oxford, 1965.

The mathematical method had previously been introduced by Jeffreys:

15. H. Jeffreys, *Proc. London Math. Soc.*, **23,** 428, 1923.

Sections 3.2 and 3.4. See References 1 and 2.

Section 3.3. See References 3 and 5. Planck's law and Rayleigh's law were both derived in 1900:

16. M. Planck, *Verhandl. deut. physik. Ges.*, **2,** 202, 237, 1900.
17. Lord Rayleigh, *Phil. Mag.*, **49,** 539, 1900.

A numerical error in Rayleigh's expression was corrected by Jeans, and Rayleigh's law is therefore often referred to as the Rayleigh-Jeans law:

18. J. H. Jeans, *Phil. Mag.*, **10,** 91, 1905.

Wien's law is older and was stated in 1896:

19. W. Wien, *Ann. Physik*, **58,** 662, 1896.

Stefan-Boltzmann's law was first deduced from experiments in 1879 by Stefan and deduced theoretically by Boltzmann in 1884:

20. J. Stefan, *Wien. Ber.*, **79,** 391, 1879.
21. L. Boltzmann, *Ann. Physik*, **22,** 31, 291, 616, 1884.

The derivation of Planck's radiation law from atomic considerations can be found in the paper in which Einstein introduced his transition probabilities:

22. A. Einstein, *Physik. Z.*, **18,** 121, 1917.

Section 3.5. An approximate formula that can be used to evaluate the partition function of the three-dimensional rigid rotator was given by Mulholland:

23. H. P. Mulholland, *Proc. Cambridge Phil. Soc.*, **24,** 280, 1928.

For an account of the history of the hydrogen rotational specific heat, see

24. A. Farkas, *Orthohydrogen, Parahydrogen and Heavy Hydrogen*, Cambridge, 1935.

The specific heat or normal hydrogen was measured by Eucken:

25. A. Eucken, *Sitz. ber. preuss. Akad. Wiss.*, **1912,** 41.

The specific heat curves of mixtures of ortho- and parahydrogen were measured by Eucken and Hiller and by Clusius and Hiller:

26. A. Eucken and K. Hiller, *Z. physik. Chem.*, **B4,** 142, 1929.
27. K. Clusius and K. Hiller, *Z. physik. Chem.*, **B4,** 158, 1929.

The theoretical explanation of the specific heat curve of normal hydrogen was given by Dennison in

28. D. M. Dennison, *Proc. Roy. Soc. (London)*, **A115,** 483, 1927.

4

Bose-Einstein and Fermi-Dirac Statistics

4.1 DEVIATIONS FROM BOLTZMANN STATISTICS

In Section 3.3 we discussed the radiation field and derived Planck's radiation law (3.308) for the energy density $\rho(\omega)$ in the radiation field. It was mentioned at that time that Planck had been obliged to introduce his quantum of action h, different from zero, since Eq. (3.308) with finite h agreed with the observational evidence. Let us examine for a moment the basic assumptions used in deriving Eq. (3.308). It was assumed that the radiation field was built up out of one-dimensional harmonic oscillators all of which possessed different frequencies. A given situation was completely described by giving the quantum numbers n_i of the stationary states in which the various oscillators were. When we wished to obtain the expression for the total energy of the system in temperature equilibrium, we remarked that it can be shown that the time average of the energy of one oscillator is equal to the system average of the energy of a system of many identical oscillators. In calculating this last average one uses the assumption of equal a priori probability for different (nondegenerate) stationary states. In other words, in order to derive Planck's radiation law—which agrees well with experimental data—we must assume that any sequence $n_1, n_2, \cdots, n_i, \cdots$ of the quantum numbers of the oscillators has the same probability. However, we know that a modern picture of the radiation field must take into account the fact that light is quantized, that is, that it occurs as light quanta. This means that the situation which we just described by the sequence $n_1, n_2, \cdots, n_i, \cdots$ in reality corresponds to a radiation field in which there are n_1 quanta with energy $\hbar\omega_1$, n_2 quanta with energy $\hbar\omega_2$, and so on. Instead of having one oscillator corresponding to a given frequency ω_i we have n_i quanta with energy $\hbar\omega_i$. If we wished to calculate the energy density of the field without having recourse to our previous considerations, our first task would be to find the a priori probability to find n_1 quanta of energy $\hbar\omega_1$, n_2 quanta of energy $\hbar\omega_2$, and so on. This probability $W(n_1,n_2,\cdots,n_i,\cdots)$* of distributing $n_1 + n_2 + \cdots$ particles over the possible energy levels $\hbar\omega_1, \hbar\omega_2, \cdots$ was calculated in Section 3.1 (Eq. (3.101)). We found there

$$W(n_i) = A \prod_i \frac{1}{n_i!}, \tag{4.101}$$

where A is a normalizing constant which we shall leave undetermined for the time being. In Section 2 of the present chapter we shall make some particular choice for A and then we shall also give the reasons for our choice. If we are dealing with a system such that we can apply Eq. (4.101), we say that it obeys *Boltzmann statistics*, which are sometimes called *classical statistics*.

A moment ago, however, we saw that in order to obtain the radiation law which agrees with experiments we must assign the value of 1 to $W(n_i)$ irrespective of the values of the n_i, or

$$W(n_i) = 1. \tag{4.102}$$

We see here a departure from the formulas that were used in the preceding chapters. If we try to use Eq. (4.101) for the a priori probability, we end up with a formula for the radiation density which is in disagreement with the observational evidence. This is often expressed in the following way. Light quanta do not obey Boltzmann statistics, but they obey *Bose-Einstein statistics*. Formula (4.102) for the a priori probability was introduced by Bose in 1924 for the case of light quanta and applied by Einstein in the same year to the case of an ideal gas.

* We shall mostly use the abbreviated notation $W(n_i)$ for $W(n_1,n_2, \cdots, n_i, \cdots)$.

It must be remarked here that the term "Bose-Einstein *statistics*" (or Boltzmann *statistics* or Fermi-Dirac *statistics*) could imply that different statistical methods are used. This is not the case. It would thus be less confusing to talk about Boltzmann *formulas*, Bose-Einstein *formulas*, or Fermi-Dirac *formulas*. However, the other nomenclature has been and is used so extensively that we shall also continue to use it, the more readily since there seems to be no danger of serious misunderstandings:

There exists a second kind of statistics that differs from the classical or Boltzmann statistics which we discussed in the preceding chapters. These statistics were introduced by Fermi and extensively discussed by Dirac in 1926. They are called the *Fermi-Dirac statistics* and are characterized by the following values of $W(n_i)$:

$$\left.\begin{aligned} W(n_i) &= 1, &\quad &\text{if all the } n_i \text{ are} \leqslant 1; \\ W(n_i) &= 0, &\quad &\text{if there is at least one } n_i > 1. \end{aligned}\right\} \tag{4.103}$$

We shall discuss presently the connection between this choice of $W(n_i)$ and Pauli's exclusion principle.

In the last section of the present chapter we shall consider briefly a few cases where Bose-Einstein and Fermi-Dirac statistics are applied and we shall also discuss why one can no longer apply classical statistics in these cases. We may mention here that Bose-Einstein statistics have been mainly applied to a discussion of the behavior of helium at low temperatures, to the case of light quanta, and to the case of various kinds of elementary excitations in solid state physics. Fermi-Dirac statistics are used in the description of the behavior of conduction electrons in metals, nucleons in nuclei, and also of some kinds of elementary excitations.

In the present chapter we shall treat the two "new" statistics, as Bose-Einstein and Fermi-Dirac statistics are sometimes called, in a way analogous to that used for Boltzmann statistics in the preceding chapters. Bose-Einstein and Fermi-Dirac statistics are also often called "quantum" statistics. The reason for this name we shall see in later sections of this chapter. The name is rather misleading, since one can quite often get a good approximation to the behavior of a system by applying Boltzmann formulas to systems where the energy levels have been computed by quantum-mechanical methods, as we did in Chapter 3, for instance. The point is that Bose-Einstein and Fermi-Dirac statistics appear as a consequence of the symmetry effects of quantum mechanics, while the diffraction effects of quantum mechanics by themselves merely slightly modify the Boltzmann formulas, usually by replacing integrals by sums, as we saw in the preceding chapter. In the limit as $h \to 0$—which is usually called the classical limit—"quantum" statistics go over into Boltzmann statistics, that is, all Bose-Einstein and Fermi-Dirac formulas go over into the corresponding Boltzmann formulas; and as in the limit as $h \to 0$ the quantum-mechanical partition function goes over into the classical one—as we saw in Section 3.4—we are, indeed, led back to the classical case, where both diffraction and symmetry effects can be neglected.

4.2 THE PROBABILITY ASPECT OF STATISTICS

We shall use here a formulation which is slightly different from the one used in Chapter 1, since we wish to apply quantum mechanics from the start.

Let us assume that we are dealing with a system consisting of N identical independent particles. The Hamiltonian of the system will be the sum of the Hamiltonians of the particles

$$\hat{\mathscr{H}}_{\text{syst}} = \sum_i \mathscr{H}_i(p_k^{(i)}, q_k^{(i)}). \tag{4.201}$$

In Eq. (4.201) k runs from 1 to s where s is the number of degrees of freedom of one particle. The $\hat{\mathscr{H}}_i$ differ only in the indices enumerating the particles, but are otherwise identical functions of their arguments.

The time-independent Schrödinger equation for the system will be

$$\hat{\mathscr{H}}_{\text{syst}}\Psi = E\Psi, \tag{4.202}$$

where E and Ψ are respectively the total energy of the system and its wave function. Since the Hamiltonian is a sum, Eq. (4.202) can be solved by the method of the separation of coordinates and we can look for solutions of Eq. (4.202) in the form

$$\Psi = \prod_i \psi(i), \tag{4.203}$$

$$E = \sum_i \varepsilon(i), \tag{4.204}$$

where $\psi(i)$ and $\varepsilon(i)$ are the eigenfunctions and eigenvalues of the single-particle Schrödinger equation

$$\hat{\mathscr{H}}_i\psi_m(i) = \varepsilon_m\psi_m(i). \tag{4.205}$$

We have written here for the eigenfunction and eigenvalue of the mth eigenstate of the ith paiticle $\psi_m(i)$ and ε_m rather than $\psi_m(q_k^{(i)})$ and $\varepsilon_m(i)$.

A stationary state of the system can be described by giving the number of particles in each stationary state corresponding to the single-particle Schrödinger equation. If there are n_1 particles in the first stationary state, n_2 particles in the second stationary state, and so on, we have from Eq. (4.203)

$$\Psi = \psi_1(1)\psi_1(2)\cdots\psi_1(n_1)\psi_2(n_1+1)\psi_2(n_1+2)\cdots\psi_2(n_1+n_2)\psi_3(\cdots)\cdots, \tag{4.206}$$

where we have numbered the particles starting from those occupying the first stationary state, and so on.

Since the particles in the system are supposed to be identical, the state of the system corresponding to the wave function (4.206) looks highly degenerate—even if we assume that there are no accidental degeneracies due to linear relations with integral coefficients between the ε_m—since all wave functions $P\psi$ given by the equation

$$P\Psi = \psi_1(k_1)\psi_1(k_2)\cdots\psi_1(k_{n_1})\psi_2(k_{n_1+1})\cdots\psi_2(k_{n_1+n_2})\psi_3(k\ldots)\cdots, \qquad (4.207)$$

where $k_1, k_2, \cdots, k_i, \cdots, k_N$ is a permutation P of the numbers $1, 2, \cdots, i, \cdots, N$, will also be eigenfunctions of the Schrödinger equation (4.202) with the same eigenvalue E given by Eq. (4.204).

In classical mechanics it was regarded as permissible to imagine labels attached to each of the N particles, even though they are identical. In that way each of them could be followed along its orbit and localized at every moment. This possibility of distinguishing between the various (identical) particles would entail that Ψ and $P\Psi$ describe two different stationary states of the system. We shall call such a stationary state a *microsituation*. Both Ψ and $P\Psi$, though they correspond to different microsituations, describe the same *macrosituation* which is characterized by the numbers $n_1, n_2, n_3, \cdots$. The a priori probability or *weight* of this macrosituation $W(n_i)$ will be given, so long as we are only considering the classical case, by the number of permutations of $1, 2, \cdots, N$ which give n_1 numbers in group 1, n_2 numbers in group 2, and so on. We computed this weight in Section 3.1 and found expression (3.101). If, indeed, we have some way of distinguishing the different microsituations, we would have for the degeneracy of the macrosituation the expression

$$W(n_i) = N!\prod_i \frac{1}{n_i!}. \qquad (4.208)$$

Such a situation will arise when we are dealing with a crystalline system where, although the particles are indistinguishable, the lattice points are not, and this means that we can use the distinguishable lattice point to distinguish the particles by stating which lattice point they occupy. In that case we can use Eq. (4.208) for the degeneracy or weight of the situation characterized by the set of numbers $n_1, n_2, n_3, \cdots$, which we shall call an n_i-distribution.

If we are dealing with a situation where we have no means of keeping track of each separate particle, we are overestimating the weight of the macrosituation by a factor $N!$ insofar as we could think of $N!$ different ways of attaching labels to the N particles. We should therefore divide expression (4.208) by $N!$ to obtain the correct weight of an n_i-distribution. In this way we avoid counting each distinguishable macrosituation $N!$ times instead of once. The weights we obtain in this way will be called, for reasons that will become obvious later in the discussion, the Boltzmann weights, and they are given by the equation

$$W_{\mathrm{Bo}}(n_i) = \prod_i \frac{1}{n_i!}. \tag{4.209}$$

The derivation of Eq. (4.209) is not very convincing and the main justification lies in the fact that it is the correct "classical" limit of the Bose-Einstein and Fermi-Dirac formulas which we shall encounter presently.

We have followed mainly Rushbrooke's arguments* in going from Eq. (4.208) to Eq. (4.209). He points out that the labeling in the case of a crystal occurs because we can label the lattice sites. Even though the atoms in the crystal can and do change places, we can describe the situation by giving the quantum state of the atom on a well-defined position in space.

Rushbrooke states that the transition from Eq. (4.208) to Eq. (4.209) in the case of a gas is *not* a quantum-mechanical effect. I would beg to differ here. So long as we are dealing with classical systems, the $N!$ microsituations are in fact different and can be distinguished from one another. In quantum-mechanical systems, however, we can no longer distinguish them. This is certainly due to the fact that we cannot follow the atoms so closely along their orbits that we can, for instance, know which atom is which after a collision. This again is a consequence of Heisenberg's famous relations. That this impossibility of following the individual particles along their orbits has a physical meaning follows, for instance, from a discussion of experiments where α-particles are scattered by α-particles.† Another case in which the indistinguishability plays an important part is in the helium atom where the two orbital electrons are identical and indistinguishable, leading to the *exchange integral* which in fact gives to some extent the "period of exchange."‡ I agree, of course, with Rushbrooke that in deriving Eq. (4.209) only part of the complete quantum-mechanical argument has been used. The exclusion principle has not been used and should have been used in a complete quantum-mechanical treatment. (Compare also Kramers' and Davydov's remarks on the interpretation of the exchange integral.)

We shall have many opportunities to discuss the consequences and importance of the omission of the factor $N!$.

So far we have not taken into account any symmetry effects. For the majority of systems with which we are dealing in nature, the degeneracy discussed a moment ago is only an apparent one, inasmuch as the wave

* G. S. Rushbrooke, *Introduction to Statistical Mechanics*, Oxford, 1949, Chap. 3, Section 1.

† Compare J. R. Oppenheimer, *Phys. Rev.*, **32,** 361, 1928; N. F. Mott, *Proc. Roy. Soc. (London)*, **A125,** 222, 1929; **A126**, 259, 1930.

‡ Compare H. A. Kramers, *Quantum Mechanics*, Amsterdam, 1957; see also A. S. Davydov, *Quantum Mechanics*, Oxford, 1965, Section 89.

function describing a system of identical particles must either be completely symmetrical or completely antisymmetrical in all the arguments corresponding to identical particles. Particles for which we need symmetrical wave functions are called *bosons* and those for which we need antisymmetrical wave functions *fermions*.

If the wave function is symmetrical in all its arguments, the only possible wave function (or *accessible* state, as Fowler calls it) Ψ_{BE} will be given by the equation

$$\Psi_{\text{BE}} = C \sum_{\text{P}} P\Psi, \tag{4.210}$$

where C is a normalization constant and where the summation is extended over all $N!$ possible permutations of the arguments.

We see that there corresponds to every n_i-distribution only *one* wave function. The weight or probability of the distribution will thus be independent of the distribution* and we have

$$W_{\text{BE}}(n_i) = 1. \tag{4.211}$$

Since Eq. (4.211) is the same as Eq. (4.102), we see the reason for the index "BE."

In the case of the antisymmetrical wave function the only possibility is given by

$$\Psi_{\text{FD}} = C \sum_{\text{P}} \varepsilon_{\text{P}} P\Psi, \tag{4.212}$$

where ε_{P} is $+1$ or -1 according as the permutation is even or odd. Expression (4.212) is often called a *Slater determinant*, since it was used extensively by Slater† in the form of a determinant,‡

$$\Psi_{\text{FD}} = C \begin{vmatrix} \psi_{l_1}(1) & \psi_{l_1}(2) \ldots\ldots\ldots & \psi_{l_1}(N) \\ \psi_{l_2}(1) & \psi_{l_2}(2) \ldots\ldots\ldots & \psi_{l_2}(N) \\ \cdot & \cdot \quad \cdot & \cdot \\ \cdot & \cdot \quad \cdot & \cdot \\ \cdot & \cdot \quad \cdot & \cdot \\ \cdot & \cdot \quad \cdot & \cdot \\ \psi_{l_N}(1) & \psi_{l_N}(2) \ldots\ldots\ldots & \psi_{l_N}(N) \end{vmatrix} \tag{4.213}$$

From Eq. (4.213) it is seen immediately that as soon as one n_i is larger than 1, $\Psi_{\text{FD}} = 0$. In the Fermi-Dirac case no two particles can be in the same state. This corresponds to the Pauli principle which was first stated for electrons.

* We equate here the weight of a distribution to the number of accessible states.

† J. C. Slater, *Phys. Rev.*, **34,** 1293, 1929.

‡ Instead of numbering the rows n_1 times 1, n_2 times, 2, and so on, as in Eq. (4.206), we have numbered them $l_1, l_2, \cdots, l_N$.

For the weight of an n_i-distribution we now have

$$\left.\begin{aligned} W_{\text{FD}}(n_i) &= 1, \qquad \sum_i n_i^{\,2} = N; \\ W_{\text{FD}}(n_i) &= 0, \qquad \sum_i n_i^2 > N, \end{aligned}\right\} \tag{4.214}$$

which is the same as Eqs. (4.103).

We note that the $W(n_i)$ given by Eqs. (4.209), (4.211), and (4.214) all have the property that they are the product of factors pertaining to the single levels.

In the remainder of this chapter we shall mainly consider cases where the energy levels are lying so densely that we can practically speak of a continuous energy spectrum. In that case we can bundle the energy levels into groups. The number of levels in the jth group will be denoted by Z_j. The number of particles with energies in that interval will be denoted by N_j. Since the energy levels are supposed to lie very densely, it will be possible to ascribe to each group a rather well-defined energy value E_j. We shall assume that we are dealing with cases where, without losing the accuracy with which E_j is defined, we can choose Z_j so large that N_j and, if necessary, also $Z_j - N_j$ are large numbers as well.

For the total energy and total number of particles we have the equations

$$\sum_j N_j E_j = E \tag{4.215}$$

and

$$\sum_j N_j = N. \tag{4.216}$$

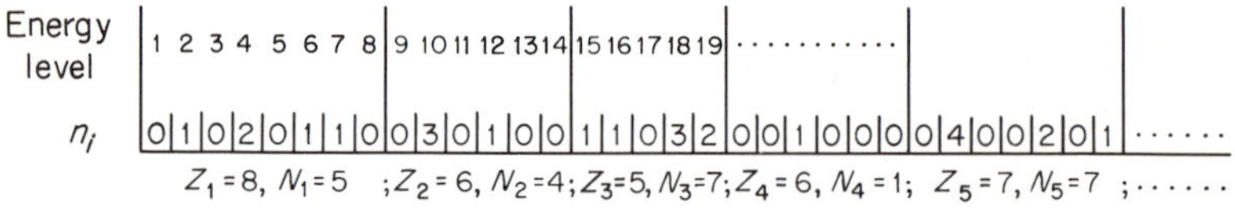

FIG. 4.1 A possible distribution of particles over energy levels that are taken together in groups.

We now wish to calculate the probability $W(N_1, N_2, \cdots, N_j, \cdots)$ $[\equiv W(N_j)]$ that there are N_1 particles in group Z_1, N_2 particles in group Z_2, and so on. Figure 4.1 depicts one possible distribution.

The $W(N_j)$ are related to the $W(n_i)$ by the equation

$$W(N_j) = \sum_{N_i = \text{Ct}} W(n_i),$$

where the summation extends over all possible n_i-distributions corresponding to a given N-distribution. Because the $W(n_i)$ are products of factors pertaining to single levels, the $W(N_j)$ will also be products of factors, this time pertaining to the different groups of levels. We can thus write generally, independently of what kind of statistics we are dealing with,

$$W(N_j) = \prod_j W_j, \tag{4.217}$$

where each of the W_j is given by the relation

$$W_j = \sum W_j(n_i), \tag{4.218}$$

where the $W_j(n_i)$ contains only those n_i referring to the ε_i in the jth group and where the summation is over all n_i combinations that add up to N_j.

We have three different cases corresponding to formulas (4.209), (4.211), and (4.214) for $W(n_i)$. First of all, we shall consider the Boltzmann case. Let us consider the group (or cell) Z_j and let the number of particles in the first level be n_1, in the second level $n_2, \cdots$, in the last level n_{Z_j}. If we keep the other n_i constant, the contribution from this group will be

$$\sum_{n_1+n_2+\cdots+n_{Z_j}=N_j} \frac{1}{n_1!\,n_2!\cdots n_{Z_j}!} = \frac{1}{N_j!}\sum \frac{N_j!}{n_1!\,n_2!\cdots n_{Z_j}!}$$
$$= \frac{Z_j^{N_j}}{N_j!},$$

where we have used Eq. (1.704) for the polynomial equation. For the $W_{j\mathrm{Bo}}$ we thus get

$$W_{j\mathrm{Bo}} = \frac{Z_j^{N_j}}{N_j!}. \tag{4.219}$$

Apart from a constant factor, expression (4.217) with W_j given by Eq. (4.219) is identical with the expression from probability calculus for the probability that N points are distributed among cells Z_j in such a way that there are N_j points in cell Z_j. We met the same expression in Section 1.7 when we discussed the distribution of representative points over volumes in velocity space.

Since $W_{\mathrm{BE}}(n_i)$ does not depend on the values of the n_i, we obtain $W_{j\mathrm{BE}}$ if we evaluate the number of different ways by which N_j particles can be distributed over Z_j levels. This number of ways is equal to $(N_j + Z_j - 1)!/N_j!(Z_j - 1)!$ and we get thus

$$W_{j\mathrm{BE}} = \frac{(N_j + Z_j - 1)!}{N_j!(Z_j - 1)!}. \tag{4.220}$$

The expression for the number of ways by which N_j particles can be distributed among Z_j levels can be obtained by a generalization of the argument given in small type in Section 3.2. We now have to write N_j as a sum of Z_j nonnegative integers. That means that we have to divide N_j dots by $Z_j - 1$ strokes. We then get immediately the final result.

If we wish to calculate $W_{j\mathrm{FD}}$, we have to evaluate the number of ways by which N_j particles can be distributed over Z_j levels without having more than one particle in any level. This number is simply the binomial coefficient

$$\binom{Z_j}{N_j}$$

We thus get

$$W_{j\mathrm{FD}} = \frac{Z_j!}{(Z_j - N_j)!\,N_j!}. \tag{4.221}$$

If $N_j \ll Z_j$, we can write Eqs. (4.220) and (4.221) in the following form

$$W_{j\mathrm{BE}} = \frac{(N_j + Z_j - 1)(N_j + Z_j - 2)\cdots(Z_j + 1)Z_j}{N_j!} \approx \frac{Z_j^{N_j}}{N_j!} = W_{j\mathrm{Bo}}$$

and

$$W_{j\mathrm{FD}} = \frac{Z_j(Z_j - 1)\cdots(Z_j - N_j + 2)(Z_j - N_j + 1)}{N_j!} \approx \prod_j \frac{Z_j^{N_j}}{N_j!} = W_{j\mathrm{Bo}}.$$

We see that in the limit where $N_j \ll Z_j$ the Bose-Einstein and the Fermi-Dirac formulas go over into the Boltzmann formulas. This can be understood as follows. If $N_j \ll Z_j$, most levels will be empty and practically all n_i will be equal to 0 or 1, which means that $W_{\mathrm{Bo}}(n_i) \cong W_{\mathrm{BE}}(n_i) \cong W_{\mathrm{FD}}(n_i) = 1$.

Let us consider for a moment two completely separated systems composed of the same kind of particles. Let the total number of particles in the two systems be $N^{(1)}$ and $N^{(2)}$ with distributions $N_j^{(1)}$ and $N_j^{(2)}$. Let $W^{(1)}$ be the probability of the $N_j^{(1)}$-distribution and $W^{(2)}$ that of the $N_j^{(2)}$-distribution. The total probability that we have the $N_j^{(1)}$-distribution in the first and the $N_j^{(2)}$-distribution in the second system will be given by the expression $W = W^{(1)} \cdot W^{(2)}$. This is a consequence of the product property (4.217) of the $W(N_j)$, and that in turn is a consequence of the product property of the $W(n_i)$. If we had used Eq. (4.208) instead of Eq. (4.209), we would not have been able to state that $W = W^{(1)} \cdot W^{(2)}$. Since the multiplication property of the $W(N_j)$ ensures that the entropy, which, as we shall see presently, is given by the equation $S = k \ln W$, will have the necessary *additive* property, the multiplication property is most desirable.

Both the fact that the choice of Eq. (4.209) for $W_{\mathrm{Bo}}(n_i)$ has as a consequence that $W_{\mathrm{Bo}}(N_j)$ is the limiting value of both $W_{\mathrm{BE}}(N_j)$ and $W_{\mathrm{FD}}(N_j)$, and the multiplication property of our $W_{\mathrm{Bo}}(N_j)$ are strong arguments in favor of the omission of the $N!$.

4.3 THE ELEMENTARY METHOD OF STATISTICS

The elementary method of statistics consists of determining the values of N_j for which $W(N_j)$ is a maximum for a given total number of particles and total energy. It is then assumed that this N_j-distribution, for which $W(N_j)$ is maximum, will be realized in an actual physical system in equilibrium.

We shall now proceed to find the N_j-distributions for which $W(N_j)$, or rather $\ln W(N_j)$, is a maximum for given values of the total energy E and total number of particles N which are related to the N_j by the equations

$$\sum_j N_j E_j = E, \tag{4.301}$$

$$\sum_j N_j = N. \tag{4.302}$$

The argument is completely analogous to that in Section 1.7. We shall assume that all Z_j, N_j, and (in the Fermi-Dirac case) $Z_j - N_j$ are large compared to unity, so that we can use Stirling's formula for the factorial in its simplest form (1.708). We then have

$$\ln W_{\mathrm{Bo}} = \sum_j N_j\left[\ln \frac{Z_j}{N_j} + 1\right], \tag{4.303}$$

$$\ln W_{\mathrm{BE}} = \sum_j \left[(N_j + Z_j) \ln \left(\frac{Z_j}{N_j} + 1\right) - Z_j \ln \frac{Z_j}{N_j}\right], \tag{4.304}$$

$$\ln W_{\mathrm{FD}} = \sum_j \left[(N_j - Z_j) \ln \left(\frac{Z_j}{N_j} - 1\right) + Z_j \ln \frac{Z_j}{N_j}\right]. \tag{4.305}$$

In order not to repeat every equation three times, we shall combine Eqs. (4.303) to (4.305) as follows

$$\ln W(N_j) = \sum_j \left[N_j\left\{\ln \left(\frac{Z_j}{N_j} + \gamma\right) + 1 - \gamma^2\right\} + \gamma Z_j \ln \left(1 + \frac{\gamma N_j}{Z_j}\right)\right], \tag{4.306}$$

where we have introduced a parameter γ which is defined by the equations

$$\gamma_{\mathrm{Bo}} = 0, \qquad \gamma_{\mathrm{BE}} = 1, \qquad \gamma_{\mathrm{FD}} = -1. \tag{4.307}$$

Since the total energy E and the total number of particles N are fixed, the conditions for a maximum of $\ln W$ are that, for any variation δN_j of the N_j, we have

$$\delta \ln W = 0, \qquad \delta N = 0, \qquad \text{and} \qquad \delta E = 0. \tag{4.308}$$

From Eq. (4.306) we now have

$$\delta \ln W = \sum_j \ln \left(\frac{Z_j}{N_j} + \gamma\right) \delta N_j, \tag{4.309}$$

while Eqs. (4.301) and (4.302) lead to

$$\delta N = \sum_j \delta N_j = 0, \tag{4.310}$$

$$\delta E = \sum_j E_j \delta N_j = 0. \tag{4.311}$$

Using Lagrange's method of undetermined multipliers, we have that for any arbitrary choice of δN_j

$$\delta \ln W + \alpha \delta N - \beta \delta E = 0. \tag{4.312}$$

Hence it follows that

$$\ln \left(\frac{Z_j}{N_j} + \gamma\right) = -\alpha + \beta E_j. \tag{4.313}$$

Equation (4.313) determines the N_j as functions of the Lagrangian multipliers α and β and of the external parameters a_k (through the E_j; compare the discussion in Section 2.6). If we wish to do so, we can use Eqs. (4.301) and (4.302) to express α and β in terms of N and E.

From Eqs. (4.313) and (4.307) we get for the N_j-distributions which make $W(N_j)$ a maximum the following expressions

$$\text{Boltzmann:} \qquad N_j = Z_j e^{\alpha} e^{-\beta E_j}, \tag{4.314}$$

$$\text{Bose-Einstein:} \qquad N_j = \frac{Z_j}{e^{-\alpha+\beta E_j} - 1}, \tag{4.315}$$

$$\text{Fermi-Dirac:} \qquad N_j = \frac{Z_j}{e^{-\alpha+\beta E_j} + 1}. \tag{4.316}$$

Equation (4.314) gives us once more the Maxwell-Boltzmann distribution. Equation (4.315) reminds us of Eq. (3.202), but with the difference that we have to put $\alpha = 0$ in order to get Eq. (3.202). This corresponds to the fact that for light quanta we have to use Bose-Einstein statistics, but without any restrictions as to the total number of particles, so that the restriction $\delta N = 0$ leading to Eq. (4.310) is not operative. This means that there is only one Lagrangian multiplier, β.

From Eqs. (4.314) to (4.316) we see that the Bose-Einstein and Fermi-Dirac formulas will go over into the Boltzmann formulas if

$$e^{-\alpha+\beta E_j} \gg 1. \tag{4.317}$$

This is equivalent, however, to the condition $N_j/Z_j \ll 1$ as can immediately be seen from Eq. (4.314). One way of satisfying condition (4.317) is by choosing α to be large and negative.

From Eqs. (4.314) and (4.316) we see that any value of α will correspond to an N_j-distribution in the case of Boltzmann or Fermi-Dirac statistics. In the case of Bose-Einstein statistics, however, we see that α is restricted to such values that

$$e^{\alpha} < e^{\beta\varepsilon_0}, \tag{4.318}$$

where ε_0 is the lowest energy level of a particle in the system.

4.4 CONNECTION WITH THERMODYNAMICS

In this section we shall discuss the connection between the probability arguments given in the previous section and thermodynamics, following much the same lines as the discussion in Sections 1.6 and 2.7.

We first of all remind ourselves that the energy levels ε_i, and hence the representative energy values E_j, may depend on external parameters,

$$\varepsilon_i = \varepsilon_i(a_1, a_2, \cdots), \tag{4.401}$$

and thus

$$E_j = E_j(a_1, a_2, \cdots). \tag{4.402}$$

We now introduce the force A_{ik} exerted by one particle in the ith single-particle eigenstate on the external parameter a_k (compare the discussion in Section 2.6),

$$A_{ik} = -\frac{\partial \varepsilon_i}{\partial a_k}. \tag{4.403}$$

The total force* A_k exerted by the system on the parameter a_k will then be given by the equation

$$A_k = \sum_i n_i A_{ik} = -\sum_i n_i \frac{\partial \varepsilon_i}{\partial a_k} = -\sum_j N_j \frac{\partial E_j}{\partial a_k}. \tag{4.404}$$

We now compare two equilibrium states of the system for which α, β, and the external parameters a_k have slightly different values, the differences being denoted by $\delta\alpha$, $\delta\beta$, and δa_k.

From Eqs. (4.309) and (4.313) we have

$$\delta \ln W_{\text{max}} = -\alpha \sum_j \delta N_j + \beta \sum_j E_j \, \delta N_j. \tag{4.405}$$

* It must be noted that, while the A_i in Section 2.6 were the forces per molecule, the A_k of Eq. (4.404) are the total forces exerted by the system.

We must emphasize here the difference between the $\delta \ln W$ and the δN_j considered at this moment and the quantities denoted by the same symbols in the preceding section. In the preceding section we were comparing for given values of N and E (or, what is equivalent, of α and β) and given values of the a_k, that is, of the ε_i or E_j, different N_j-distributions and hence different W-values, where one distribution corresponded to a maximum of W and the varied one to a value of W which was slightly off the maximum. Now, however, we are comparing different equilibrium distributions, but for varying values of the α, β, and a_k. In the remainder of this section we shall understand by N_j the values corresponding to the equilibrium distribution for given values of α, β, and a_k and by W(or $W_{\max}$) the maximum (equilibrium) value for the same values of α, β, and a_k.

Denoting by $\delta\!\!\!^{-} A$ the work done by the system on the external parameters when these parameters change from a_k to $a_k + \delta a_k$, and using Eq. (4.404), we have

$$\delta\!\!\!^{-} A = \sum_k A_k \delta a_k = -\sum_j \sum_k N_j \frac{\partial E_j}{\partial a_k} \delta a_k = -\sum_j N_j \delta E_j. \tag{4.406}$$

From Eqs. (4.302) and (4.301) we have also

$$\delta N = \sum_j \delta N_j, \tag{4.407}$$

$$\delta E = \sum_j E_j \delta N_j + \sum_j N_j \delta E_j = \sum_j E_j \delta N_j - \delta\!\!\!^{-} A. \tag{4.408}$$

We define now a quantity $\delta\!\!\!^{-} Q$ by the equation

$$\delta\!\!\!^{-} Q = \sum_j E_j \delta N_j. \tag{4.409}$$

From Eqs. (4.408) and (4.409) it follows that

$$\delta\!\!\!^{-} Q = \delta E + \delta\!\!\!^{-} A, \tag{4.410}$$

or: $\delta\!\!\!^{-} Q$ is the heat added to the system, that is, the increase of energy not accounted for by macroscopically measurable work (compare the definition of $\delta\!\!\!^{-} A$).

Combining Eqs. (4.405) and (4.407) to (4.410), we have

$$\delta \ln W_{\max} = -\alpha \delta N + \beta \delta\!\!\!^{-} Q = -\alpha \delta N + \beta(\delta E + \delta\!\!\!^{-} A). \tag{4.411}$$

In connection with Eq. (4.313) we mentioned that the N_j were functions of α, β, and the a_k, or if we preferred it, of N, E, and the a_k. In Eq. (4.411) we have, indeed, used N, E, and the a_k as independent variables, since these seem to be the natural variables to use in this case. From thermodynamics we know* that the thermodynamic potential for which N, E, and the a_k are

* See, for instance, D. ter Haar, and H. Wergeland, *Elements of Thermodynamics*, Reading, Mass., 1966.

the natural independent variables is the entropy, and we are thus led to identify $\ln W_{max}$, apart from possible multiplicative and additive constants, with the entropy S. We can also see this by considering two equilibrium states of the system for which $\delta N = 0$. In that case we see from Eq. (4.411) that β is an integrating factor of δQ, which means that it must be proportional to the reciprocal of the absolute temperature T and that $\beta \delta Q$ must be proportional to δS. Thus we have

$$\beta = \frac{1}{kT}, \tag{4.412}$$

$$S = k \ln W_{max}. \tag{4.413}$$

The constant k that occurs in Eqs. (4.412) and (4.413) is again Boltzmann's constant. This can be proved, for instance, by comparing the equation of state following from Eq. (4.411) with the perfect gas law (see Eq. (4.613)).

The complete proof that k is really Boltzmann's constant is slightly more complicated. It involves the consideration of a system consisting of the system under consideration combined with a "gas thermometer." One then first proves that β is the same for the two systems and, secondly, one assumes that the gas in the "thermometer" is so diluted that it can be treated as a perfect gas. We refer to the literature for a more extensive discussion.*

From Eqs. (4.411) and (4.413) we see that the entropy is not only a function of the a_k and β (or T), but also of α. We know from thermodynamics that the entropy is an extensive or additive quantity, that is, that for two systems (1) and (2), which are independent, the total entropy will be equal to $S^{(1)} + S^{(2)}$ This property follows from Eq. (4.413), since the W_{max} are multiplicative (compare the discussion at the end of Section 4.2).

Let us now consider the physical meaning of α. For the variation of the (Helmholtz) free energy $F(=E - ST)$ of the system we have from Eqs. (4.411) to (4.413)

$$\delta F = \delta(E - ST) = -S\delta T + \frac{\alpha}{\beta}\delta N - \delta A. \tag{4.414}$$

We thus get

$$\frac{\alpha}{\beta} = \left(\frac{\partial F}{\partial N}\right)_{T,a_k} = \mu, \tag{4.415}$$

where μ is the free energy per particle, or the thermal potential per particle. For α we thus have

$$\alpha = \beta\mu = \frac{\mu}{kT}. \tag{4.416}$$

* See, for example, R. C. Tolman, *The Principles of Statistical Mechanics*, Oxford, 1938, p.85.

That μ can also be interpreted as the partial thermal potential follows easily when we consider a system where the volume V is one of the external parameters. We shall assume for the sake of simplicity that it is actually the only one, so that $\delta A = P\delta V$.

Introducing the thermal potential* or Gibbs free energy G by the equation

$$G = F + PV = E - ST + PV, \tag{4.417}$$

we have for its variation

$$\delta G = \delta F + P\delta V + V\delta P = -S\delta T + \frac{\alpha}{\beta}\delta N + V\delta P, \tag{4.418}$$

whence

$$\mu = \left(\frac{\partial G}{\partial N}\right)_{T,p}. \tag{4.419}$$

Before concluding this section, we wish to introduce a new thermodynamical quantity, called by Kramers the q-potential. This q-potential is dimensionless and is defined by the equation

$$q = \ln W_{\max} + \alpha N - \beta E = \frac{ST + \mu N - E}{kT}. \tag{4.420}$$

Using Eq. (4.411) we get for its variation

$$\delta q = N\delta\alpha - E\delta\beta + \beta \ A. \tag{4.421}$$

We see that in this case α, β, and the a_k are the natural choices of independent variables. From Eqs. (4.421) and (4.406) we get the equations

$$N = \frac{\partial q}{\partial \alpha}, \tag{4.422}$$

$$E = -\frac{\partial q}{\partial \beta}, \tag{4.423}$$

and

$$A_k = \frac{1}{\beta}\frac{\partial q}{\partial a_k}. \tag{4.424}$$

If we are dealing with a homogeneous system of which the volume is one of the external parameters, we have

$$\mu N = G = E - ST + PV, \tag{4.425}$$

* The thermal potential is also sometimes called the chemical potential or simply the thermodynamic potential.

and hence from Eq. (4.420),

$$q = \frac{PV}{kT}. \tag{4.426}$$

4.5 THE DARWIN-FOWLER METHOD

Before we apply the formulas derived in the two preceding sections, we shall give a different derivation of Eqs. (4.314) to (4.316), using the so-called Darwin-Fowler method. The reason is partly that it is a method which has been widely used in statistical mechanics literature and partly that it uses mathematical techniques which can be applied to other problems. We shall see in Chapter 5 that there are strong physical arguments to use ensemble theory to derive Eqs. (4.314) to (4.316) for the case where the Z_j are not large compared to unity, and inasmuch as we accept these arguments there is no reason to use the Darwin-Fowler method, and to consider it merely as a mathematical trick. Boltzmann* emphasized that there are two ways of defining the equilibrium situation: either by identifying it with the most probable distribution—as we did in Section 4.3, for instance—or by identifying it with the average distribution. As long as we are dealing with systems consisting of a large number of particles, the most probable situation will be so much more probable that it is also the average situation.† Furthermore, by taking the energy levels in groups, which was necessary in order that we could use Stirling's formula for the factorial, we are in fact calculating the average and not the most probable situation.‡

We shall now look for the average situation rather than the most probable one, restricting ourselves to the case of a system in which there is only one kind of independent particles. Defining $W(n_i)$ as in Section 4.2, we first

* L. Boltzmann, *Wien. Ber.*, **76**, 373, 1877.

† This is a sketchy statement which we do not wish to make more rigorous at this point.

‡ See E. Schrödinger, *Berliner*, *Ber*. **1925**, 434. The most probable situation is found by writing for the entropy

$$S = k \ln W_{\max}, \tag{A}$$

where $W_{\max}$ should be the maximum value of $W(n_i)$. The average situation, on the other hand corresponds to the equation

$$S = k \ln \sum W, \tag{B}$$

as follows after a moment's reflection. In Section 4.3 we lumped all levels into groups where all levels in one group had about the same energy E_j, and again using Eq. (A), but now with W being the probability for a situation where N_j particles are in the jth group, we calculated the entropy. We now must take into account the following two facts. (a) The maximum of $W(N_j)$ is so steep that, apart from a negligible constant, we may in Eq. (B) substitute $W_{\max}(N_j)$ for the sum. (b) $W_{\max}(N_j)$ itself is, however, a sum of several $W(n_i)$ (see Eq. (4.217) and (4.218). We see thus that by using the $W(N_j)$ instead of the $W(n_j)$ we can reduce Eq. (B) to Eq. (A).

of all remind ourselves that $W(n_i)$ can always be written in the form

$$W(n_i) = \prod_i \gamma(n_i), \tag{4.501}$$

where the actual form of the $\gamma(n_i)$ depends on the nature of the particles and is different for the three cases of Boltzmann, Bose-Einstein, or Fermi-Dirac statistics.

We now wish to calculate the average values of the $\overline{n_i}$ under the supplementary conditions of given total number of particles and given total energy,

$$N = \sum n_i, \tag{4.502}$$

$$E = \sum n_i \varepsilon_i. \tag{4.503}$$

We have clearly for the average values $\overline{n_i}$ the equation

$$\overline{n_i} = \frac{1}{G} \sum{}'' n_i W(n_k), \tag{4.504}$$

where

$$G = \sum{}'' W(n_k). \tag{4.505}$$

The summations extend over all possible combinations of the n_k compatible with the conditions (4.502) and (4.503); this is indicated by the two primes on the summation signs.

In order to be able, if necessary, to consider fluctuations and correlations, we are also interested in expressions of the type

$$\overline{n_i{}^p n_j{}^q} = \frac{1}{G} \sum{}'' n_i{}^p n_j{}^q W(n_k). \tag{4.506}$$

The first problem is the evaluation of G. This is done by introducing a generating function* $F(x,y;\varepsilon_i)$ by the equation

$$F(x,y;\varepsilon_i) = \sum W(n_k) x^{n_1+n_2+\cdots} y^{n_1\varepsilon_1+n_2\varepsilon_2+\cdots}, \tag{4.507}$$

where the summation is over all values of the n_k without any restrictions. In order to obtain G from $F(x,y;\varepsilon_i)$ we must pick out of the sum those terms of which the exponent of x is equal to N and the exponent of y is equal to E. This is done by using the theory of complex functions so that one has

$$G = \left(\frac{1}{2\pi i}\right)^2 \oint dx \oint dy \; x^{-N-1} y^{-E-1} F(x,y;\varepsilon_i), \tag{4.508}$$

* This method goes back to Laplace, who introduced it in his *Théorie Analytique des Probabilités*.

where both integrations are counterclockwise along a closed contour around the origin in the complex planes of x and y, respectively.*

In a similar way one can show that

$$\overline{n_i{}^p n_j{}^q} = \frac{1}{G}\left(\frac{1}{2\pi i}\right)^2 \oint dx \oint dy\; x^{-N-1} y^{-E-1} \left(\frac{1}{\ln y}\frac{\partial}{\partial \varepsilon_i}\right)^p \left(\frac{1}{\ln y}\frac{\partial}{\partial \varepsilon_j}\right)^q F(x,y;\varepsilon_i). \tag{4.509}$$

We may remark here that if we use Eq. (4.501) for $W(n_k)$, $F(x,y;\varepsilon_i)$ can be written as

$$F(x,y;\varepsilon_i) = \prod_i f(x\,y^{\varepsilon_i}), \tag{4.510}$$

where

$$f(z) = \sum_{n=0}^{\infty} \gamma(n) z^n. \tag{4.511}$$

Equations (4.508) and (4.509) are exact. In some cases one can evaluate the integrals and thus obtain exact expressions for the n_i. However, in most cases these integrals cannot be evaluated and we must have recourse to approximation methods.

Method of Steepest Descents.† One such approximate method is the method of steepest descents, which is used to obtain an approximate evaluation of integrals of the kind

$$I = \int_A^B \chi(z) e^{tf(z)}\, dz, \tag{4.512}$$

where t is large, real, and positive, and $f(z)$ is an analytic function of the complex variable z $(= x + iy)$. The real and imaginary parts of $f(z)$ will be called φ and ψ

$$f = \varphi + i\psi, \tag{4.513}$$

and since $f(z)$ is an analytic function, φ and ψ both have to satisfy Laplace's equation

$$\frac{\partial^2 \varphi}{\partial x^2} + \frac{\partial^2 \varphi}{\partial y^2} = \frac{\partial^2 \psi}{\partial x^2} + \frac{\partial^2 \psi}{\partial y^2} = 0. \tag{4.514}$$

In the integral on the right-hand side of Eq. (4.512) the largest contribution

* The contours are supposed to lie within the circles of convergence of F. It must be noted that x and y are both complex variables and not the real and imaginary parts of a complex variable.

† The method of steepest descents is due to P. Debye (*Math. Ann.*, **67**, 535, 1909). For an account, see H. and B. S. Jeffreys, *Methods of Mathematical Physics*, Cambridge, 1946, Sections 17-03 and 17-04; or E. Schrödinger, *Statistical Thermodynamics*, Cambridge, 1948, p. 29.

will come from regions where φ is large. From the theory of complex functions it follows that φ has nowhere an absolute maximum, but that there are points where

$$\frac{\partial \varphi}{\partial x} = \frac{\partial \varphi}{\partial y} = 0. \tag{4.515}$$

Since

$$\frac{\partial \varphi}{\partial x} = \frac{\partial \psi}{\partial y} \quad \text{and} \quad \frac{\partial \varphi}{\partial y} = -\frac{\partial \psi}{\partial x} \tag{4.516}$$

it then follows that

$$\frac{\partial \psi}{\partial x} = \frac{\partial \psi}{\partial y} = 0 \quad \text{and} \quad f'(z) = 0, \tag{4.517}$$

where f' stands for df/dz. These points are called *saddle points*. It is possible to draw through such a saddle point at least two curves along which φ is constant. Between those curves, φ will alternately be larger and smaller than at the saddle point; in other words, there are at least two hills and two valleys.

Let us now consider a path in the complex plane. The line element of such a path will be denoted by ds and the angle between ds and the positive x axis by θ. We then have

$$\frac{\partial \varphi}{\partial s} = \cos\theta \frac{\partial \varphi}{dx} + \sin\theta \frac{\partial \varphi}{\partial y}. \tag{4.518}$$

If we are looking for paths along which $|\partial\varphi/\partial s|$ is maximum, θ has to be solved from the equation

$$0 = -\sin\theta \frac{\partial \varphi}{\partial x} + \cos\theta \frac{\partial \varphi}{\partial y}. \tag{4.519}$$

From Eq. (4.516) it then follows that on these paths $\partial\psi/\partial s = 0$ or ψ is constant. Such a path will be called a *line of steepest descent*. In each valley there will be such a line. The path over which we shall now integrate from A to B will be the following one. From A along a line of constant φ to the line of steepest descent, then via this line over the saddle point to the valley in which B is situated and finally again along a line of constant φ to B. (We do not consider here the case where A and B are in the same valley.)

Let z_0 be the coordinate of the saddle point. In the neighborhood of z_0, $f(z)$ can be expanded in the form (compare Eq. (4.517))

$$f(z) = f(z_0) + \tfrac{1}{2}(z - z_0)^2 f''(z_0) + \cdots. \tag{4.520}$$

We take the path in the direction of the valley such that $(z - z_0)^2 f''(z_0)$ is real and negative, because we *descend* along a line of constant ψ. If we introduce a new variable ζ by putting

$$f(z) - f(z_0) = -\tfrac{1}{2}\zeta^2, \tag{4.521}$$

Eq. (4.512) takes the form

$$I = e^{tf(z_0)} \int_A^B \chi(z) e^{-\frac{1}{2} t\zeta^2} \frac{dz}{d\zeta} d\zeta. \tag{4.522}$$

Introducing

$$z - z_0 = re^{-i\alpha}, \tag{4.523}$$

we have

$$\frac{dz}{d\zeta} = e^{i\alpha} | f''(z_0) |^{-1/2}, \tag{4.524}$$

where we have made use of the fact that $(z - z_0)^2 f''(z_0)$ is real and negative.

Before introducing expression (4.524) into Eq. (4.522), we will consider the integral

$$J = \int_{-\infty}^{+\infty} g(z) e^{-\frac{1}{2} a^2 z^2} dz, \tag{4.525}$$

where $g(z)$ is analytic and bounded on the real axis, and a is real and large.

The function $g(z)$ can be expanded as follows:

$$g(z) = a_0 + a_1 z + a_2 z^2 + \cdots = \sum_{n=0}^{\infty} a^n z^n. \tag{4.526}$$

Introducing this expansion into J, we get

$$J = \sum_{n=0}^{\infty} \int_{-\infty}^{+\infty} a_n z^n e^{-\frac{1}{2} a^2 z^2} dz, \tag{4.527}$$

or,

$$J = \sqrt{2\pi} \sum_{n=0}^{\infty} \frac{(2n-1)! a_{2n}}{(n-1)! 2^n a^{2n+1}}. \tag{4.528}$$

Since a is supposed to be large, we break off after only a few terms. Taking the first term only and remembering the meaning of a_0, we have, asymptotically,

$$J \approx \sqrt{2\pi} \frac{g(0)}{a}. \tag{4.529}$$

Now introducing expression (4.624) into Eq. (4.522) and using Eq. (4.525), we get

$$I \approx e^{tf(z_0)} \chi(z_0) e^{i\alpha} \left| \frac{2\pi}{tf''(z_0)} \right|^{1/2} \tag{4.530}$$

The other terms in the expansion are difficult to obtain and are very often not needed.

We now return to Eqs. (4.508) and (4.509). It can be shown that the integrands of the integrals under consideration possess a steep minimum on the real axis which lies inside the circle of convergence. The path of integration can then be taken through this minimum. In order to have our integrals in a suitable form, we write

$$F(x,y;\varepsilon_i)x^{-N-1} = e^{Ng(x)}. \tag{4.531}$$

Concentrating on the integration over x, we are concerned with the integral

$$I(x) = \frac{1}{2\pi i} \oint e^{Ng(x)}, \tag{4.532}$$

which is approximately given by the equation (the path is along a circle around the origin so that $\alpha = \frac{1}{2}\pi$)

$$I(x) = e^{Ng(x_0)} \sqrt{\frac{1}{2\pi N g''(x_0)}}, \qquad g''(x) = \frac{\partial^2 g}{\partial x^2} \tag{4.533}$$

where x_0 satisfies the equation

$$\frac{\partial g}{\partial x} = 0, \qquad \text{or} \qquad x\frac{\partial F}{\partial x} - (N+1)F = 0. \tag{4.534}$$

The last factor in the expression for $I(x)$ is only of the order $N^{1/2}$, while the first factor contains a factor N in the exponent. We may therefore to a first approximation neglect the second factor* and write

$$I(x) = e^{Ng(x_0)} = F(x_0,y;\varepsilon_i){x_0}^{-N-1}. \tag{4.535}$$

In a similar way the integration over y can be performed, and we finally get

$$G \cong {x_0}^{-N-1}{y_0}^{-E-1}F(x_0,y_0;\varepsilon_i), \tag{4.536}$$

where x_0 and y_0 are the roots of Eq. (4.534) and the equation

$$y\frac{\partial F}{\partial y} - (E+1)F = 0. \tag{4.537}$$

Since the method of the steepest descents provides only a reasonable approximation, if N and E are large, we may neglect the 1 in Eqs. (4.534) and (4.537) and have thus

$$N = \frac{x}{F}\frac{\partial F}{\partial x}. \tag{4.538}$$

$$E = \frac{y}{F}\frac{\partial F}{\partial y}. \tag{4.539}$$

* Since in averages like the ones in Eq. (4.509) we are concerned with the ratio of two integrals, the neglect of this factor is completely negligible.

By the same method we get for $\overline{n_i^p n_j^q}$,

$$\overline{n_i^p n_j^q} = \frac{1}{G} x_0^{-N-1} y_0^{-E-1} \left(\frac{1}{\ln y_0}\frac{\partial}{\partial \varepsilon_i}\right)^p \left(\frac{1}{\ln y_0}\frac{\partial}{\partial \varepsilon_j}\right)^q F(x_0, y_0; \varepsilon_i)$$

$$= \frac{1}{F} (\ln y_0)^{-p-q} \left(\frac{\partial}{\partial \varepsilon_i}\right)^p \left(\frac{\partial}{\partial \varepsilon_j}\right)^q F, \tag{4.540}$$

where we used Eq. (4.536) for G.

If we use for F Eq. (4.510), we see immediately that $\overline{n_i^p n_j^q}$ is zero unless either p or q is equal to zero, and we then have

$$\overline{n_i^p} = (\ln y_0,^{-p} \frac{1}{f_i}\frac{\partial^p f_i}{\partial \varepsilon_i^p}, \qquad f_i = f(x_0 y_0^{\varepsilon_i}), \tag{4.541}$$

where f is given by Eq. (4.511).

Equations (4.536), (4.541), (4.538), and (4.539) together give us all the necessary information which we shall now apply to the case of a system of independent particles, obeying either Boltzmann, Bose-Einstein, or Fermi-Dirac statistics. The difference between the statistics lies in the expressions for the $\gamma(n_i)$. We have from Eqs. (4.209), (4.211), and (4.214)

$$\text{Boltzmann:} \qquad \gamma(n_i) = \frac{1}{n_i!}; \tag{4.542}$$

$$\text{Bose-Einstein:} \qquad \gamma(n_i) = 1; \tag{4.543}$$

$$\text{Fermi-Dirac:} \qquad \left.\begin{aligned} \gamma(n_i) &= 1, \quad n_i = 0 \quad \text{or} \quad 1, \\ &= 0, \quad n_i > 1. \end{aligned}\right\} \tag{4.544}$$

And we get from Eq. (4.511) for the functions f,

$$f_{\text{Bo}} = \sum_{n=0}^{\infty} \frac{z^n}{n!} = e^z, \tag{4.545}$$

$$f_{\text{BE}} = \sum_{n=0}^{\infty} z^n = \frac{1}{1-z}, \tag{4.546}$$

$$f_{\text{FD}} = \sum_{n=0}^{1} z^n = 1 + z. \tag{4.547}$$

We then get from Eq. (4.510) for F

$$F_{\text{Bo}} = \exp[x \sum_i y^{\varepsilon_i}], \tag{4.548}$$

$$F_{\text{BE}} = \prod_i \frac{1}{1 - x y^{\varepsilon_i}} \tag{4.549}$$

$$F_{\text{FD}} = \prod_i (1 + x y^{\varepsilon_i}). \tag{4.550}$$

The x_0 and y_0 follow from Eqs. (4.538) and (4.539), or

$$N = \sum_i \frac{\partial \ln f_i}{\partial \ln x_0} \tag{4.551}$$

$$E = \sum_i \frac{\partial \ln f_i}{\partial \ln y_0}. \tag{4.552}$$

Using Eqs. (4.541) with $p = 1$, we see that the Eqs. (4.551) and (4.552) can be written in the following translucent form

$$N = \sum_i n_i, \tag{4.553}$$

$$E = \sum_i n_i \varepsilon_i. \tag{4.554}$$

From Eq. (4.541) we have

$$\overline{n_i} = \frac{x_0}{f_i} \frac{\partial f_i}{\partial x_0}, \tag{4.555}$$

or, using Eqs. (4.545) to (4.547),

$$(\overline{n_i})_{\mathrm{Bo}} = x_0 y_0^{\varepsilon_i}, \tag{4.556}$$

$$(\overline{n_i})_{\mathrm{BE}} = \frac{x_0 y_0^{\varepsilon_i}}{1 - x_0 y_0^{\varepsilon_i}}, \tag{4.557}$$

$$(\overline{n_i})_{\mathrm{FD}} = \frac{x_0 y_0^{\varepsilon_i}}{1 + x_0 y_0^{\varepsilon_i}}. \tag{4.558}$$

If we compare expressions (4.556) to (4.558) with Eqs. (4.314) to (4.316), with $Z_j \equiv 1$, $N_j \equiv n_i$, we see that we have the same equations, provided

$$y_0 = e^{-\beta}, \tag{4.559}$$

$$x_0 = e^{\alpha}. \tag{4.560}$$

These equations give us the physical meaning of x_0 and y_0. They can be derived directly, as was done by Darwin and Fowler in their original papers.

4.6 THE PERFECT BOLTZMANN GAS

We shall consider in the present and in the subsequent sections the case of a perfect gas. In order to simplify the expression for the energy levels, we assume that the gas is enclosed in a cube of edge length L and of volume $V(=L^3)$.

If m is the mass of one particle, the energy levels are given by the equation

$$\varepsilon = \frac{\varepsilon_0}{3}(k_1{}^2 + k_2{}^2 + k_3{}^2), \qquad \varepsilon_0 = \frac{3h^2}{8mL^2}, \tag{4.601}$$

where k_1, k_2, and k_3, which are positive integers, are the quantum numbers corresponding to the three translational degrees of freedom. The quantity ε_0 is the lowest energy level corresponding to $k_1 = k_2 = k_3 = 1$.

Each stationary state corresponds to a point in the three-dimensional k_1,k_2,k_3-space which is situated in the positive octant. Introducing k by the equation

$$k^2 = k_1{}^2 + k_2{}^2 + k_3{}^2, \tag{4.602}$$

we find for the number of energy levels dZ with k between k and $k + dk$ the expression

$$dZ = \tfrac{1}{8} \cdot 4\pi k^2 \, dk, \tag{4.603}$$

where we have assumed that the particle has no spin. If the particle had spin S, expression (4.604) should be multiplied by $(2S + 1)$. Using Eqs. (4.601) and (4.602), we can write this expression for dZ in the form*

$$dZ = 2\pi\left(\frac{2m}{h^2}\right)^{3/2} V\sqrt{\varepsilon}\, d\varepsilon. \tag{4.604}$$

If we consider a Boltzmann gas, we have from Eqs. (4.303) and (4.313)

$$\ln W_{\text{Bo}} = \sum_j N_j(-\alpha + \beta E_j + 1) = -(\alpha - 1)N + \beta E, \tag{4.605}$$

where we have dropped the index "max" on the W and where the N_i are once more the equilibrium values.

For the q-potential we get

$$q_{\text{Bo}} = \ln W + \alpha N - \beta E = N = \sum_j N_j, \tag{4.606}$$

or

$$q_{\text{Bo}} = e^\alpha \sum_j Z_j e^{-\beta E_j} = e^\alpha Z(\beta), \tag{4.607}$$

where $Z(\beta)$ which, apart from the factor K', is the same as $Z_\mu{}^{(\text{qu})}$ of Eq. (3.413) is the *partition function* which is a function of the β (and of the external parameters),

$$Z(\beta) = \sum_j Z_j e^{-\beta E_j}. \tag{4.608}$$

For the groups of energy levels Z_j we can in the case of a gas take energy intervals dZ, and then all sums will go over into integrals.

Using expression (4.604) for dZ, we get for the partition function

$$Z = 2\pi\left(\frac{2m}{h^2}\right)^{3/2} V \int_0^\infty \varepsilon^{1/2} e^{-\beta\varepsilon}\, d\varepsilon = \left(\frac{2\pi m}{\beta h^2}\right)^{3/2} V, \tag{4.609}$$

* It has to be remarked that Eq. (4.604) remains valid even if the volume V is not a cube. See R. Courant and D. Hilbert, *Methods of Mathematical Physics*, Vol. I, New York, 1957.

and for the q-potential

$$q_{\text{Bo}} = \left(\frac{2\pi m}{\beta h^2}\right)^{3/2} Ve^{\alpha}. \tag{4.610}$$

From Eq. (4.422) we get

$$N = \frac{\partial q_{\text{Bo}}}{\partial \alpha} = q_{\text{Bo}}, \tag{4.611}$$

which is the same as Eq. (4.606).

If we assume that V is the only external parameter so that $\delta A = P\delta V$, we have from Eq. (4.424) (see also Eq. (4.426)),

$$\beta P = \frac{\partial q_{\text{Bo}}}{\partial V} = \frac{q_{\text{Bo}}}{V} = \frac{N}{V}, \tag{4.612}$$

or

$$PV = \frac{N}{\beta} = NkT, \tag{4.613}$$

which is once more the perfect gas law.

The energy E of the system follows from Eq. (4.423)

$$E = -\frac{\partial q_{\text{Bo}}}{\partial \beta} = \frac{3}{2}\frac{q_{\text{Bo}}}{\beta} = \frac{3}{2}NkT, \tag{4.614}$$

corresponding to the equipartition of the kinetic energy, giving a contribution $\frac{1}{2}kT$ for each degree of freedom.

The partial thermal potential divided by kT, α, is given by

$$\alpha = \ln \frac{q_{\text{Bo}}}{Z} = \ln P - \frac{5}{2}\ln T - \mathfrak{C} - \frac{3}{2}\ln M, \tag{4.615}$$

where M is the molecular weight of the gas and $\mathfrak{C}$ a constant which is connected with the so-called chemical constants (see Section 9.2) and which is given by the equation

$$\mathfrak{C} = \ln\left[k\left(\frac{2\pi m_{\text{H}} k}{h^2}\right)^{3/2}\right], \tag{4.616}$$

where m_{H} is the molecular weight unit.

Finally we get for the entropy from Eq. (4.420), using Eqs. (4.611) and (4.614),

$$S = kq_{\text{Bo}} - k\alpha N + \frac{E}{T} = Nk\left(\frac{5}{2} - \alpha\right), \tag{4.617}$$

or, using Eq. (4.615)

$$S = k[\tfrac{5}{2}\ln T - \ln P + \mathfrak{C} + \tfrac{3}{2}\ln M + \tfrac{5}{2}], \tag{4.618}$$

the so-called Sackur-Tetrode equation.*

4.7 THE PERFECT BOSE-EINSTEIN GAS

We now turn to quantum gases. First of all, we shall investigate under what circumstances we can safely apply Boltzmann statistics.

Since the energy eigenvalues given by Eq. (4.601) are all positive and since $\beta\varepsilon_0$ is very small compared to unity for all attainable temperatures, as we saw in Section 3.5, the condition for the applicability of Boltzmann statistics (Eq. (4.317)) reduces to the condition

$$e^{-\alpha} \gg 1. \tag{4.701}$$

If inequality (4.701) is satisfied, the total number of particles will be given to a fair approximation by the same equation for the three statistics,

$$N = \sum_j N_j \approx \sum_j Z_j e^{\alpha - \beta E_j},$$

or

$$N \cong e^{\alpha}\left(\frac{2\pi mkT}{h^2}\right)^{3/2} V, \tag{4.702}$$

where we have used Eqs. (4.314) to (4.316), (4.701), (4.608), and (4.609).

We thus get for $e^{-\alpha}$ approximately

$$e^{-\alpha} \cong \left(\frac{2\pi mk}{h^2}\right)^{3/2} T^{3/2}\left(\frac{N}{V}\right)^{-1}. \tag{4.703}$$

From Eqs. (4.703) and (4.701) we see first of all that, in the limit $h \to 0$, Boltzmann statistics may be applied. We can thus understand why Boltzmann statistics are often called the classical statistics. Secondly we see that in the limit of high temperatures or low densities condition (4.701) is satisfied and Boltzmann statistics can be applied.

We should like to write Eq. (4.703) in a slightly different form. The average energy per particle—which is in the present case the same as the average kinetic energy—is in the classical limit equal to $\frac{3}{2}kT$ (compare Eq. (4.614). It follows from the usual quantum-mechanical arguments that

* O. Sackur, *Ann. Physik*, **36,** 958, 1911; **40,** 67, 1912. H. Tetrode, *Ann. Physik*, **38, 434,** 1912 (corrections in *Ann. Physik*, **39,** 255, 1912); *Proc. Kon. Ned. Akad. Wet.* (Amsterdam), **17,** 1167, 1915.

the De Broglie wavelength λ_{DBr} of the particle will be given by the equation

$$\lambda_{\text{DBr}} = \frac{h}{p} = \frac{h}{\sqrt{2m\varepsilon}} = \left(\frac{h^2}{3mkT}\right)^{1/2}, \tag{4.704}$$

where p is the momentum of the particle.

Combining Eqs. (4.703) and (4.704), we have

$$e^\alpha \cong \left(\frac{3}{2\pi}\right)^{3/2} \lambda_{\text{DBr}}{}^3 \frac{N}{V} \equiv \zeta. \tag{4.705}$$

We see from Eq. (4.705) that condition (4.701) is equivalent to the condition that the De Broglie wavelength of the particles is small compared to the average distance apart in the gas.

The quantity ζ introduced by Eq. (4.705) is essentially what Sommerfeld* calls the *degeneracy parameter*. We shall see presently that it is the natural parameter in which to expand for the case when the perfect gas is nearly classical, while in the case of extreme quantum behavior, when $\zeta \gg 1$, the natural expansion parameter is an inverse power of ζ. Of course, Eqs. (4.703) and (4.705) are only valid in the case when $e^\alpha \ll 1$, and they actually represent the first terms in a series expansion (in ζ).

We now turn to a perfect Bose-Einstein gas, and to simplify our formulas we assume that the particles have zero spin. From Eqs. (4.304) and (4.315) we have

$$\ln W_{\text{BE}} = \sum_j N_j(-\alpha + \beta E_j) - \sum_j Z_j \ln(1 - e^{\alpha - \beta E_j}). \tag{4.706}$$

The q-potential follows from Eq. (4.420) and is given by the equation

$$q_{\text{BE}} = -\sum_j Z_j \ln(1 - e^{\alpha - \beta E_j}). \tag{4.707}$$

In most cases we can again use an integral instead of the sum, and then we can apply Eq. (4.604) and get

$$q_{\text{BE}} = -2\pi\left(\frac{2m}{h^2}\right)^{3/2} V \int_0^\infty \sqrt{\varepsilon} \ln(1 - e^{\alpha - \beta\varepsilon})\, d\varepsilon, \tag{4.708}$$

or

$$q_{\text{BE}} = \left(\frac{2\pi m}{\beta h^2}\right)^{3/2} V g(\alpha), \tag{4.709}$$

where

$$g(\alpha) = \sum_{n=1}^{\infty} \frac{e^{n\alpha}}{n^{5/2}}. \tag{4.710}$$

* A. Sommerfeld, *Z. Physik*, **47**, 1, 1928.

We have for $g(\alpha)$, as long as $e^\alpha \leqslant 1$,

$$g(\alpha) = \frac{-2}{\sqrt{\pi}} \int_0^\infty \sqrt{x} \ln (1 - e^{\alpha - x})\, dx$$

$$= \frac{2}{\sqrt{\pi}} \sum_{n=1}^{\infty} \frac{1}{n} \int_0^\infty \sqrt{x} e^{n\alpha - nx}\, dx,$$

whence Eq. (4.710) follows.

For the total number of particles we have from Eq. (4.422)

$$N = \frac{\partial q_{\mathrm{BE}}}{\partial \alpha} = \left(\frac{2\pi m}{\beta h^2}\right)^{3/2} V g'(\alpha) = q_{\mathrm{BE}} \frac{g'(\alpha)}{g(\alpha)}, \tag{4.711}$$

where

$$g'(\alpha) = \frac{dg(\alpha)}{d\alpha} = \sum_{n=1}^{\infty} \frac{e^{n\alpha}}{n^{3/2}}. \tag{4.712}$$

The pressure and total energy are given by the equations

$$P = \frac{q_{\mathrm{BE}}}{\beta V}, \tag{4.713}$$

$$E = -\frac{\partial q_{\mathrm{BE}}}{\partial \beta} = \frac{3}{2} \frac{q_{\mathrm{BE}}}{\beta}. \tag{4.714}$$

We see that from Eqs. (4.713) and (4.714) it follows that

$$P = \frac{2}{3} \frac{E}{V}. \tag{4.715}$$

This equation could have been derived much more easily. From Eq. (4.601) we see that the energy of a particle is proportional to $V^{-2/3}$. As $-\partial\varepsilon/\partial V$ is the pressure exerted by one particle, it follows that this partial pressure is equal to $2\varepsilon/3V$ and Eq. (4.715) follows immediately. It also follows from this derivation that this equation must hold irrespective of the kind of statistics. Indeed, from Eqs. (4.613) and (4.614) it follows in the Boltzmann case, and it will also be found to hold in the Fermi-Dirac case.

From Eqs. (4.711) and (4.713) we get

$$PV = NkT \frac{g(\alpha)}{g'(\alpha)}. \tag{4.716}$$

Since in the Bose-Einstein case Eq. (4.318) must always be satisfied, we can always expand in powers of e^α. If we expand $g(\alpha)/g'(a)$ in powers of e^α we get from Eq. (4.716)

$$PV = NkT\left[1 - \frac{e^\alpha}{2^{5/2}} + \cdots\right]. \tag{4.717}$$

We observe here deviations from the perfect gas law. These deviations tend to zero when α goes to $-\infty$.

Introducing now the degeneracy parameter ζ through Eq. (4.705), we get from Eq. (4.711) the following series expansion for e^α,

$$e^\alpha = \zeta\left(1 - \frac{\zeta}{2^{3/2}} + \cdots\right), \tag{4.718}$$

and we see that, indeed, Eq. (4.703) follows in the limit as $\zeta \to 0$.

From Eqs. (4.717) and (4.718), we then get for the equation of state

$$PV = NkT\left(1 - \frac{\zeta}{2^{5/2}} + \cdots\right). \tag{4.719}$$

Using Eqs. (4.714), (4.711), and (4.718), we get the following expansion for the energy,

$$E = \frac{3}{2}NkT\left(1 - \frac{\zeta}{2^{5/2}} + \cdots\right). \tag{4.720}$$

Equation (4.720) could, of course, also have been obtained from Eqs. (4.715) and (4.719).

The entropy follows from Eq. (4.420):

$$S = k(q - \alpha N + \beta E). \tag{4.721}$$

Using Eqs. (4.713), (4.718), (4.719), and (4.720), we get for S the following expansion

$$S = Nk\left[\frac{5}{2} - \ln\zeta - \frac{\zeta}{2^{7/2}} + \cdots\right]. \tag{4.722}$$

We now wish to calculate the specific heats C_v and C_P and their ratio $\kappa = C_P/C_v$. For C_v we have from Eqs. (4.705) and (4.720)

$$C_v = \left(\frac{\partial E}{\partial T}\right)_V = \frac{3}{2}Nk\left[1 + \frac{\zeta}{2^{7/2}} + \cdots\right]. \tag{4.723}$$

In order to obtain C_P we have to express δQ in terms of δT under the condition that the pressure is kept constant. This can be done in the following way. First of all we use Eq. (4.715) to get

$$\delta Q = \delta E + P\delta V = \delta E + \frac{2}{3}\frac{E}{V}\delta V = \frac{5}{3}\delta E, \tag{4.724}$$

since from Eq. (4.715) it follows that in the case of constant pressure $\delta \ln E = \delta \ln V$.

From Eqs. (4.720) and (4.705) we have

$$\delta E = \frac{3}{2} Nk\left[1 - \frac{\zeta}{2^{5/2}} + \cdots\right]\delta T - \frac{3}{2} NkT \frac{\delta\zeta}{2^{5/2}} + \cdots, \tag{4.725}$$

and

$$\delta \ln \zeta = -\tfrac{3}{2}\delta \ln T - \delta \ln V. \tag{4.726}$$

Combining Eqs. (4.724) to (4.726), we finally get for C_P the expansion

$$C_P = \left(T \frac{\partial S}{\partial T}\right)_P = \frac{5}{2} Nk\left[1 + \frac{3\zeta}{2^{7/2}} + \cdots\right], \tag{4.727}$$

and for the ratio of the specific heats

$$\kappa = \frac{C_P}{C_v} = \frac{5}{3}\left[1 + \frac{\zeta}{2^{5/2}} + \cdots\right]. \tag{4.728}$$

All series converge at sufficiently high temperatures, and everywhere the first term gives the expression for the perfect Boltzmann gas.

We mentioned in Section 4.3 that in the Bose-Einstein case α should be restricted to values smaller than $\beta\varepsilon_0$. Let us consider for a moment the case where α approaches this limiting value. We shall assume that α is so close to $\beta\varepsilon_0$ that we have

$$-\alpha + \beta\varepsilon_0 \ll 1 \qquad \text{and} \qquad -\alpha + \beta\varepsilon_0 \ll \beta(\varepsilon_1 - \varepsilon_0), \tag{4.729}$$

where ε_1 is the lowest level but one (it follows from Eq. (4.601) that $\varepsilon_1 = 2\varepsilon_0$).

If we use Eqs. (4.216) and (4.319) to calculate the total number of particles, we have

$$N = \sum_j \frac{Z_j}{e^{-\alpha+\beta E_j} - 1}. \tag{4.730}$$

We shall ignore for the moment the fact that in our derivations we have used the assumption $Z_j \gg 1$ and we shall assume all Z_j to be equal to 1, so that the E_j are the energy levels of the particles. Using the inequalities (4.729), Eq. (4.730) then reduces to

$$N = \frac{1}{e^{-\alpha+\beta\varepsilon_0} - 1} + \frac{1}{e^{-\alpha+\beta\varepsilon_1} - 1} + \cdots \tag{4.731}$$

or

$$N \approx \frac{1}{-\alpha + \beta\varepsilon_0}. \tag{4.732}$$

Since the different terms on the right-hand side of Eq. (4.730) represent the number of atoms occupying the various energy levels, Eq. (4.732) expresses that as $\alpha \to \beta\varepsilon_0$ all atoms tend to occupy the lowest level. This phenomenon is the so-called *Einstein condensation.* It is a typical quantum effect, since for $h \to 0$, α will go to $-\infty$.

We cannot at this point enter into a detailed discussion of condensation phenomena which will be discussed in Vol. II.* However, we shall briefly describe the way the isotherms of a perfect Bose-Einstein gas behave, showing that they have a horizontal part.

It is convenient to rewrite the equations we have derived a moment ago. We introduce the activity z by the equation

$$z = e^{\alpha}, \tag{4.733}$$

and can then write Eqs. (4.713) and (4.709) in the form

$$q = \beta PV = \frac{V}{v_0} f(z), \tag{4.734}$$

where

$$f(z) = \sum_{n=1}^{\infty} \frac{z^n}{n^{5/2}}, \tag{4.735}$$

while v_0 is, apart from a numerical factor of order of magnitude unity, the cube of the average De Broglie wavelength (compare Eq. (4.704)),

$$v_0 = \left(\frac{\beta h^2}{2\pi m}\right)^{3/2} \tag{4.736}$$

Introducing the specific volume $v = V/N$, we find the following relation between v and z from Eq. (4.711):

$$N = z\frac{\partial q}{\partial z} = \frac{V}{v_0} zf'(z), \qquad \left(f'(z) = \sum_{n=1}^{\infty} \frac{z^n}{n^{3/2}}\right), \tag{4.737}$$

or

$$\frac{1}{v} = \frac{1}{v_0} zf'(z) \tag{4.738}$$

The power series on the right-hand side of Eqs. (4.734) and (4.738) converge for $z \leqslant 1$, but are no longer convergent for $z > 1$.† As long as the specific

* See also Chapter 9 of the first edition of this book where an extensive bibliography is also given.

† It has been shown by Opechowski (*Physica*, **4**, 722, 1937) that the following expansions are valid when z lies in the neighborhood of 1:

$$f(z) = 2.36(-\ln z)^{3/2} + 1.34 + 2.61 \ln z - 0.73(\ln z)^2 + \cdots,$$

$$zf'(z) = -3.54(-\ln z)^{1/2} + 2.61 - 1.46 \ln z - 0.10(\ln z)^2 + \cdots.$$

volume v is larger than a critical value v_c given by the equation

$$\frac{1}{v_c} = \frac{1}{v_0} f'(1) = \frac{2.61}{v_0}, \tag{4.739}$$

we can use Eqs. (4.734) and (4.738) to obtain the equation of state. However, one has to consider with some special care the case where $v < v_c$.

Let us now consider the isotherms. They will consist of two parts. The first part, where $v \geqslant v_c$, is given by Eq. (4.734),

$$P = \frac{kT}{v_0} f(z), \qquad v \geqslant v_c, \tag{4.740}$$

where z as a function of v follows from Eq. (4.738). The second part of the isotherm is given by the equation

$$P = \frac{kT}{v_0} f(1), \qquad v < v_c. \tag{4.741}$$

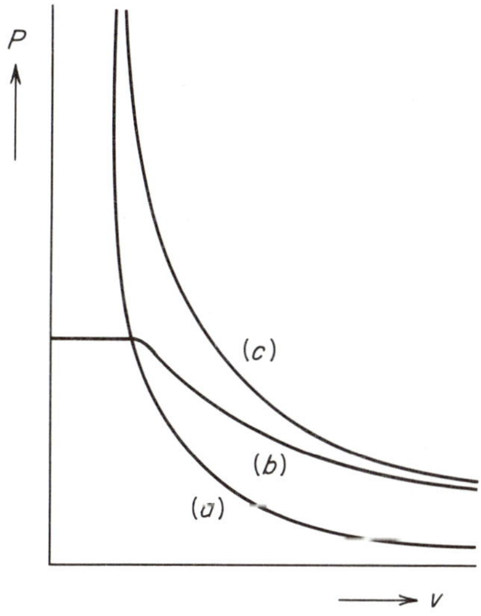

FIG. 4.2 The isotherms of a perfect boson gas: a is the curve on which all condensation points lie; b the isotherm of a boson gas; and c the isotherm of a Boltzmann gas at the same temperature.

Curve b of Fig. 4.2 depicts one of these isotherms. We have drawn in the same figure for comparison the isotherm of a perfect Boltzmann gas at the same temperature (curve c). Curve a in Fig. 4.2 is finally the curve on which all condensation points lie. The equation of this curve is obtained by eliminating

T from Eqs. (4.734) and (4.738) and putting $z=1$, and we have in this way

$$Pv^{5/3} = \frac{h^2}{2\pi m} \frac{f(1)}{[f'(1)]^{5/3}} \tag{4.742}$$

Equations (4.740) and (4.741) are exactly valid only in the limiting case as $N \to \infty$. They can be proved most easily by first considering the case of finite N and afterward letting N go to infinity. We shall briefly sketch the argument. We must make use of the fact that although Eqs. (4.734) and (4.738) have been derived by replacing sums over energy levels by integrals, the energy spectrum will really be discrete as long as the volume remains finite, and instead of Eqs. (4.734) and (4.738) we should really use Eqs. (4.707) and (4.422) with all Z_j put equal to 1:

$$q = -\sum_s \ln (1 - e^{\alpha - \beta\varepsilon_s}), \tag{4.743}$$

$$N = \sum_s \frac{1}{e^{-\alpha+\beta\varepsilon_s} - 1}. \tag{4.744}$$

If v approaches v_c, α will approach 0, and it is dangerous to replace the sums in Eqs. (4.743) and (4.744) by integrals. We shall, for the sake of simplicity, shift the zero of the energy levels in such a way that the lowest energy level, ε_0, is zero. Since α will be in the neighborhood of zero, the first term in the sum on the right-hand side of Eq. (4.744) will become much more important than the other terms and we shall therefore write Eq. (4.744) as

$$N = \frac{1}{e^{-\alpha} - 1} + \sum_{s \neq 0} \frac{1}{e^{-\alpha+\beta\varepsilon_s} - 1}. \tag{4.745}$$

Replacing the sum over all energy levels but the lowest by an integral and using the fact that the first term on the right-hand side of Eq. (4.745) represents the average number, n_0, of particles in the lowest energy level (compare Eq. (4.315)), we can write Eq. (4.745) in the form

$$N = n_0 + \frac{V}{v_0} zf'(z). \tag{4.746}$$

Introducing the specific volume v $(=V/N)$ and its critical value v_c given by Eq. (4.739), we have

$$1 = \frac{n_0}{N} + \frac{vzf'(z)}{v_c f'(1)}. \tag{4.747}$$

As long as $v > v_c$, Eq. (4.747) can easily be satisfied and, indeed, only a small fraction of the particles will occupy the lowest level so that Eqs. (4.747) and (4.738) are identical. However, if $v < v_c$, Eq. (4.747) must replace Eq. (4.738). In that case we should expect z to be about equal to one—which can be verified by a careful analysis—and using for n_0 Eq. (4.315) and using the fact that $z \cong 1$, we can write Eq. (4.747) in the form

$$1 = \frac{1}{N(1-z)} + \frac{v}{v_c}, \qquad v < v_c, \tag{4.748}$$

or

$$z = 1 - \frac{1}{N}\frac{v_c}{v_c - v}, \tag{4.749}$$

and we see that, indeed, in the limit as $N \to \infty$, z will be equal to one for all values of v smaller than v_c and Eq. (4.741) follows.

We shall not give here a discussion of the isotherm either for $v < v_c$ or for v in the neighborhood of v_c,* or a discussion of the possible relation between the Einstein condensation and the λ-transition of liquid helium.†

Before leaving the case of the Einstein condensation we want to see what happens when, instead of decreasing the specific volume while keeping the temperature constant, we lower the temperature while keeping the density (or v) constant. Condensation will then set in as soon as the temperature becomes equal to a critical temperature T_0 given by the equation (compare Eqs. (4.739) and (4.736))

$$T_0 = \frac{h^2}{2\pi mk}[2.61v]^{-2/3} = \frac{h^2}{2\pi mk}\left(\frac{\rho}{2.61m}\right)^{2/3}, \tag{4.750}$$

or

$$T_0 = 115\rho^{2/3}M^{-5/3}\ {}^\circ\mathrm{K}, \tag{4.751}$$

where ρ is the density of the system in g cm^{-3} and where M is the molecular weight of the gas.

Let us briefly summarize the behavior of pressure, energy, and specific heat as functions of the temperature for given values of N and V.

First consider the pressure, from which the energy follows immediately from Eq. (4.715). As long as the temperature is above T_0 we have Eq. (4.740), or

$$P = CT^{5/2}f(z), \qquad T > T_0, \tag{4.752}$$

where

$$C = \left(\frac{2\pi mk}{h^2}\right)^{3/2} k, \tag{4.753}$$

and where z is still a function of T by Eq. (4.738).

At temperatures below T_0, when z will be equal to one, we have

$$P = CT^{5/2}f(1), \qquad T \leqslant T_0. \tag{4.754}$$

* See, for instance, D. ter Haar, *Proc. Roy. Soc.* (*London*), **A212**, 552, 1952.

† See, for example, Chapter 9 of the first edition of this book. This point will also be discussed in Vol. II.

The specific heat per particle, c_v, follows from Eqs. (4.752), (4.754), and (4.740) and from the relation

$$c_v = \frac{1}{N}\frac{\partial E}{\partial T}, \tag{4.755}$$

and we have

$$c_v = \frac{3}{2} v \frac{\partial P}{\partial T}. \tag{4.756}$$

At temperatures below T_0 Eq. (4.756) leads to the equation

$$c_v = \tfrac{15}{4}\, CvT^{3/2} f(1), \qquad T \leqslant T_0. \tag{4.757}$$

The value of c_v on the condensation curve, where $z = 1$ and $T = T_0$, follows from Eqs. (4.757) and (4.750) and we have

$$c_v = \frac{15}{4} k \frac{f(1)}{f'(1)} = 1.926k, \tag{4.758}$$

which is actually more than the classical value of $1.5k$. From Eq. (4.756) it follows that c_v is a continuous function of T, though $\partial c_v/\partial T$ is discontinuous. Figure 4.3 depicts P and c_v as functions of T.

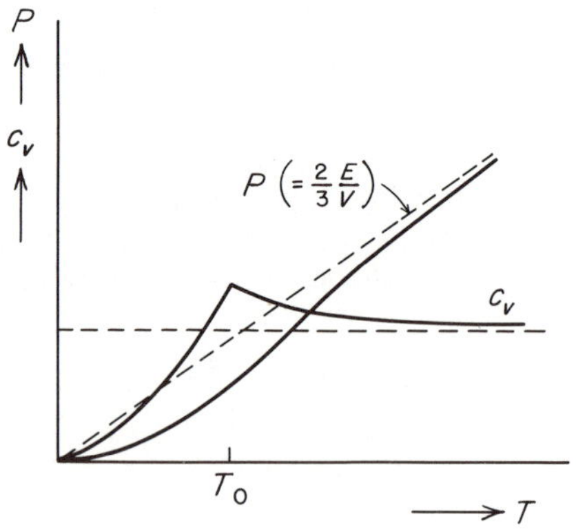

FIG. 4.3 Pressure and specific heat of a perfect boson gas.

To what extent can the Einstein condensation be compared to the condensation of an imperfect gas? In the latter case, part of the system will form the liquid and thus form a different phase. This is not the case when we are considering a perfect Bose-Einstein gas. The particles that are too many, that is, those particles in excess of Nv/v_c will all be found in the lowest energy level (compare Eq. (4.747)), as was first pointed out by Einstein.*

4.8 THE PERFECT FERMI-DIRAC GAS

The discussion in this section develops parallel to that given in the preceding section as long as $e^\alpha \ll 1$. We shall first give the results without any further comments. Instead of Eqs. (4.706) to (4.714) we have now†

$$\ln W_{\mathrm{FD}} = \sum_j N_j(-\alpha + \beta E_j) + \sum_j Z_j \ln (1 + e^{\alpha - \beta E_j}), \tag{4.801}$$

$$q_{\mathrm{FD}} = \sum_j Z_j \ln [1 + e^{\alpha - \beta E_j}], \tag{4.802}$$

$$q_{\mathrm{FD}} = 4\pi\left(\frac{2m}{h^2}\right)^{3/2} V \int_0^\infty \sqrt{\varepsilon} \ln [1 + e^{\alpha - \beta\varepsilon}]\, d\varepsilon, \tag{4.803}$$

$$q_{\mathrm{FD}} = 2\left(\frac{2\pi m}{\beta h^2}\right)^{3/2} V h(\alpha), \tag{4.804}$$

$$h(\alpha) = -\sum_{n=1}^{\infty} \frac{(-e^\alpha)^n}{n^{5/2}}, \tag{4.805}$$

$$h'(\alpha) = -\sum_{n=1}^{\infty} \frac{(-e^\alpha)^n}{n^{3/2}}, \tag{4.806}$$

$$N = q_{\mathrm{FD}} \frac{h'(\alpha)}{h(\alpha)}, \tag{4.807}$$

$$P = \frac{q_{\mathrm{FD}}}{\beta V}, \tag{4.808}$$

$$E = \frac{3}{2} \frac{q_{\mathrm{FD}}}{\beta}. \tag{4.809}$$

From Eqs. (4.808) and (4.809) it follows again that Eq. (4.715) is satisfied. Introducing once more a quantity ζ by Eq. (4.705), but with $(3/4\pi)^{3/2}$ instead of $(3/2\pi)^{3/2}$ in the second member of that equation to take into account the spin of the fermions, we now get instead of Eqs. (4.718) to (4.720), (4.722),

* It seems to me that the distinction which London makes between a condensation in momentum space in the case of the Einstein condensation and a condensation in coordinate space in the case of an imperfect gas is not justified, since the condensed phase of an imperfect gas also does not contribute to the pressure and thus also corresponds to a phase condensed in momentum space. Compare also the discussion by Becker (*Z. Physik*, **128,** 120, 1950), who points out that the Einstein condensation can also be considered to be a condensation in coordinate space.

† We have assumed that we are dealing with spin-$\frac{1}{2}$ particles. This leads to extra factors 2 in Eqs. (4.803), (4.804), and so on.

(4.727), and (4.728), the expansions

$$e^{\alpha} = \zeta\left[1 + \frac{\zeta}{2^{3/2}} + \cdots\right], \tag{4.810}$$

$$PV = NkT\left[1 + \frac{\zeta}{2^{5/2}} + \cdots\right], \tag{4.811}$$

$$E = \frac{3}{2}NkT\left[1 + \frac{\zeta}{2^{5/2}} + \cdots\right], \tag{4.812}$$

$$S = Nk\left[\frac{5}{2} - \ln\zeta + \frac{\zeta}{2^{7/2}} + \cdots\right], \tag{4.813}$$

$$C_v = \frac{3}{2}Nk\left[1 - \frac{\zeta}{2^{7/2}} + \cdots\right], \tag{4.814}$$

$$C_V = \frac{5}{2}Nk\left[1 - \frac{3\zeta}{2^{7/2}} + \cdots\right], \tag{4.815}$$

$$\kappa = \frac{5}{3}\left[1 - \frac{\zeta}{2^{5/2}} + \cdots\right]. \tag{4.816}$$

All series converge as long as $e^{\alpha} \leqslant 1$. We see that in first approximation Bose-Einstein and Fermi-Dirac statistics differ only in the *sign* of the deviation from the Boltzmann statistics.

We shall now discuss the case where $e^{\alpha} \gg 1$. This is called the case of *strong degeneracy* and the gas is called a *degenerate quantum gas*. As $T \to 0$, we shall find that $\alpha \to \infty$, and we shall say that at the absolute zero the degeneracy is complete. Let us first discuss the behavior at absolute zero.

From Eqs. (4.803) and (4.422) we get for the total number of particles

$$N = \frac{\partial q}{\partial \alpha} = 4\pi\left(\frac{2m}{h^2}\right)^{3/2} V \int_0^{\infty} \frac{\sqrt{\varepsilon}\, d\varepsilon}{1 + e^{-\alpha+\beta\varepsilon}}. \tag{4.817}$$

From this equation we see that the number of particles with energies between ε and $\varepsilon + d\varepsilon$ is given by the equation

$$dN = f(\varepsilon)\, d\varepsilon = 4\pi\left(\frac{2m}{h^2}\right)^{3/2} V \frac{\sqrt{\varepsilon}\, d\varepsilon}{1 + e^{-\alpha+\beta\varepsilon}}. \tag{4.818}$$

This equation is actually the same as Eq. (4.316) if we use for Z_j the expression (4.604) (with an extra spin-factor 2) for the number of levels dZ with energies between ε and $\varepsilon + d\varepsilon$. We note that, whereas $f(\varepsilon)$ varies as $\sqrt{\varepsilon}$ for small values of ε, for high energies it varies as $\sqrt{\varepsilon}\, e^{-\beta\varepsilon}$: the Fermi distribution (4.818) has a "Maxwellian tail" (compare Eq. (1.104) and take into account that for a spherical symmetric distribution $d^3\mathbf{c}$ can be written as $4\pi c^2\,dc = 4\pi\sqrt{(2\varepsilon/m^3)}\,d\varepsilon$). We have depicted in Fig. 4.4 the function $f(\varepsilon)$ both for a finite temperature and for $T = 0$, when the Maxwellian tail is absent. It is often useful to consider instead of $f(\varepsilon)$ a function $F(\varepsilon)$, sometimes called the Fermi function, which differs from $f(\varepsilon)$ by a factor $dZ/d\varepsilon$:

$$F(\varepsilon) = \frac{1}{e^{-\alpha+\beta\varepsilon} + 1}. \tag{4.819}$$

This function and the negative of its derivative with respect to ε are shown in Fig. 4.4 for a finite temperature. The dotted line in Fig. 4.4 shows $F(\varepsilon)$ for $T = 0$; in that case, $-\partial F/\partial\varepsilon$ becomes a delta function.

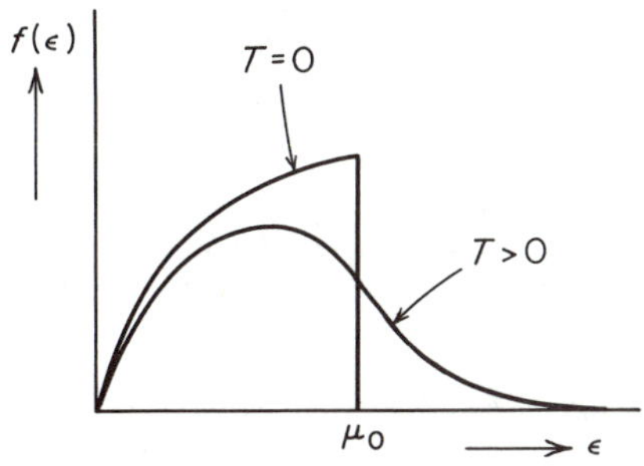

FIG. 4.4 The distribution function $f(\varepsilon)$ of Eq. (4.818) at $T = 0$ and at a finite temperature.

We now introduce by Eq. (4.415) the partial chemical potential $\mu = \alpha/\beta$. Its value for $T = 0$ is denoted by μ_0. In order to determine the value of μ_0, we split the integral in Eq. (4.817) into two parts and write

$$N = 4\pi\left(\frac{2m}{h^2}\right)^{3/2} V\left\{\int_0^{\mu}\frac{\sqrt{\varepsilon}\,d\varepsilon}{1 + e^{-\beta(\mu-\varepsilon)}} + \int_{\mu}^{\infty}\frac{\sqrt{\varepsilon}\,d\varepsilon}{1 + e^{\beta(\mu-\varepsilon)}}\right\}. \tag{4.820}$$

This equation is valid for any value of T (or β). In the limit as $T \to 0$, or $\beta \to \infty$, the second integral vanishes and the first one is equal to $\frac{2}{3}{\mu_0}^{3/2}$. We have thus in that limit

$$N = \frac{8\pi}{3}\left(\frac{2m}{h^2}\right)^{3/2} V{\mu_0}^{3/2}, \tag{4.821}$$

or

$$\mu_0 = \frac{h^2}{2m}\left(\frac{3N}{8\pi V}\right)^{2/3}. \tag{4.822}$$

We note, by the way, that in the limit as $T = 0$ the functions $f(\varepsilon)$ ($= dN/d\varepsilon$) and $F(\varepsilon)$ have a discontinuity at $\varepsilon = \mu_0$ (see Figs. 4.4 and 4.5).

The energy of the system follows from Eqs. (4.803) and (4.423) and is given by the equation

$$E = -\frac{\partial q}{\partial \beta} = 4\pi\left(\frac{2m}{h^2}\right)^{3/2} V \int_0^\infty \frac{\varepsilon\sqrt{\varepsilon}\, d\varepsilon}{e^{-\alpha+\beta\varepsilon} + 1}. \tag{4.823}$$

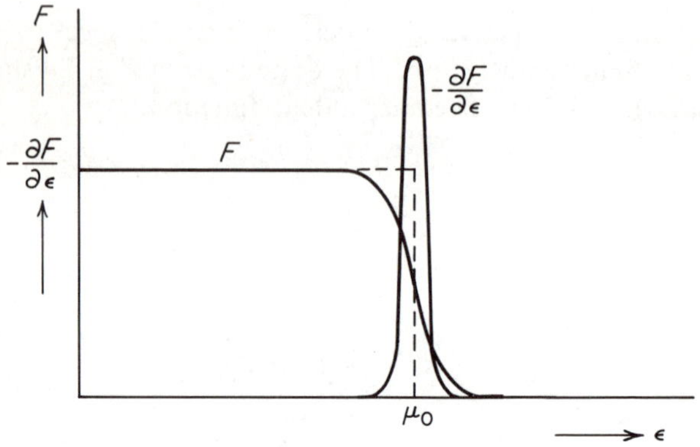

FIG. 4.5 The Fermi function $F(\varepsilon)$ of Eq. (4.819), and its derivative at a finite temperature. The dotted curve corresponds to $T = 0$.

At $T = 0$ we have

$$E_0 = \frac{8\pi}{5}\left(\frac{2m}{h^2}\right)^{3/2} V {\mu_0}^{5/2} = \frac{3}{5}\frac{h^2}{2m} N \left(\frac{3N}{8\pi V}\right)^{2/3}, \tag{4.824}$$

or

$$E_0 = \tfrac{3}{5} N \mu_0. \tag{4.825}$$

Equations (4.824) or (4.825) give us the so-called zero-point energy of a Fermi-Dirac gas. We could also have calculated E_0 by using the Pauli principle and filling up the lowest energy levels until all particles had been assigned.

Combining Eqs. (4.715) and (4.824), we find for the *zero-point pressure*

$$P_0 = \frac{2}{3}\frac{E_0}{V} = \left(\frac{3}{8\pi}\right)^{2/3} \frac{h^2}{5m} \left(\frac{N}{V}\right)^{5/3}, \tag{4.826}$$

and we see that P_0 varies as the $\frac{5}{3}$ power of the density.

That this zero-point pressure is not always negligible can be seen when we consider the "gas" of the conduction electrons in a metal. The density N/V will be of the order of 10^{24} cm^{-3}, and with $m \cong 10^{-27}$ g we have

$$P_0 \sim 10^6 \text{ atm.} \tag{4.827}$$

With the same values of N/V and m, we find that

$$\mu_0 \approx 35 \text{ eV}, \tag{4.828}$$

corresponding to a temperature μ_0/k of about 4.10^5 °K.

We note that we could also have obtained Eq. (4.826) by combining Eqs. (4.803) and (4.426), performing an integration by parts and taking the limit $T \to 0$. It follows from the occurrence of h in the expression for P_0 that the zero-point pressure, like the Einstein condensation, is a typical quantum effect.

We should at this point mention that μ is often called the *Fermi level*,* especially in metal and semiconductor theory.† This Fermi level is the *electrochemical potential* and is equal to the partial thermal potential μ as can be seen as follows. Addition of δN electrons will result in a change δF of the free energy equal to $\mu\,\delta N$ (compare Eq. (4.414)). Consider now two systems A and B at electrical potentials V_A and V_B. An adiabatic transfer of δN electrons with charge e from A to B will result in a change δF of the free energy of the combined system which, on the one hand, is equal to $e(V_B - V_A)$ and, on the other hand, is equal to $\mu_B - \mu_A$. This shows that μ is the electrochemical potential.

We note in passing that, as it follows from thermodynamics‡ that in the case of equilibrium between two systems which can exchange particles the partial thermal potentials in the two systems must be equal, we find that two metals or semi-conductors in contact should have the same value for their Fermi levels in equilibrium. When we use the term Fermi level, μ is often denoted by E_F. The name "Fermi level" derives from the fact that at absolute zero all levels up to E_F will be occupied (compare Eq. (4.820)). The particles occupying these levels up to E_F are sometimes called the *Fermi sea*.

From the result of Eq. (4.828) it is clear that for the case of the conduction electrons in a metal for any temperature the inequality $\beta\mu \gg 1$ will hold,

* See, for example, W. Shockley, *Electrons and Holes in Semiconductors*, New York, 1950, p. 231 and Chap. 16; J. C. Slater, *Quantum Theory of Matter*, New York, 1951, Chap. 12.

† Sometimes the Fermi level is called the Fermi energy; this is rather confusing because this term is also used for the average energy of the electrons in the system (see Slater, *op. cit.*, p. 420).

‡ See, for example, D. ter Haar and H. Wergeland, *Elements of Thermodynamics*, Reading, Section VI.1.

which means that we are then always dealing with a strongly degenerate gas. We can therefore again expect that we might be able to give series expansions for the various quantities of interest in terms of inverse powers of $\beta\mu$ or of $\beta\mu_0$. We shall use as our expansion parameter a quantity η defined by the equation

$$\eta = \frac{1}{\beta\mu_0} = \frac{2mkT}{h^2}\left(\frac{8\pi V}{3N}\right)^{2/3}\left[=\left(\frac{8}{9\pi}\right)^{1/3}\zeta^{-2/3}\right], \tag{4.829}$$

where ζ is defined by Eq. (4.705) including the spin factor 2.

To find the expansions we use the general theorem that

$$\int_0^\infty G(\varepsilon)\frac{\partial F}{\partial \varepsilon}\,d\varepsilon = -G(\mu) - \frac{\pi^2}{6\beta^2}\frac{\partial^2 G}{\partial \varepsilon^2}\bigg|_{\varepsilon=\mu} + \cdots \tag{4.830}$$

Equation (4.830) can be proved as follows.*

First of all, we introduce a new variable x by the equation

$$x = \beta(\varepsilon - \mu), \tag{4.831}$$

so that we have

$$F(\varepsilon) \to F(x) = \frac{1}{e^x + 1}. \tag{4.832}$$

The integral now becomes

$$\int_{-\beta\mu}^{\infty} G(x)\frac{\partial F}{\partial x}\,dx \equiv I. \tag{4.833}$$

We now expand $G(x)$ in a Taylor series:

$$G(x) = G(0) + G'(0)x + \tfrac{1}{2}G''(0)x^2 + \cdots. \tag{4.834}$$

Defining the quantities I_i by the equations

$$I_i = \frac{1}{i!}\int_{-\beta\mu}^{\infty}\frac{\partial F}{\partial x}x^i\,dx, \tag{4.835}$$

we find from Eqs. (4.833) to (4.835)

$$I = \sum_{n=0}^{\infty} G^{(n)}(0)I_n. \tag{4.836}$$

The problem is thus reduced to finding the I_i. We now first of all note that as $\beta\mu \gg 1$, we can replace the lower limit in all integrals I_i by $-\infty$; the errors introduced in this way are of the order of magnitude of $e^{-\beta\mu}$ and thus totally negligible in the case of interest to us. We further note that $\partial F/\partial x$ is an even

* A. Sommerfeld and H. Bethe, *Handb. d. Phys.*, **24₂**, 346, 1933.

function of x, so that all I_i with i odd vanish, while for the I_i with i even, we have*

$$I_i = -\frac{2}{i!}\int_0^\infty \frac{\varepsilon^i e^{-\varepsilon}}{(1+e^{-\varepsilon})^2}\,d\varepsilon = -\frac{2}{i!}\int_0^\infty \varepsilon^i \sum_{n=1}^\infty (-1)^n n e^{-n\varepsilon}\,d\varepsilon$$

$$= 2\sum_{n=1}^\infty \frac{(-1)^{n+1}}{n^i} = \frac{2\pi^i(2^{i-1}-1)}{i!}B_i, \tag{4.837}$$

where the B_i are the Bernoulli numbers. Changing back from x to ε as variable now gives us Eq. (4.830).†

Applying Eq. (4.830) we get for N from Eq. (4.817)

$$N = \frac{8\pi}{3}\left(\frac{2m}{h^2}\right)^{3/2} V\mu^{3/2}\left[1 + \frac{\pi^2}{8}(\beta\mu)^{-2} + \cdots\right], \tag{4.838}$$

and hence for μ:

$$\mu = \mu_0\left[1 - \frac{\pi^2}{12}\eta^2 + \cdots\right]. \tag{4.839}$$

Using Eqs. (4.830) and (4.839), we then get for E, P, C_v, and S:

$$E = E_0\left[1 + \frac{5\pi^2}{12}\eta^2 + \cdots\right], \tag{4.840}$$

$$P = P_0\left[1 + \frac{5\pi^2}{12}\eta^2 + \cdots\right], \tag{4.841}$$

$$C_v = \frac{\partial E}{\partial T} = \frac{\pi^2}{2}Nk\eta + \cdots, \tag{4.842}$$

$$S = \int_0^T \frac{C_v\,dT}{T} = \frac{\pi^2}{2}Nk\eta + \cdots. \tag{4.843}$$

In Fig. 4.6 we have drawn E, P, and C_v as functions of the temperature. The dotted lines indicate the behavior of a Boltzmann gas.

It is easy to see the physical meaning of Eqs. (4.840) and (4.842). If we study Figs. 4.4 and 4.5, we see that only a fraction of the particles is excited when we go from absolute zero to a finite temperature. Calculating the slope of $F(\varepsilon)$ at the point of inflection and drawing a straight line through the inflection point at a slope twice that of the tangent to $F(\varepsilon)$ in that point, we find that the line will intersect the ε axis at a distance $4kT$ from μ_0. Replacing the shaded

* See, for instance, I. N. Bronshtein and K. A. Semendyayev, *A Guidebook to Mathematics*, Oxford, 1963, p. 354. (Note that there is a misprint in their formula 20: 2^{2k} should be 2^{2k-1}.)

† We note that in the limit as $\beta \to \infty$, when only the first term on the right-hand side of (4.830) has to be taken into account, $-\partial F/\partial\varepsilon$ acts, indeed, as a δ-function.

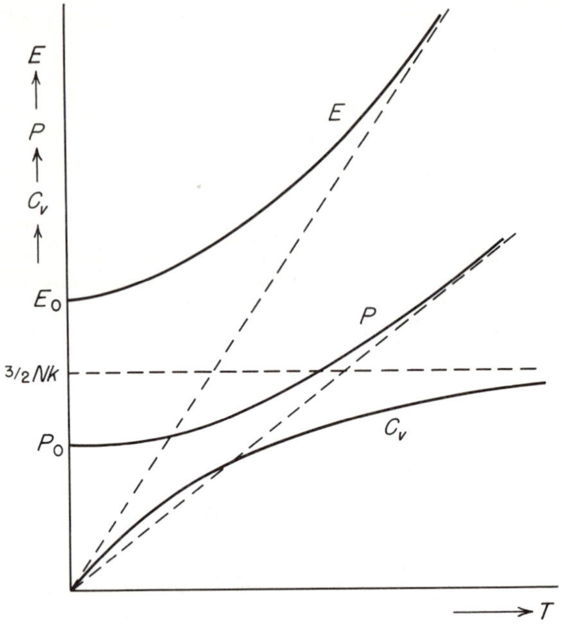

FIG. 4.6 The energy, pressure, and specific heat of a perfect fermion gas.

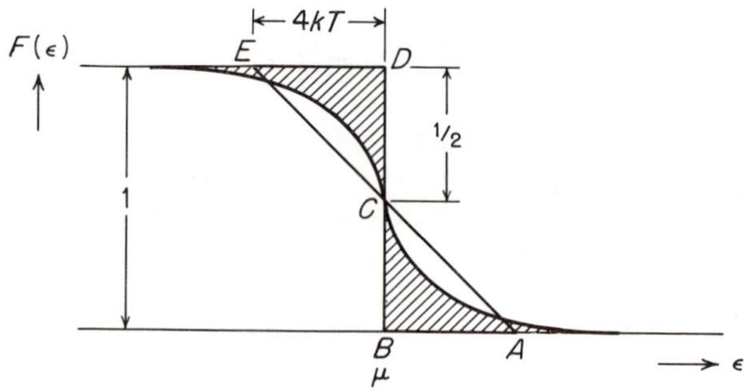

FIG. 4.7 The Fermi function at a finite temperature. The shaded areas correspond to the electrons missing from the totally degenerate Fermi sea and being above the Fermi level, respectively.

areas in Fig. 4.7 by the triangles ABC and CDE, we find that the total number of particles involved in the thermal properties of the system will be of the order of $(kT/\mu)N$. As the energy of each of these particles is increased roughly by an amount kT, we find for the total increase in energy an amount of the order of $(kT/\mu)^2 N\mu$ in agreement with Eq. (4.842). If instead of the shaded areas we take the triangles, we find for the numerical factor $\frac{8}{3}$ which is, indeed, extremely close to the factor $\frac{3}{5} \cdot (5\pi^2/12)$ of Eq. (4.842). As only a fraction kT/μ of the particles is involved in the thermal properties, we can understand that the specific heat is less than the classical value Nk by a factor of that order of magnitude.

We note also that it follows from Eq. (4.841) that the quantity $\partial P/\partial T$, which is constant in the classical case, is now proportional to T.

4.9 APPLICATIONS OF THE BOSE-EINSTEIN OR FERMI-DIRAC STATISTICS

In the preceding sections we discussed the behavior of perfect gases. The question we wish to consider in this section is the following one: "When can we expect deviations from Boltzmann statistics?" Let us confine ourselves to gaseous systems, or systems which to a first approximation may be treated as gases. We saw (Eq. (4.701)) that the condition for the applicability of Boltzmann statistics was that e^α is small compared to one. We can calculate e^α for various gases at their boiling point, using Eq. (4.703), and the result is given in Table 4.1.

TABLE 4.1

Gas	Boiling Point	e^α
Argon	87.4 °K	2.10^{-6}
Neon	27.2 °K	10^{-4}
Hydrogen	20.3 °K	7.10^{-3}
Helium_4	4.2 °K	0.13
Helium_3	3.2 °K	0.4

We see that the only gas for which deviations from classical statistics could conceivably be detected should be helium. In the case of He_4 there seem to be some indications that the equation of state follows more closely that of a Bose-Einstein gas than that of a Fermi-Dirac gas, but the measurements are still far from conclusive. We should expect Bose-Einstein behavior from He_4, since the nucleus is built up out of 4 nucleons and there are 2 orbital electrons. Since e^α in the case of He_3 is even larger due to the smaller mass of the He_3 atom and its lower boiling point, one might hope to get a definite confirmation that He_3 obeys Fermi-Dirac statistics through an accurate measurement of its equation of state.

We must remark here that helium has much more often been connected with Bose-Einstein statistics because of the peculiar behavior of the liquid. The transition from helium I to helium II resembles in some respects the Einstein condensation, as we mentioned in Section 4.7. Another case where Bose-Einstein statistics must be applied is the radiation field, as we discussed at the beginning of this chapter.

There are also a number of cases where Fermi-Dirac statistics have to be applied. From Eq. (4.703) it follows that a small mass or a high density will make it necessary to abandon Boltzmann statistics. The first case where Fermi-Dirac statistics are applied is that of the "gas" of the conduction electrons in a metal. Assuming an average spacing of 1 ÅU for the atoms in the metal and one conduction electron per atom, we have at room temperatures $e^{\alpha} \sim 10^5$. Many properties of metals can be understood if we apply Fermi-Dirac statistics to this case.

A second case where Fermi-Dirac statistics must be applied is the nuclear "gas." To a first rough approximation the nucleons inside a nucleus may be treated as a perfect Fermi-Dirac gas, or rather a mixture of two perfect Fermi-Dirac gases, that of the protons and that of the neutrons. We have here a high value of e^{α} due to the very high nuclear density ($\rho > 10^{12}$ g cm^{-3}). In the interior of some stars such high densities will occur that, notwithstanding the very high stellar temperatures, the gas is degenerate. Both Fermi-Dirac and Bose-Einstein degeneracy may occur.

Finally, we must mention here that the various quasi-particles that enter so prolifically into modern theoretical physics usually obey well-defined statistics.

The question of what particles obey what statistics is a very profound and difficult one. In the bibliographical notes at the end of this chapter we have given some references to a discussion of this problem, and we shall only briefly mention a few aspects of it at this point. First of all, it must be emphasized that it is far from certain that all particles occurring in nature are necessarily either fermions or bosons. As a first, rough approximation one can say that all elementary particles that have integral spin, such as the photon, the pion, or the kaon, are bosons, while those that have half-odd-integral spin, such as the electron, the neutrino, the muon, the proton, and the neutron, are fermions. The problem becomes more difficult when we come to compound particles or to quasi-particles. As long as the binding energy of the constituents in a compound particle is high compared to other energies in the problem, a compound particle containing an even number of fermions will obey Bose-Einstein statistics (examples are the α-particle and the He_4 atom), while those containing an odd number of fermions will obey Fermi-Dirac statistics (for example, the He_3 atoms).*

* For a discussion of this problem, see P. Ehrenfest and J. R. Oppenheimer, *Phys. Rev.* **37**, 333, 1931.

Applications of the Fermi-Dirac and Bose-Einstein statistics to various physical systems will be discussed further in Vol. II, or the reader may refer to the first edition and other literature mentioned in the bibliographical notes.

PROBLEMS

1. Prove the equivalence of Eqs. (4.213) and (4.212).
2. Gentile* introduced in 1940 the so-called intermediate statistics in which the maximum-allowed number of particles in any one energy level is d. Derive the equilibrium value of the number of particles N_j in the jth group of levels, containing Z_j levels.
3. Use the method of Sections 4.2 and 4.3 to express the ratio $c_A c_B / c_{AB}$ in terms of Z_A, Z_B, and Z_{AB} for the case of the following dissociative equilibrium:

$$\mathrm{A} + \mathrm{B} \rightleftarrows \mathrm{AB},$$

where the c's are the concentrations. Assume that there are no interactions between the particles and that there is an energy of formation of AB equal to χ, that is, that the ground state of AB lies lower by an amount χ than the ground state of A + B.
4. Use the Darwin-Fowler method to derive the result obtained in the preceding problem.†
5. Write down the integral representation for the factorial,

$$z! = \int_0^\infty e^{-t} t^z \, dt,$$

and introduce a new variable v through the equations

$$(1 + u)e^{-u} = e^{-\frac{1}{2}v^2}, \qquad t = z(1 + u).$$

Then use the method of steepest descents to find Stirling's asymptotic formula

$$z! \approx \sqrt{2\pi z}\left(\frac{z}{e}\right)^z.$$

In most cases we can in the expression for $\ln z!$ drop the factor $\sqrt{(2\pi z)}$.

* G. Gentile, *Nuovo Cimento*, **17**, 493, 1940. See also: A. Sommerfeld, *Ber. Dtsch. Chem. Ges.*, **75**, 1988, 1942. H. Wergeland, *Proc. Roy. Norwegian Acad, Sc.*, **17**, 51, 1944, G. Schubert, *Z. Naturf.*, **1**, 113, 1946, and H. Müller, *Ann. Physik*, **7**, 420, 1950.

† See, for example, R. H. Fowler, *Statistical Mechanics*, Cambridge, 1929. pp. 104ff.

6. Consider a perfect gas in a volume V without an external field of force. Divide the volume into m equal cells and denote by n_k the number of particles in the kth cell; use the Darwin-Fowler method to find an expression for the average value $\overline{n_k}$ of n_k.
7. Find the equation of the curve a in Fig. 4.2.
8. Find the way the number of bosons in the ground state depends on the temperature for temperatures below the condensation temperature T_0.
9. Prove that a two-dimensional or one-dimensional boson gas does not undergo an Einstein condensation.
10. Consider a perfect gas of bosons, each of which can be either in a ground state or in an excited state Δ above the ground state. Assume that $\Delta/kT_0^{(0)} \gg 1$ ($T_0^{(0)}$ is the condensation temperature for the case where $\Delta = \infty$). Find an approximate expression for the condensation temperature of this gas, and explain qualitatively the result obtained.
11. Find an equation for the isentropics in the P-V diagram for a perfect monatomic quantum gas.
12. Prove that the specific heat C_v for a perfect two-dimensional gas of fermions is identical with that for a perfect two-dimensional boson gas.*
13. Find the next terms in the series expansions in ζ for the various quantities given in Sections 4.7 and 4.8.
14. Find the first few terms in a power series expansion in the temperature for the energy of a two-dimensional gas of fermions.
15. Find the next terms in the series expansions in η for the various quantities given in Section 4.8.
16. Prove that the tangent at the point of inflection in Fig. 4.5 cuts the ε axis at $\mu + 2kT$.
17. Prove that, if we replace the shaded areas in Fig. 4.7 by the areas ABC and CDE, the total energy of the system becomes $E_0 + \frac{8}{3}N\mu(\beta\mu)^{-2}$.
18. Evaluate the compressibility of a strongly degenerate, perfect Fermi-Dirac gas.
19. Show that the specific heat of a strongly degenerate, perfect fermion gas is given by

$$C_v \approx \tfrac{1}{3}\pi^2 k^2 T \rho(\mu_0),$$

where $\rho(\varepsilon)$ is the single-particle density of states.

● The next few problems deal with metals and semiconductors; first we shall briefly review the basic ideas upon which the elementary theory of metals and semiconductors is based. We can only give an outline of

* R. M. May, *Phys. Rev.*, **135**, A1515, 1964.

these ideas and must refer to the literature* for details and for complications such as overlapping bands. The theory given here refers essentially to one-dimensional solids, but possesses all the characteristic features that we need for our present applications. We also refer to the literature for a discussion of such concepts as the effective mass and the role of interactions between the electrons.

We must now enter into a slightly detailed discussion of the band structure of the energy levels in crystals. If we are concerned with the movement of electrons in a crystal and if we treat the electrons as noninteracting with each other, the first problem is to find the characteristic energy values of a particle moving in a one-dimensional potential energy field with periodic structure. One can show that—at any rate, if one is dealing with an infinite crystal—the energy spectrum consists of "bands."† This means that there exist energy values E_0, E_1, E_2, $\cdots$ such that all energies satisfying the inequalities

$$E_0 \leqslant E \leqslant E_1, \qquad E_2 \leqslant E \leqslant E_3, \qquad E_4 \leqslant E \leqslant E_5, \cdots \tag{A}$$

are characteristic values, while none of the energies satifying the inequali ities

$$E < E_0, \qquad E_1 < E < E_2, \qquad E_3 < E < E_4, \cdots \tag{B}$$

are characteristic values. The energy values satisfying Eq. (A) are called *allowed* energies and the energy values satisfying Eq. (B) are called *forbidden* energies; one also speaks of *allowed* and *forbidden bands.*

Let us consider the situation at absolute zero. There are two possibilities. The first possibility is that the electrons fill up the lowest bands completely, but only half fill the last band that is at all occupied, and we are dealing with a metal. In that case any amount of energy can remove electrons from a level occupied at absolute zero. The second possibility is that the electrons exactly fill up the lowest bands, but that the next allowed band is completely empty. This is the case of an *intrinsic semiconductor* or an *insulator.* The difference between these two types of materials is only quantitative and depends on the energy gap $\Delta\varepsilon$ between the band last filled and the first empty band. In insulators this gap is of the order of 5 to 10 eV, while in intrinsic semiconductors it is of the order of 1 eV. The gap is, for instance, 0.4 eV in the case of tellurium and 0.76 eV in the case of germanium.

Sometimes the structure of a crystal is not completely periodic due, for

* See, for example, A. H. Wilson, *The Theory of Metals*, Cambridge, 1953; F. Seitz, *The Modern Theory of Solids*, New York, 1940; A. J. Dekker, *Solid State Physics*, Englewood Cliffs, 1957.

† We refer to a paper by Kramers (*Physica*, **2,** 483, 1935) for a general discussion of this behavior.

instance, to impurities in the crystal. In that case it is possible that there are a few energy levels that lie in the forbidden band (see Fig. 4.8). These levels can either lie close to the conduction band (Fig. 4.8(1)), that is, the first band in which there are empty levels, or close to the valence band (Fig. 4.8(2)). In the first case they are called *donor levels*. Only a relatively

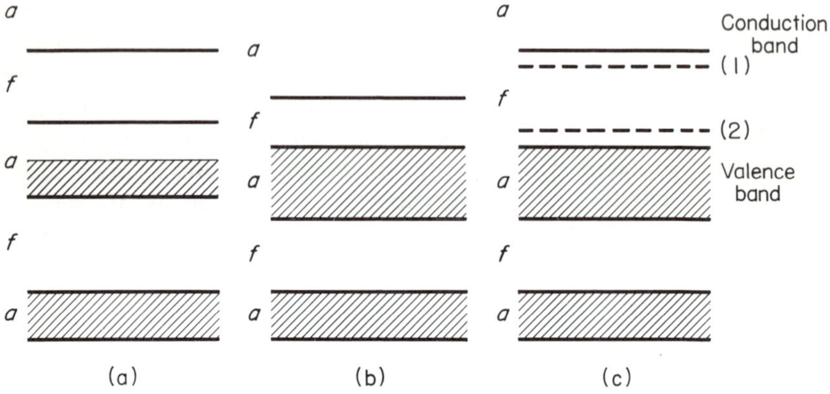

FIG. 4.8 The band structure of a metal (a), *an intrinsic semiconductor* (b), *and an impurity semiconductor* (c).

small amount of energy will lift electrons from a donor level into the conduction band. Such a case will occur when germanium, for instance, is doped with an element from the fifth column in the periodic table, such as phosphor. In the second case they are called *acceptor levels*. Only a relatively small amount of energy will lift electrons from the valence band into an acceptor level, leading to hole conduction. We must note that usualy the wave functions corresponding to the donor and acceptor levels are localized. Indeed, in Fig. 4.8, in which we have indicated the cases of a metal (*a*), an intrinsic semiconductor (*b*), an impurity semiconductor (*c*), (*a* and *f* indicate, respectively, the allowed and the forbidden bands), we have given two-fold information. First of all, one can read from the figure the values of the characteristic energies. However, this could be done from a figure such as is given by Fig. 4.9 without leading to confusion.* Secondly, one gives information about the wave functions corresponding to the various values of the energy: if the shaded

* One sees from Fig. 4.9 that the energy levels corresponding to the impurity levels, although often lying very closely—much more closely in fact than is indicated in Fig. 4.9 —are different and not necessarily all equal as one might expect from Fig. 4.8.

area, indicating the levels occupied at absolute zero, or the empty area of the allowed energy bands extend all the way, it is implied that the corresponding wave function has an appreciable amplitude throughout that region of space, while the extent of the lines indicating the impurity levels is limited, corresponding to the localized nature of the corresponding wave functions. This kind of information is especially useful when

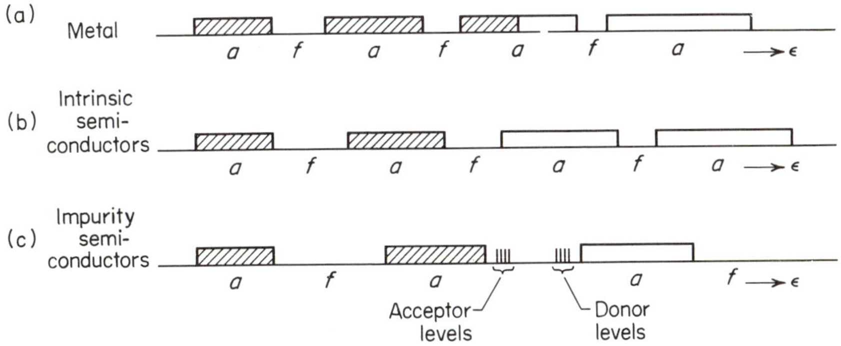

FIG. 4.9 The energy spectrum of a metal (a), *an intrinsic semiconductor* (b), *and an impurity semiconductor* (c).

one considers the band structure near the surface of a solid or near the interface of two solids in contact (see, for instance, Dekker's textbook). For the discussion in the problems of this chapter we shall use simplified models, both of the metal and of the semiconductor. Our model of a metal is the following one (see Fig. 4.10). We shall assume that the conduction electrons are moving as free and independent particles in a constant negative potential, $-\chi$, produced by the metallic ions. Our problem is thus first of all to derive an expression for the energy levels of an electron when its potential energy has the form given in Fig. 4.10.* Actually, we need only the energy level density, and it turns out that, if we take as the zero of energy the value of the potential energy inside the metal, this density is the same as that of an electron in a volume V with rigid walls, as long as we are only interested in bound levels. We have thus for the number of levels, dZ, with energies between ε and $\varepsilon + d\varepsilon$, where ε is

* We are not concerned with any detailed features of the potential energy and have therefore assumed U to be constant throughout the metal and to change discontinuously to zero at the surface of the metal. In actual cases there will be a smooth transition from $-\chi$ to 0, but the simplified potential of Fig. 4.10 is sufficient for our purposes.

reckoned from the bottom of the potential well, expression (4.604), multiplied by 2 to take the electron spin into account.

We saw in Section 4.8 that for the gas of conduction electrons we are dealing with a strongly degenerate gas, so that to a first approximation all levels up to a distance μ_0 from the bottom of the potential well are

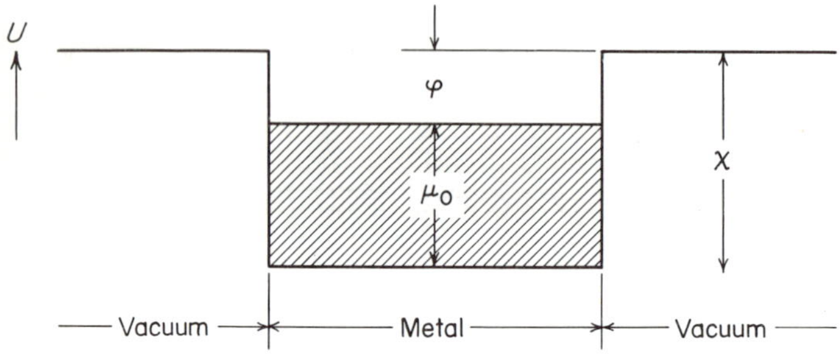

FIG. 4.10 The electron Fermi sea in a metal at absolute zero; φ is the work function.

occupied. This is again indicated by shading in Fig. 4.10. The work function φ is defined by the equation

$$\varphi = \chi - \mu_0. \tag{C}$$

The work function is the energy which one must give to an electron at the top of the "Fermi sea" in order to get it out of the metal.

The model we shall use for an impurity semiconductor is one due to Wilson.* In this model (see Fig. 4.11) we assume that we may neglect the influence of the occupied bands. Furthermore, we assume that there is a conduction band for $\varepsilon \geqslant 0$ which is completely unoccupied at absolute zero and that there are N_b impurity levels, all situated at $\varepsilon = -\Delta\varepsilon$, which are completely occupied at absolute zero. This can be expressed by writing for the energy level density

$$Z(\varepsilon) = N_b \delta(\varepsilon + \Delta\varepsilon), \qquad \varepsilon < 0,$$

where $\delta(\varepsilon)$ is the Dirac δ-function. For the energy level density in the conduction band we use again expression (4.604), multiplied by 2 to take the electron spin into account.

* A. H. Wilson, *Proc. Roy. Soc. (London)*, **A133,** 458, 1931; **A134,** 277, 1931.

Finally, we shall use for an intrinsic semiconductor a model in which the energy level density in the valence band is given by the expression

$$dZ = 4\pi\left(\frac{2m_{\text{h}}}{h^2}\right)^{3/2} V\sqrt{-\varepsilon}\, d\varepsilon, \qquad \varepsilon < 0, \tag{D}$$

where we have assumed the top of the valence band to be at $\varepsilon = 0$, while

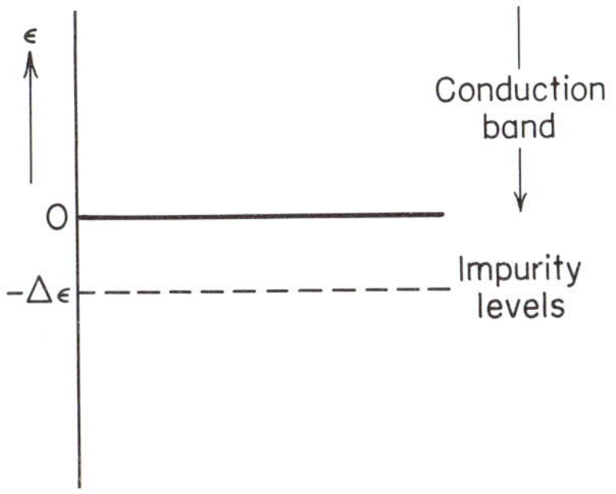

FIG. 4.11 Wilson's model of an impurity semiconductor in which there are a number of impurity levels at a depth $\Delta\varepsilon$ *below the conduction band.*

the energy level density in the conduction band is given by the expression

$$dZ = 4\pi\left(\frac{2m_{\text{el}}}{h^2}\right)^{3/2} V\sqrt{\varepsilon - \Delta}\, d\varepsilon, \qquad \varepsilon > \Delta, \tag{E}$$

where the bottom of the conduction band is at $\varepsilon = \Delta > 0$. ●

20. Find for an intrinsic semiconductor (1) the number of electrons in the conduction band as a function of the temperature, and (2) the Fermi level of the semiconductor as a function of $m_{\text{h}}/m_{\text{el}}$, assuming $\beta\Delta \gg 1$.
21. Find for an impurity semiconductor (1) the number of electrons in the conduction band, and (2) the Fermi level of the semiconductor, assuming $\beta\Delta\varepsilon \gg 1$. Find also the distribution of the electrons in the conduction band over the various energies.
22. Consider an impurity semiconductor with a large number of impurity levels and a small gap. Take as an example $N_{\text{b}}/V = 10^{18}\ \text{cm}^{-3}$ and $\Delta\varepsilon = 10^{-3}$ eV or 10^{-5} eV. Determine the temperature range in which one cannot treat the gas of conduction electrons as a classical gas, and discuss why at both sufficiently high and at sufficiently low temperatures the gas becomes classical.

23. Find an expression for the Richardson current,* that is, the current of the so-called thermionic emission which arises because at finite temperatures there will be a certain number of electrons with sufficient energy to leave the metal. Assume that $\beta\varphi \gg 1$, and that the electron gas is strongly degenerate.

24. A metal is illuminated by light of frequency ω. Find the photocurrent, assuming that the components of the electron momentum parallel to the metal surface remain unchanged by the absorption of a photon.

25. Consider the Lorentz model of the transport properties of metals (see Problem 4 in Chapter 2). Taking for f_0 the Fermi-Dirac distribution instead of the Maxwell distribution as was done in Problems 4 to 7 in Chapter 2, evaluate (1) the electrical conductivity, (2) the thermal conductivity, (3) the isothermal Hall coefficient, (4) the Wiedemann-Franz ratio, and (5) magnetoresistance of a metal.

26. Calculate the spin paramagnetic susceptibility of a system of free electrons (1) at $T = 0$, and (2) at a finite temperature, assuming that $\beta\mu \gg 1$ (Pauli paramagnetism†).

27. Given that the energy levels of an electron moving freely in a uniform magnetic field **B** along the z axis are given by the equation‡

$$E = \left(n + \frac{1}{2}\right)\hbar\omega + \frac{\hbar^2 k_z^{\,2}}{2m}, \qquad \omega = eB/m, \tag{F}$$

and that the number of levels for a given value of n and with k_z in the interval k_z, $k_z + dk_z$ is equal to

$$\frac{VeB\, dk_z}{2\pi^2\hbar},$$

and using the summation formula

$$\sum_{n=0}^{m} g_{n+\frac{1}{2}} \approx \int_0^{m+1} g(x)\, dx - \frac{1}{24}\,[g'(x)]_0^{\,m+1},$$

find an expression for the orbital diamagnetism of a strongly degenerate free electron gas in a volume V, assuming that $\beta\mu_B B \ll 1$, where μ_B is the Bohr magneton (Landau diamagnetism§).

* O. W. Richardson, *Phil. Mag.*, **23**, 594, 1912.

† W. Pauli, *Z. Physik*, **41**, 81, 1926.

‡ See, for instance, D. ter Haar, *Men of Physics: L. D. Landau*, Vol. I, Oxford, 1965, p. 31.

§ L. D. Landau, *Z. Physik*, **64**, 629, 1930; *Collected Papers*, Oxford and New York, 1965, No. 4, p. 31.

28. Use Eq. (1.717) as the definition of Boltzmann's H-function, and expression (4.306) for ln W; prove that

$$\frac{dH}{dt} \leqslant 0,$$

if the average number of transitions $N_{ij\to i'j'}$ per unit time where the groups Z_i and Z_j each lose one particle, while at the same time the groups Z_i and Z_j each gain one particle, is given by the equation

$$N_{ij\to i'j'} = A_{ij\to i'j'}N_iN_j(Z_{i'} + \gamma N_{i'})(Z_{j'} + \gamma N_{j'}),$$

if we may assume that

$$A_{ij\to i'j'} = A_{i'j'\to ij}.$$

29. Find the condition that $dH/dt = 0$ from the result of the preceding problem; derive Eqs. (4.314) to (4.316).

BIBLIOGRAPHICAL NOTES

General references for this chapter are

1. G. E. Uhlenbeck, *Over Statistische Methoden in de Theorie der Quanta*, Thesis, Leiden, 1927.
2. P. Jordan, *Statistische Mechanik auf Quantentheoretischer Grundlage*, Braunschweig, 1933.
3. E. Schrödinger, *Statistical Thermodynamics*, Cambridge, 1948.

Section 4.1. The Bose-Einstein statistics were first introduced by Bose using a method only slightly different from the elementary method:

4. S. N. Bose, *Z. Physik*, **26,** 178, 1924.

This paper had been translated by Einstein, who added the following note: "Boses Ableitung der Planckschen Formel bedeutet nach meiner Meinung einen wichtigen Fortschritt. Die hier benutzte Methode liefert auch die Quantentheorie des idealen Gases, wie ich an anderer Stelle ausführen will."

This application to the case of a perfect gas appeared the same year:

5. A. Einstein, *Berliner Ber.*, **1924,** 261; **1925,** 3, 18.

Following the introduction of the exclusion principle by Pauli in 1925 in the paper

6. W. Pauli, *Z. Physik*, **31,** 776, 1925,

Fermi introduced it into the statistical discussion:

7. E. Fermi, *Z. Physik*, **36,** 902, 1926.

The connection between statistics and wave mechanics was investigated by Dirac:

8. P. A. M. Dirac, *Proc. Roy. Soc.* (*London*), **A112,** 661, 1926.

Section 4.2. The fact that the omission of the factor $N!$ is essential in order that S is an extensive quantity was noted by Gibbs. Compare, for instance,

9. J. W. Gibbs, *Elementary Principles in Statistical Mechanics* (Vol. II of his Collected Works), New Haven, 1948, Chap. 15.

Compare also

10. L. Nordheim, *Z. Physik*, **27**, 65, 1924.
11. G. S. Rushbrooke, *Introduction to Statistical Mechanics*, Oxford, 1949, pp. 36ff.

For the connection between accessible states and probabilities see References 1 and 8. A discussion about the splitting up of the $N!$ possible wave functions according to their symmetry properties was given by Wigner:

12. E. Wigner, *Z. Physik*, **40**, 492, 1926.
13. E. Wigner, *Z. Physik*, **40**, 883, 1927.

The reason why some particles obey Bose-Einstein and others Fermi-Dirac statistics lies rather deep. One can show that if the annihilation and creation operators corresponding to a description where "second quantization" (P. Jordan and O. Klein, *Z. Physik*, **45**, 751, 1927; P. Jordan and E. Wigner, *Z. Physik*, **47**, 631, 1928; compare also H. A. Kramers *Quantum Mechanics*, Amsterdam, 1957. Section 72; G. Wentzel, *Quantum Theory of Fields*, New York, 1949, Section 22; A. S. Davydov, *Quantum Mechanics*, Oxford, 1965, Chapters 14 and 15) is introduced, follow the symmetric commutation rules, the corresponding particles obey Bose-Einstein statistics, while they obey Fermi-Dirac statistics if annihilation and creation operators are anticommuting.

Pauli and Belinfante gave in 1940 (*Physica*, **7**, 177, 1940) the following general principles (we quote their summary and have inserted further references): "Es wird untersucht, in wie weit Spekulationen über die Statistik willkürlicher, hypothetischer Teilchen in einer relativistisch invarianten Theorie möglich sind, wenn alle oder ein Teil der folgenden drei Postulate vorausgesetzt werden: (I) die Energie ist positiv; (II) Observable an verschiedenen Raum-Zeitpunkten mit raumartiger Verbindungslinie sind kommutativ; (III) es gibt zwei äquivalente Beschreibungen der Natur, in welchen die Elementarladungen entgegengesetzte Vorzeichen haben und in welchen einander entsprechende Feldgrössen sich bei Lorentz-Transformationen in gleicher Weise transformieren. [Cf. H. A. Kramers, *Proc. Kon. Ned. Akad. Wet.* (Amsterdam), **40**, 814, 1937.]

"Der eine von uns [W. Pauli, *Phys. Rev.*, **58**, 716, 1940] hatte bereits gezeigt, dass allgemein bei ganzzahligem Spin aus (II) allein Einstein-Bose-Statistik, bei halbzahligem Spin aus (I) allein Fermi-Dirac-Statistik folgt. Ferner hat der andere von uns [F. J. Belinfante, *Physica*, **6**, 870, 1939] gezeigt, dass für eine gewisse Klasse von Teilchen, (die alle bisher in der Natur beobachteten umfasst, und die dadurch charakterisiert ist, dass sie höchtensdurch *einen* Undor [cf. F. J. Belinfante, *Physica*, **6**, 849, 1939] von gegebener Stufe beschrieben wird), aus dem durch ein spezielles Transformationsgesetz spezialisierten Postulat (III) für ganzen Spin E.-B.-Statistik, für halbzahligen Spin F.-D.-Statistik gefolgert werden kann. In der vorliegenden Note wird in den typischen Fällen von Spin 0 and Spin $\frac{1}{2}$ durch Beispiele gezeigt, dass im allgemeinen Fall mehrerer Undoren von gleicher Stufe aus (III) nicht mehr eindeutig auf die Statistik der Teilchen geschlossen werden kann, während (II) bzw. (I) für ganzen bzw. halben Spin hiezu stets hinreichend bleiben. In den speziellen Fällen des Skalarfeldes, des Vektorfeldes und des Dirac-Elektrons, wo nur ein einziger Undor gegebener Stufe in die Theorie eingeht, folgt dagegen das Transformationsgesetz der Ladungskonjugierung eindeutig, so dass hier (III) zur Festlegung der Statistik ausreicht."

See also:

14. N. Burgoyne, *Nuovo Cimento*, **8,** 607, 1958.
15. G. Lüders and B. Zumino, *Phys. Rev.*, **110,** 1450, 1958.
16. G. F. Dell 'Antonio, *Ann. Physics*, **16,** 153, 1961.
17. A. Messiah and W. Greenberg, *Phys. Rev.*, **136**, B248, 1964.
18. M. D. Girardeau, *Phys. Rev.*, **139**, B500, 1965.

For a discussion of the spin-statistics connection for strange particles, see, for instance,

19. R. Gatto, *Phys. Letters*, **5,** 56, 1963.

Finally we refer to the 1963 Brandeis Theoretical Physics Summer School lectures by M. Dresden in which an extensive discussion is given of the possible existence of parastatistics, that is, statistics which are neither Bose-Einstein nor Fermi-Dirac statistics (compare Problem 2 of Chapter 4):

20. M. Dresden, "The Existence and Significance of Parastatistics," *State University of Iowa Report No. SUI* 64-2, 1964.

Recently, Greenberg and Messiah have shown that no particle known in June 1965, could obey parastatistics:

21. O. W. Greenberg and A. M. L. Messiah, *Phys. Rev.*, **138,** B1155, 1965.

For a discussion of various $W(n_i)$ see also

22. P. Ehrenfest and G. E. Uhlenbeck, *Z. Physik*, **41,** 24, 1927.

Sections 4.3 and 4.4. We have used here notes of lectures given by H. A. Kramers in Leiden during the winter 1944–1945 for a small circle of physicists. In these lectures, Kramers introduced the q-potential.

Section 4.5. The Darwin-Fowler method was first developed in the following papers:

23. C. G. Darwin and R. H. Fowler, *Phil. Mag.*, **44,** 450, 1922.
24. C. G. Darwin and R. H. Fowler, *Phil. Mag.*, **44,** 823, 1922.
25. C. G. Darwin and R. H. Fowler, *Proc. Cambridge Phil. Soc.*, **21,** 262, 1922.
26. R. H. Fowler, *Phil. Mag.*, **45,** 1, 1923.
27. R. H. Fowler. *Phil. Mag.*, **45,** 497, 1923.
28. C. G. Darwin and R. H. Fowler, *Proc. Cambridge Phil Soc.*, **21,** 391, 1923.
29. C. G. Darwin and R. H. Fowler, *Proc. Cambridge Phil. Soc.*, **21,** 730, 1923.

30. R. H. Fowler, *Proc. Cambridge Phil. Soc.*, **22,** 861, 1925.
31. R. H. Fowler, *Phil. Mag.*, **1,** 845, 1926.
32. R. H. Fowler, *Proc. Roy. Soc.* (*London*), **A113,** 432, 1926.

See also References 1, 3 and

33. R. H. Fowler, *Statistical Mechanics*, Cambridge, 1936.

Sections 4.6 to 4.8. These sections are applications of the formulas developed in the earlier sections. Compare also References 5 and 7 and

34. E. H. Kennard, *Kinetic Theory of Gases*, New York, 1938, pp. 397 ff.

For a discussion of the Einstein condensation see References 5, and

35. D. ter Haar, *Elements of Statistical Mechanics*, 1st ed., New York, 1954 (and references given there on p. 224).
36. F. London, *Phys. Rev.*, **54,** 947, 1938.
37. R. H. Fowler and H. Jones, *Proc. Cambridge Phil. Soc.*, **34,** 573, 1938.
38. H. Wergeland and K. Hove-Storhoug, *Proc. Norwegian Acad. Sc.*, **15,** 131, 135, 181, 1943.
39. R. Becker, *Z. Physik*, **128,** 120, 1950.
40. D. ter Haar, *Proc. Roy. Soc.* (*London*), **A212,** 552, 1952.

Section 4.9. For a discussion of the applicability of quantum statistics to gases see

41. D. ter Haar, *Am. J. Phys.*, **17,** 399, 1949.

For an application of Bose-Einstein or Fermi-Dirac statistics to the equation of state, see

42. J. de Boer, *Contribution to the Theory of Compressed Gases*, Thesis, Amsterdam, 1940.
43. J. de Boer, J. van Kranendonk, and K. Compaan, *Physica*, **16,** 545, 1950.

For the statistical theory of metals see

44. A. Sommerfeld and H. Bethe, *Handb. d. Phys.*, $\mathbf{24_2}$**,** 333, 1933.

For a discussion of statistics of nuclear particles see

45. H. A. Bethe, *Revs. Modern Phys.*, **9,** 69, 1937.
46. H. Wergeland, *Fra Fysikkens Verden*, **1945,** 223.

For astrophysical applications see

47. S. Chandrasekhar, *Introduction to the Study of Stellar Structure*, Chicago, 1939, where further references can be found.
48. L. Biermann, *Z. Naturf.*, **3a,** 481, 1948.

Several applications to the theory of metals and semiconductors, to nuclear physics, and to astrophysics can be found in Part C of Reference 35.

5

Classical Ensembles

5.1 THE Γ-SPACE; ENSEMBLES

In the preceding chapters we discussed systems that consisted of independent particles and which, moreover, existed isolated from the rest of the universe. We thus neglected in our treatment various points. Firstly, except when discussing Boltzmann's H-theorem, where collisions—that is, interactions between particles—play an essential role, we neglected interactions within the system. Secondly, we neglected interactions within the surroundings. Thirdly, we neglected any experimental limitations to the amount of information that can be obtained about a physical system. The first point will be dealt with in the present section, where we shall discuss how one can treat systems of interacting particles.

The second point is important in connection with the introduction of the concept of temperature. This concept entails necessarily a means of fixing this temperature, which in turn involves the introduction of a large thermostat in which the system is contained. However, we then no longer have an isolated system and we must expect that due to the interaction between system and thermostat, or *temperature bath*, as we shall often call the thermostat, the energy of the system is no longer constant. Indeed, there exists a certain *complementarity* between energy and temperature.* We shall see later (Sections 5.4 and 5.13) that, if we fix the temperature, the energy may have various values which spread around a most probable value—but in most physical cases the actual dispersion is so small as to be negligible—and that, on the other hand, if we are dealing with a really isolated system, we must expect that the temperature measured in the system will be fluctuating.

The third point which has not yet been considered with sufficient care is the fact that there are very serious limitations to the amount of information we can obtain about a physical system that we are considering. We have mentioned, but not sufficiently stressed, that in practice only a few quantities such as the pressure, the total kinetic energy, the total linear momentum, and the total angular momentum of the system can be measured. In the previous chapters our discussion was such that one might get the impression that it would be possible, if we wished to do so, to determine the distribution function to any degree of detailed accuracy and hence calculate all average values in which we might be interested. Since only a few data will be available, we are confronted with the problem of determining the distribution function under these circumstances. This question will be discussed in some detail in Sections 5.13 and 5.14.

Let us discuss first the question of interaction between the particles in the system. We shall briefly consider under what circumstances our neglect of the interactions would be permissible. In the Bibliographical Notes to Chapter 1 we said that a perfect gas, that is, a gas where one can neglect the interactions, would be a system where the total potential energy (or interaction energy) is at all times negligibly small compared to the total kinetic energy. Using the language of quantum mechanics, we can put it slightly differently. The total energy of an isolated system may be given within certain limits, that is, we may know that the energy of the system lies in the interval $E - \delta$, $E + \delta$. If it is still to have any sense at all to talk about a system with a fixed energy, δ must be small compared to E.

The fact that the energy of a system can never be completely fixed is, of course, an essential point in quantum mechanics. In order to fix our ideas we may, for example, take for δ the natural line width of the energy of the system. Or, if Δt is the smallest time interval during which observations can

* It must be emphasized that this complementarity is a classical one and does not involve the quantum of action.

be made, we can take δ in accordance with the Heisenberg relations to satisfy the equation

$$\delta \sim \frac{h}{\Delta t}.$$

If, as is mostly the case, the system consists of many particles, the spacing ΔE of the energy levels of the particles — treated as independent entities — will be very small (compare the remarks in Section 3.5) and we will have $\Delta E \ll \delta$. Now if the interaction between the particles is so weak that the total interaction energy U is small compared to δ, we can still treat the system as consisting of independent particles. However, in order that transitions between various states can occur at all, U must be larger than ΔE, since otherwise practically no transitions would occur. We have thus the inequalities

$$\Delta E \ll U \ll \delta \ll E. \tag{5.0101}$$

One way to introduce the interaction between the particles in the system is by treating the whole system as one large molecule. If each particle has s degrees of freedom and if N is the number of particles in the system, the total number of degrees of freedom for the whole system will be sN. The state of the system will be completely determined by $2sN$ coordinates, such as

$$\begin{array}{l}
q_1{}^1, q_2{}^1, q_3{}^1, \cdots, q_s{}^1; p_1{}^1, p_2{}^1, p_3{}^1, \cdots, p_s{}^1; \\
q_1{}^2, q_2{}^2, q_3{}^2, \cdots, q_s{}^2; p_1{}^2, p_2{}^2, p_3{}^2, \cdots, p_s{}^2; \\
\cdots\cdots\cdots\cdots\cdots\cdots\cdots\cdots\cdots\cdots\cdots\cdots \\
\cdots\cdots\cdots\cdots\cdots\cdots\cdots\cdots\cdots\cdots\cdots\cdots \\
q_1{}^N, q_2{}^N, q_3{}^N, \cdots, q_s{}^N; p_1{}^N, p_2{}^N, p_3{}^N, \cdots, p_s{}^N,
\end{array}$$

where the $p_k{}^j$ and $q_k{}^j$ are the canonical momenta and coordinates of the jth particle.* Let $\mathcal{T}(p,q)$ be the total kinetic energy and $U(q)$ the total potential energy of the system, where p and q stand respectively for all sN p's and all sN q's. We shall only consider conservative systems so that neither $\mathcal{T}$ nor U will depend explicitly on the time, and we shall assume that the forces in the system are velocity-independent. The total energy $\mathcal{H}$ is given by the equation

$$\mathcal{H} = \mathcal{T} + U = \mathcal{H}(p,q). \tag{5.0102}$$

It is advantageous to number the p's and q's continuously from 1 to sN, and we shall assume that this has been done unless otherwise stated. Since

* It may be remarked here that the $2sN$ coordinates need not be the coordinates and momenta of the individual particles. We shall only consider canonical variables. In that case the sN q's can be any sN quantities which completely determine the position of all particles in the system. The sN p's are determined from the kinetic energy (compare Eq. (2.205)).

the p's and q's are canonically conjugate variables, we have (compare also Eqs. (2.207))

$$\dot{p}_i = -\frac{\partial \mathscr{H}}{\partial q_i}, \quad \dot{q}_i = \frac{\partial \mathscr{H}}{\partial p_i}; \qquad i = 1, 2, \cdots, sN. \tag{5.0103}$$

To each system there corresponds a point, the so-called *representative point*, in the $2sN$-dimensional phase-space. Following the Ehrenfests' notation, this space will be called *Γ-space* where Γ stands for *g*as.

From theoretical mechanics it is known that there are $2sN$ functions $\varphi_i(p,q)$ corresponding to Eqs. (5.0103) with the following properties. Of the $2sN$ functions, one is the total energy ε of the system, one is — apart from a constant — the time t, and the others are constant:

$$\left.\begin{aligned} \varphi_1(p,q) &= \varepsilon(p,q) = c_1, \\ \varphi_2(p,q) &= c_2, \\ &\cdots\cdots\cdots\cdots\cdots \\ \varphi_{2sN-1}(p,q) &= c_{2sN-1}, \\ \varphi_{2sN}(p,q) &= c_{2sN} + t, \end{aligned}\right\} \tag{5.0104}$$

where the c_i are the $2sN$ integration constants. The c_i ($i = 1,2,\cdots,sN$) or the corresponding φ_i are called integrals or constants of motion.

That one can put the integration of Eqs. (5.0103) in the form (5.0104) follows if one divides the first $2sN - 1$ equations of motion by the last one. The resulting $2sN - 1$ equations do not contain the time and give rise to the first $2sN - 1$ equations of (5.0104). One is then left with one equation and from that it is possible after substitution to determine t by a quadrature, and the last of the equations (5.0104) results.

As time goes on, the representative point of the system will describe an orbit in Γ-space. The q_i change their values because of the velocities of the particles, and the p_i becuase of the external and internal forces acting on the particles. The orbit will be situated on the hypersurface with equation $\varepsilon(p,q) = c_1$. This hypersurface will be called an *energy surface*.

Instead of considering one system it is often advantageous and even unavoidable, as we shall discuss in Section 5.13, to consider a large number of similar systems that all contain N particles, each of which has s degrees of freedom and which all exist under the same external circumstances. In other words, all of these systems will possess the same Hamiltonian (5.0102), but their representative points will in general not be the same. Let us assume that there are so many systems that it is possible to speak of the density $D(p,q)$ with which the representative points are distributed in Γ-space. Such a collection of similar systems was called by Gibbs an *ensemble*.

Since the representative points are moving in phase space, the density will in general be a function of time. Let the density, that is, the number of representative points per unit volume of Γ-space, at time t be denoted by $D(p,q,t)$. In considering the change of D with time it is important to distinguish between its rate of change at a fixed point in Γ-space and the rate of change at a point moving along an orbit in Γ-space. The first of these — which we may call the local rate of change — will be denoted by $\partial D/\partial t$, while for the second, or convected rate of change as it may be called, we use the symbol dD/dt. The relation between the two is given by the equation

$$\frac{dD}{dt} = \frac{\partial D}{\partial t} + \sum_i \left(\frac{\partial D}{\partial p_i} \dot{p}_i + \frac{\partial D}{\partial q_i} \dot{q}_i \right). \tag{5.0105}$$

Let us consider for a moment the volume element $d\Omega$ in Γ-space corresponding to coordinates in the intervals p_i, $p_i + dp_i$ and q_i, $q_i + dq_i$, so that we have

$$d\Omega = \prod_i dp_i \, dq_i. \tag{5.0106}$$

Elementary volumes in Γ-space, $d\Omega$, will sometimes be called, with Gibbs, elements of *extension in phase*.

At a later moment, $t + \delta t$, the representative points which at t were inside $d\Omega$ will have moved along their orbits and their coordinates which were p_i, q_i at t will now be $p_i + \dot{p}_i\delta t$, $q_i + \dot{q}_i\delta t$. Let $d\Omega'$ be the extension in phase at $t + \delta t$. We shall show presently that

$$\frac{dD}{dt} = 0. \tag{5.0107}$$

This means that the *density in phase*, as D is called by Gibbs, remains constant in the neighborhood of a point as it moves along its orbit, or that we may treat the points in Γ-space as an incompressible fluid. A result of this will be that we have

$$d\Omega = d\Omega', \tag{5.0108}$$

expressing the fact that the extension in phase occupied by a collection of representative points is constant in time. Equations (5.0107) and (5.0108) are two different forms of *Liouville's theorem*.

We can prove Eq. (5.0107) as follows by considering the right-hand side of Eq. (5.0105). (Compare the derivation of Eq. (2.507).) Consider the extension in phase $d\Omega$ at t and at $t + \delta t$. The number of representative points N in $d\Omega$ at t will be given by the equation

$$N = D d\Omega. \tag{5.0109}$$

At $t + \delta t$ we have

$$N + \delta N = \left(D + \frac{\partial D}{\partial t}\,\delta t\right) d\Omega. \tag{5.0110}$$

The difference δN arises from the fact that the number of points entering $\delta\Omega$ during δt may be different from the number of points leaving $d\Omega$ during this time interval. The number of points entering $d\Omega$ during δt through the "face" at q_i will be given by the expression

$$D\dot{q}_i \frac{d\Omega}{dq_i}\,\delta t, \tag{5.0111}$$

while the number of points leaving $d\Omega$ during δt through the "opposite face" at $q_i + dq_i$ is given by the expression

$$\left(D + \frac{\partial D}{\partial q_i}\,dq_i\right)\left(\dot{q}_i + \frac{\partial \dot{q}_i}{\partial q_i}\,dq_i\right)\frac{d\Omega}{dq_i}\,\delta t. \tag{5.0112}$$

In expressions (5.0111) and (5.0112) $d\Omega/dq_i$ stands for the "area" of the "face," that is, $d\Omega$ divided by dq_i, and not for a derivative.

Combining expressions (5.0111) and (5.0112), summing over all $2sN$ coordinates (q_i and p_i), and neglecting higher-order terms, we get

$$\delta N = -\sum_i \left[D\left(\frac{\partial \dot{q}_i}{\partial q_i} + \frac{\partial \dot{p}_i}{\partial p_i}\right) + \left(\frac{\partial D}{\partial q_i}\,\dot{q}_i + \frac{\partial D}{\partial p_i}\,\dot{p}_i\right)\right] d\Omega\,\delta t,$$

or, using Eqs. (5.0109) and (5.0110),

$$\frac{\partial D}{\partial t} = -\sum_i \left(\frac{\partial D}{\partial q_i}\,\dot{q}_i + \frac{\partial D}{\partial p_i}\,\dot{p}_i\right), \tag{5.0113}$$

since from Eqs. (5.0103) it follows that

$$\frac{\partial \dot{q}_i}{\partial q_i} + \frac{\partial \dot{p}_i}{\partial p_i} = 0. \tag{5.0114}$$

From Eqs. (5.0105) and (5.0113), Eq. (5.0107) follows.

Equation (5.0113) can also be written in the following form:

$$\frac{\partial D}{\partial t} = \{\mathscr{H}, D\}, \tag{5.0115}$$

which expresses the rate of change of D in terms of the *Poisson bracket* of D and the Hamiltonian.

The Poisson bracket $\{a,b\}$ is defined by the equation

$$\{a,b\} = \sum_i \left[\frac{\partial a}{\partial q_i}\frac{\partial b}{\partial p_i} - \frac{\partial a}{\partial p_i}\frac{\partial b}{\partial q_i}\right].$$

An ensemble for which $\partial D/\partial t = 0$ is called a *stationary ensemble* and in the following we shall restrict ourselves mainly to that kind of ensemble. From Eq. (5.0115) and the definition of a stationary ensemble it follows that an ensemble is stationary if

$$\{\mathscr{H},D\} = 0. \tag{5.0116}$$

It is shown in theoretical mechanics that Eq. (5.0116) is satisfied if and only if

$$D = D(\varphi_1,\varphi_2, \cdots, \varphi_{2sN-1}), \tag{5.0117}$$

where the φ_i are the functions discussed above (Eqs. (5.0104)).

Instead of considering stationary ensembles with a density given by the general equation (5.0117), we shall restrict ourselves to those stationary ensembles for which

$$D = D(\varepsilon), \tag{5.0118}$$

where $\varepsilon = \varphi_1$ is the energy of the system. The reasons for this restriction will be discussed in Section 5.13.

Instead of working with D we shall mostly use the normalized *coefficient of probability* ρ, which is given by the equation

$$\rho = \frac{D}{n}, \tag{5.0119}$$

where n is the total number of systems in the ensemble and given by the equation

$$n = {}_{\Gamma}\!\!\int D\, d\Omega, \tag{5.0120}$$

and where in Eq. (5.0120) and henceforth the symbol ${}_{\Gamma}\!\int$ will indicate integration over the whole of the available part of Γ-space.*

We shall often refer to ρ as the *density of the ensemble*, which should not be confused with the density in phase D.

With Gibbs we also introduce a function η, the *index of probability*, given by the equation

$$\eta = \ln \rho. \tag{5.0121}$$

As a consequence of Eq. (5.107) we have

$$\frac{d\rho}{dt} = \frac{d\eta}{dt} = 0, \tag{5.0122}$$

* In most cases, some parts of Γ-space will not be available due to external circumstances or to our choice of coordinates. In the case of a gas enclosed in a finite volume, the x_i,y_i,z_i can range only over a finite interval, while a q_i which is an angle is restricted to the interval 0 to 2π.

while for stationary ensembles

$$\frac{\partial \rho}{\partial t} = \frac{\partial \eta}{\partial t} = 0. \tag{5.0123}$$

We shall mainly be concerned with ensembles for which Eq. (5.0118) holds, and in that case we have

$$\rho = \rho(\varepsilon), \qquad \eta = \eta(\varepsilon). \tag{5.0124}$$

For future reference we remark that ρ must be a single-valued, real, and nonnegative function of the p's and q's which in virtue of Eqs. (5.0119) and (5.0120) satifies the *normalization equation*

$$\int_\Gamma \rho \, d\Omega = 1. \tag{5.0125}$$

Any function $G(p,q)$ of the p_i and q_i is called a *phase function* and its average value $\bar{G}$, where the average is taken over the ensemble, is defined by the equation

$$\bar{G} = \int_\Gamma G\rho \, d\Omega. \tag{5.0126}$$

If we are dealing with a stationary ensemble, $\bar{G}$ will be a constant, that is, not dependent on time, provided G does not contain the time explicitly.

We want to indicate briefly and very incompletely how one can derive the Boltzmann transport equation (2.507) from the Liouville equation in the form

$$\frac{d\rho^{(N)}}{dt} = 0 \tag{5.0127}$$

where we have emphasized that we are dealing with a system consisting of N particles—all of which we assume to be identical—by the addition of a superscript on ρ.

We now consider the case where the only degrees of freedom are those of the center of the mass; that is, we are considering the case of point particles without internal degrees of freedom. In that case we can write Eq. (5.0127) in the form

$$\frac{\partial \rho^{(N)}}{\partial t} + \sum_j (\nabla_{pj}\mathscr{H} \cdot \nabla_j \rho^{(N)}) - \sum_j (\nabla_j \mathscr{H} \cdot \nabla_{pj} \rho^{(N)}) = 0 \tag{5.0128}$$

or

$$\frac{\partial \rho^{(N)}}{\partial t} + \sum_j \left[\frac{1}{m}(\mathbf{p}_j \cdot \nabla_j \rho^{(N)}) + (\mathbf{F}_j^{\text{ext}} + \mathbf{F}_j^{\text{int}} \cdot \nabla_{pj} \rho^{(N)})\right] = 0 \tag{5.0129}$$

where we have used the canonical equations of motion (5.0103) and the fact that for point particles the kinetic energy is a sum of terms $\mathbf{p}_j^2/2m$ while the potential energy consists of two parts, one part corresponding to external forces $\mathbf{F}_j^{\text{ext}}$, and the other part corresponding to the interparticle forces, $\mathbf{F}_j^{\text{int}}$. To simplify our notation we have indicated by ∇_p the vector operator with components $\partial/\partial p_x$, $\partial/\partial p_y$, and $\partial/\partial p_z$.

We now introduce the single-particle distribution function $f(\mathbf{r},\mathbf{p})$ by the equation

$$f(\mathbf{r},\mathbf{p}) = \int \rho^{(N)}\, d^3\mathbf{r}_2 \cdots d^3\mathbf{r}_N d^3\mathbf{p}_2 \cdots d^3\mathbf{p}_N, \tag{5.0130}$$

We first of all note that it is immaterial over which $N-1$ of the N particles we integrate, because all particles are identical. Secondly, we note that we are here in contradistinction to Section 2.5, considering f as a function of $\mathbf{r}$ and $\mathbf{p}$ rather than of $\mathbf{r}$ and $\mathbf{c}$. Finally, we note that it follows from Eqs. (5.0125) and (5.0130), where the integration is over all allowable values of $\mathbf{r}_2, \cdots, \mathbf{p}_N$, that $f(\mathbf{r},\mathbf{p})$ is normalized.

To find the equation to be satisfied by $f(\mathbf{r},\mathbf{p})$ we must integrate Eq. (5.0129) over $N-1$ of the particles. Before doing that, we shall introduce one further assumption: that the forces between the particles are two-particle forces so that

$$\mathbf{F}_j^{\text{int}} = \sum_{k(\neq j)} \mathbf{F}_{jk}^{\text{int}}.$$

Because of the identical nature of the particles and the fact that f will vanish as $|\mathbf{p}| \to \infty$ and at the walls of the vessel containing the system, we find after integrating

$$\frac{\partial f_1}{\partial t} + \frac{1}{m}(\mathbf{p}_1 \cdot \nabla_1 f_1) + (\mathbf{F}_1^{\text{ext}} \cdot \nabla_{p1} f_1) = -\int (\mathbf{F}_{12}^{\text{int}} \cdot \nabla_{p1} f_{12}^{(2)})\, d^3\mathbf{r}_2\, d^3\mathbf{p}_2, \tag{5.0131}$$

where $f_{12}^{(2)}$ is the two-particle distribution function defined by the equation

$$f_{12}^{(2)} = \int \rho^{(N)}\, d^3\mathbf{r}_3 \cdots d^3\mathbf{r}_N d^3\mathbf{p}_3 \cdots d^3\mathbf{p}_N. \tag{5.0132}$$

We note that Eq. (5.0131) already begins to look like the Boltzmann equation (2.507). Kirkwood* has shown how a time-averaging procedure can reduce the right-hand side of Eq. (5.0131) to the collision integral of Eq. Eq. (2.507). Since the analysis is too complicated to be reproduced here, we merely draw attention to the fact that integrating over time will produce something like the difference between $f_{12}^{(2)}$ before and after a collision, while if there were no correlations between the particles we would expect that $f_{12}^{(2)}$ would be simply the product $f_1 f_2$.

5.2 STATIONARY ENSEMBLES

In the present section we shall briefly introduce three kinds of stationary ensembles which we shall use in the subsequent discussion. We mentioned that

* J. G. Kirkwood, *J. Chem. Phys.*, **15**, 72, 1947. See also N. N. Bogolyubov, *Studies in Stat. Mech.*, **1**, 1, 1962. E. G. D. Cohen and T. H. Berlin, *Physica*, **26**, 717, 1960.

we would restrict ourselves to ensembles with densities given by Eq. (5.0124), that is, densities depending on the energy only.

Since the density must satisfy Eq. (5.0125), we cannot use the following two expressions for ρ, which are the two simplest possibilities,*

$$\rho = \text{const}, \text{ or } \rho = \text{const} \times \varepsilon. \qquad (5.0201)$$

However, we can try a linear dependence of η on ε,

$$\eta = \beta(\psi - \varepsilon) \qquad (5.0202)$$

and we have for the density

$$\rho = e^{\beta(\psi - \varepsilon)}, \qquad (5.0203)$$

where ψ and β are constants. Gibbs called the ensemble with density given by Eq. (5.0203) a *canonical ensemble*. In order to distinguish it from the microcanonical ensemble to be introduced presently, we shall call it a *macrocanonical ensemble*. The quantity β is related to the temperature, as we shall see in the next section. The inverse of β which Gibbs denoted by Θ is called the *modulus* of the macrocanonical ensemble. The quantity ψ is independent of the p_i and q_i, but depends on β and the external parameters a_i (called by Gibbs the *external coordinates*) on which the energy ε depends, since in virtue of the normalization condition (5.125) it is determined by the equation

$$e^{-\beta\psi} = {}_{\Gamma}\!\int e^{-\beta\varepsilon}\, d\Omega. \qquad (5.0204)$$

We may remark here that there is a strong resemblance between the density of a macrocanonical ensemble on the one hand and the distribution function of the Maxwell-Boltzmann distribution (Eq. (2.107) on the other. It looks as if a macrocanonical ensemble can be considered to be a system of "molecules" that interact weakly and therefore tend toward the Maxwell-Boltzmann distribution. Indeed, it can be shown (see Section 5.13) that a system in a thermostat will show the average behavior of a system belonging to a macrocanonical ensemble.

Another reason why macrocanonical ensembles play a special role among the stationary ensembles is that if two of them with the same value of β are coupled together, the resulting ensemble is again a macrocanonical ensemble. We shall return to this point in Section 5.6.

Apart from the macrocanonical ensembles we shall introduce two other stationary ensembles, namely, the microcanonical and the F-ensemble.

* It is, of course, possible that ρ satisfies Eq. (5.0201) for a restricted range of energy values; an example is given by the microcanonical or by the F-ensemble to be introduced presently.

The *microcanonical ensemble* (sometimes called an *ergodic ensemble*) consists of systems that all have the same energy ε_0; its density is given by the equation

$$\rho = A\delta(\varepsilon - \varepsilon_0), \tag{5.0205}$$

where A is a constant to be determined from the normalization condition and $\delta(x)$ is Dirac's δ-function.

The microcanonical ensemble corresponds to a surface density $\sigma(p,q)$ which is given by the equation

$$\sigma(p,q) = \frac{1}{Q(p,q)} \tag{5.0206}$$

where $Q(p,q)$ is given by the equation

$$Q = \sqrt{\sum_{i=1}^{sN}\left[\left(\frac{\partial \mathscr{H}}{\partial p_i}\right)^2 + \left(\frac{\partial \mathscr{H}}{\partial q_i}\right)^2\right]} \tag{5.0207}$$

The proof of Eq. (5.0206) is as follows. According to Liouville's theorem the density in Γ-space is invariant if one follows the representative points along their orbits in Γ-space. Consider now a volume element lying between two neighboring energy surfaces (corresponding to E_0 and $E_0 + \delta E$). The volume element will retain its extension if we follow its points along their orbits. Since the extension is given by the expression $d\mathscr{S}\cdot\delta N$, where $d\mathscr{S}$ is a surface element on the energy surface and δN the distance between the two energy surfaces, and since δN will be proportional to $1/Q$, we see that $d\mathscr{S}$ will vary as Q. In this way the number of points inside a fixed surface element ($=\sigma\cdot d\mathscr{S}$) will remain constant in time if we choose σ according to Eq. (5.0206).

Sometimes it is convenient to define the microcanonical ensemble by equation

$$\begin{aligned} \rho &= A, \qquad \varepsilon_0 \leqslant \varepsilon \leqslant \varepsilon_0 + \Delta, \\ \rho &= 0, \qquad \varepsilon < \varepsilon_0 \text{ or } \varepsilon > \varepsilon_0 + \Delta, \end{aligned} \tag{5.0208}$$

instead of by Eq. (5.0205). In this case the ensemble is sometimes called an *energy-shell ensemble*. The microcanonical ensemble is really a limiting case of an energy-shell ensemble.

The density of the *F-ensemble* is given by the equation

$$\rho = BF(\varepsilon), \tag{5.0209}$$

where B is to be determined by the normalization condition and where $F(x)$ is a function defined by the equations

$$\left.\begin{aligned} F(x) &= 1, \qquad x \leqslant \varepsilon_0; \\ F(x) &= 0, \qquad x > \varepsilon_0. \end{aligned}\right\} \tag{5.0210}$$

In Fig. 5.1 we have drawn ρ as a function of ε for the three ensembles mentioned in this section.

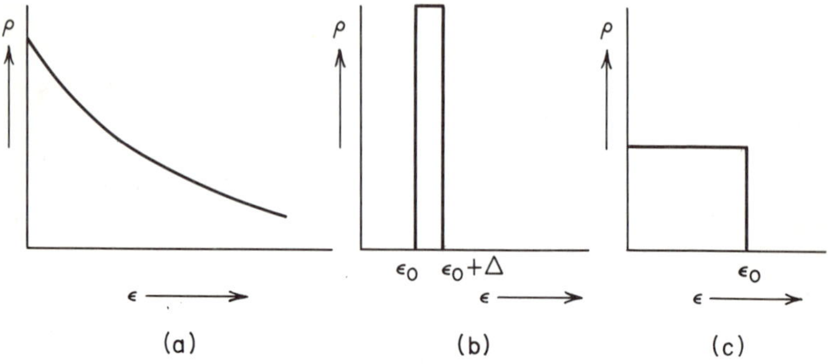

FIG. 5.1 Ensemble densities: (a) *macrocanonical ensemble,* (b) *energy shell ensemble,* (c) *F*-ensemble.

5.3 THE MACROCANONICAL ENSEMBLE

We mentioned in the preceding section the following theorem. *The macroscopic behavior of a system at equilibrium at a temperature T will be described correctly by taking the average behavior of a system in a macrocanonical ensemble with a β that is related to T by the equation*

$$\beta^{-1} = kT. \tag{5.0301}$$

That this is a sensible way of describing the behavior of a system in temperature equilibrium is due to the following two facts:

(a) The fluctuations of most quantities from the average values are negligibly small.
(b) Those fluctuations which are not small are so fast that they cannot be measured with our instruments.

Fluctuations, that is, deviations from the average value, will be discussed in the next section, when we shall have an opportunity to see how far conditions (a) and (b) are satisfied.

Let us now prove our basic theorem. From the normalization of ρ we have the equation

$$e^{-\beta\psi} = \int_{\Gamma} e^{-\beta\varepsilon}\, d\Omega. \tag{5.0302}$$

Remembering that ε depends on the external parameters a_i we shall, as in Section 2.6, introduce the generalized forces A_i by the equations

$$A_i = -\frac{\partial \varepsilon}{\partial a_i}. \tag{5.0303}$$

If the external parameters a_i are varied from a_i to $a_i + \delta a_i$, we have for the variation $\delta\varepsilon$ of the energy

$$\delta\varepsilon = -\sum A_i \delta a_i. \tag{5.0304}$$

Let us now consider a new macrocanonical ensemble derived from the given one by a variation of β and the external parameters. Since the normalization condition must also hold for this new ensemble, we have

$$-e^{-\beta\psi}\delta(\beta\psi) = \int_\Gamma e^{-\beta\varepsilon}[-\varepsilon\delta\beta - \beta\delta\varepsilon]\, d\Omega,$$

that is,

$$\delta(\beta\psi) = \int_\Gamma e^{\beta(\psi-\varepsilon)}[\varepsilon\delta\beta + \beta\delta\varepsilon]\, d\Omega,$$

or

$$\delta(\beta\psi) = \bar{\varepsilon}\delta\beta + \beta\overline{\delta\varepsilon} = \bar{\varepsilon}\delta\beta - \beta\sum_i \bar{A}_i\delta a_i, \tag{5.0305}$$

where we have used Eq. (5.0126) in the form

$$\bar{G} = \int_\Gamma G\, e^{\beta(\psi-\varepsilon)}\, d\Omega. \tag{5.0306}$$

Since

$$\delta(\beta\bar{\varepsilon}) = \bar{\varepsilon}\delta\beta + \beta\delta\bar{\varepsilon}, \tag{5.0307}$$

we can write Eq. (5.0305) in the form

$$\delta[\beta(\bar{\varepsilon} - \psi)] = -\delta\bar{\eta} = \beta(\delta\bar{\varepsilon} - \overline{\delta\varepsilon}). \tag{5.0308}$$

If arguments similar to those given in Section 4.7 (compare also Section 2.7), are used, it follows that

(a) $\overline{\delta\varepsilon}$ corresponds to the work done on the system when the external parameters change form a_i to $a_i + \delta a_i$;

(b) $\delta\varepsilon - \overline{\delta\varepsilon}$ is thus the heat added to the system, δQ, that is, the increase of energy not accounted for by macroscopically measurable work, this last quantity, δA being equal to $-\overline{\delta\varepsilon}$;

(c) hence β, being an integrating factor of δQ, must be inversely proportional to the absolute temperature T, or

$$\beta^{-1} = kT,$$

which proves Eq. (5.0301);

(d) $-\bar{\eta}$, apart possibly from an additive constant, corresponds to the entropy S divided by k, or

$$S = k\beta(\bar{\varepsilon} - \psi) + \text{const} = -k\bar{\eta} + \text{const}; \tag{5.0309}$$

(e) that ψ, apart possibly from an additive multiple of the temperature, is the (Helmholtz) free energy F.

We shall fix the additive constants by putting

$$e^{\beta F} = N!\, h^{sN} e^{\beta\psi}, \tag{5.0310}$$

where h is Planck's constant. From Eq. (5.0302) we then have

$$e^{-\beta F} = \frac{1}{N!} \int_\Gamma e^{-\beta\varepsilon} \frac{d\Omega}{h^{sN}}, \tag{5.0311}$$

and we see that both sides of Eq. (5.0311) are dimensionless.

The factor $N!$ enters in order that the free energy and the entropy will be extensive quantities and the factor h^{sN} ensures the correct dimensions and together they ensure that the formulas given here are, indeed, the classical limits of the corresponding quantum-mechanical equations.

From this discussion it follows that given ψ as a function of β and the a_i, or given β and given ε as a function of the external parameters—in which case ψ can be computed from Eq. (5.0302)—we have obtained an expression for the free energy of a system at equilibrium at temperature T for the given values of the a_i. It is well known that once the free energy is given as a function of T and the external parameters, the thermodynamical state of the system is determined. Most thermodynamical quantities can be derived from the free energy by differentiation. As δF and $\delta\psi$ are the same, we should expect relations between the average values of phase functions and partial derivatives of ψ. Indeed, from Eq. (5.0305) we easily have

$$\bar{\varepsilon} = \frac{\partial(\beta\psi)}{\partial\beta}, \tag{5.0312}$$

which is the Gibbs-Helmholtz equation, and

$$\bar{A}_i = -\frac{\partial\psi}{\partial a_i}. \tag{5.0313}$$

If a_i corresponds to the volume V, Eq. (5.0313) gives us

$$P = -\frac{\partial\psi}{\partial V}, \tag{5.0314}$$

from which the equation of state can be derived once ψ is computed.

One often denotes the right-hand side of Eq. (5.0311) by Z_Γ and this *partition function* plays a similar role to that of Z_μ of Chapter 2. Expressing

$\bar{\varepsilon}$, $\bar{A}_i$, and P in terms of Z_Γ we have, instead of Eqs. (5.0312) to (5.0314),

$$\bar{\varepsilon} = -\frac{\partial \ln Z_\Gamma}{\partial \beta}, \tag{5.0315}$$

$$\bar{A}_i = \frac{\partial \ln Z_\Gamma}{\beta \partial a_i}, \tag{5.0316}$$

$$P = \frac{\partial \ln Z_\Gamma}{\beta \partial V}. \tag{5.0317}$$

Other averages can be computed by straightforward integration using Eq. (5.0306). In that way we can, for instance, calculate the average value of the kinetic energy $\mathscr{T}$. This kinetic energy is a homogeneous quadratic expression in the P_i and hence we have (compare Eq. (2.211))

$$\mathscr{T} = \frac{1}{2} \sum_{i=1}^{sN} p_i \frac{\partial \mathscr{T}}{\partial p_i}. \tag{5.0318}$$

By using the same methods that led to Eq. (2.212) we find

$$\bar{\mathscr{T}} = \frac{sN}{2\beta} = \frac{1}{2} sNkT, \tag{5.0319}$$

expressing the *equipartition* of kinetic energy. Each degree of freedom contributes $1/2\beta = \frac{1}{2}kT$ to the average kinetic energy.

Let us now consider an ensemble of systems each containing N independent particles in a volume V. Let ε_j be the total energy of the jth particle ($j = 1, 2, \cdots, N$) and $p_k{}^j$ and $q_k{}^j$ the kth generalized momentum and coordinate pertaining to the jth particle. We then have

$$\varepsilon = \sum_{j=1}^{N} \varepsilon_j \tag{5.0320}$$

and

$$d\Omega = \prod_j d\omega_j, \tag{5.0321}$$

where

$$d\omega_j = \sum_{k=1}^{s} dp_k{}^j dq_k{}^j. \tag{5.0322}$$

We can now evaluate Z_Γ as

$$Z_\Gamma = \frac{1}{N!} \int_\Gamma e^{\beta \Sigma \varepsilon_j} \prod_j \frac{d\omega_j}{h^s} = \left[\frac{e}{N} \int e^{-\beta \varepsilon_j} \frac{d\omega_j}{h^s} \right]^N = Z_\mu{}^N, \tag{5.0323}$$

where Z_μ is given by Eq. (2.701) with $K = e/Nh^s$ and where we have used Stirling's formula for the factorial (Eq. (1.708)).

If the system consists of free monatomic particles, Z_Γ can be evaluated directly. The energy is given by the equation

$$\varepsilon = \sum_j \frac{\mathbf{p}_j^2}{2m} + \sum_j F(x_j,y_j,z_j), \tag{5.0324}$$

where $F(x,y,z)$ is the function defined by Eq. (2.721),

$$\left.\begin{aligned} F(x,y,z) &= 0, && \text{if } x,y,z \text{ lies inside } V; \\ F(x,y,z) &= \infty, && \text{if } x,y,z \text{ lies outside } V. \end{aligned}\right\} \tag{5.0325}$$

The integration over Γ-space is elementary and leads to the result (compare Eqs. (2.722) and (5.0323))

$$Z_\Gamma = \frac{1}{N!} V^N \left(\frac{2\pi m}{\beta h^2}\right)^{3N/2}. \tag{5.0326}$$

Using Eqs. (5.0315) and (5.0317), we have

$$\bar{\varepsilon}(=\overline{\mathcal{T}}) = -\frac{\partial \ln Z_\Gamma}{\partial \beta} = \frac{3N}{2\beta}, \tag{5.0327}$$

$$P = \frac{\partial \ln Z_\Gamma}{\beta \partial V} = \frac{N}{\beta V}. \tag{5.0328}$$

The last equation leads once more, as expected, to the perfect gas law.

5.4 FLUCTUATIONS IN A MACROCANONICAL ENSEMBLE

In the preceding section formulas were derived for the average values of the total energy and the generalized forces. If we wish to ascertain what the chances are of finding these average values realized when we pick a system at random from the macrocanonical ensemble, we must compute the fluctuations of these quantities. At the same time we shall also find formulas for the correlations between the various quantities.

The easiest way of obtaining formulas for the fluctuations is to start from the normalizing equation in the form

$$\int_\Gamma e^{\beta(\psi-\varepsilon)}\, d\Omega = 1. \tag{5.0401}$$

Taking the derivative with respect to β, we get

$$0 = \int_\Gamma \frac{\partial}{\partial \beta}[\beta(\psi-\varepsilon)] e^{\beta(\psi-\varepsilon)}\, d\Omega = \int_\Gamma \left[\frac{\partial \beta\psi}{\partial \beta} - \varepsilon\right] e^{\beta(\psi-\varepsilon)}\, d\Omega, \tag{5.0402}$$

or

$$\bar{\varepsilon} = \frac{\partial(\beta\psi)}{\partial\beta}, \tag{5.0403}$$

in accordance with Eq. (5.0312).

Similarly, by taking the derivative with respect to a_i in Eq. (5.0401), we get

$$0 = \int_\Gamma \frac{\partial}{\partial a_i} [\beta(\psi - \varepsilon)]\, e^{\beta(\psi-\varepsilon)}\, d\Omega = \beta \int_\Gamma \left(\frac{\partial\psi}{\partial a_i} - \frac{\partial\varepsilon}{\partial a_i}\right) e^{\beta(\psi-\varepsilon)}\, d\Omega, \tag{5.0404}$$

or, using Eq. (5.0303),

$$\bar{A}_i = -\frac{\partial\psi}{\partial a_i}. \tag{5.0405}$$

Differentiating Eq. (5.0402) with respect to β gives us

$$\int_\Gamma \left[\frac{\partial^2(\beta\psi)}{\partial\beta^2} + \left\{\frac{\partial(\beta\psi)}{\partial\beta} - \varepsilon\right\}^2\right] e^{\beta(\psi-\varepsilon)}\, d\Omega = 0, \tag{5.0406}$$

or, using Eq. (5.0403),

$$\overline{(\varepsilon - \bar{\varepsilon})^2} = -\frac{\partial^2(\beta\psi)}{\partial\beta^2} = -\frac{\partial\bar{\varepsilon}}{\partial\beta}. \tag{5.0407}$$

Similarly, differentiating Eq. (5.0404) with respect to a_j, we get

$$\int_\Gamma \left[\frac{\partial^2\psi}{\partial a_i\, \partial a_j} - \frac{\partial^2\varepsilon}{\partial a_i\, \partial a_j} + \beta\left(\frac{\partial\psi}{\partial a_i} - \frac{\partial\varepsilon}{\partial a_i}\right)\left(\frac{\partial\psi}{\partial a_j} - \frac{\partial\varepsilon}{\partial a_j}\right)\right] e^{\beta(\psi-\varepsilon)}\, d\Omega = 0, \tag{5.0408}$$

or, using Eqs. (5.0405) and (5.0303),

$$\overline{(A_i - \bar{A}_i)(A_j - \bar{A}_j)} = \frac{1}{\beta}\left[\overline{\frac{\partial^2\varepsilon}{\partial a_i\, \partial a_j}} - \frac{\partial^2\psi}{\partial a_i\, \partial a_j}\right], \tag{5.0409}$$

which is valid both for $i = j$ and for $i \neq j$.

Finally we get by differentiating either Eq. (5.0402) with respect to a_i or Eq. (5.0404) with respect to β

$$\overline{(A_i - \bar{A}_i)(\varepsilon - \bar{\varepsilon})} = \frac{\partial^2\psi}{\partial a_i\, \partial\beta}. \tag{5.0410}$$

We shall see later on that in many cases ψ and its derivatives, and thus also $\bar{\varepsilon}$ and $\bar{A}_i$, will be proportional to the number of particles in the system, N. In that case it follows from Eqs. (5.0407), (5.0409), and (5.0410) that all quantities such as

$$\frac{\overline{(\varepsilon - \bar{\varepsilon})^2}}{\bar{\varepsilon}^2}, \quad \frac{\overline{(A_i - \bar{A}_i)(A_j - \bar{A}_j)}}{\bar{A}_i\bar{A}_j}, \quad \frac{\overline{(A_i - \bar{A}_i)(\varepsilon - \bar{\varepsilon})}}{\bar{A}_i\bar{\varepsilon}}$$

will be of the order of $1/N$ and hence negligibly small if N is sufficiently large.

It sometimes happens, however, that fluctuations can be very large. At an ideal wall, for instance, the force $(-\partial U/\partial q)$ will be infinite (compare Fig. 1.5 where we have a similarly idealized potential energy) and the pressure will fluctuate violently. It is then, however, practically always possible to show that the fluctuations of the time average B_i defined by the equation

$$B_i = \frac{1}{\tau}\int_t^{t+\tau} A_i\,dt, \tag{5.0411}$$

are small, that is,

$$\frac{\overline{(B_i - \bar{B}_i)^2}}{\bar{B}_i^2} \sim \frac{1}{N} \ll 1. \tag{5.0412}$$

The fact that any actual measurement of the pressure will take a finite time means that it is more in accordance with physical reality to consider the B_i than the A_i (compare also the discussion in Section 5.8, and especially that in Section 8.2).

5.5 THE ENTROPY IN A MACROCANONICAL ENSEMBLE

In Section 5.3 we saw that the entropy S of a system was related to the average value $\bar{\eta}$ of the index of probability through Eq. (5.0309). There are two other ways to correlate the entropy with quantities relating to the ensemble. We shall reserve for later discussion the fact, which may seem strange at first sight, that it is possible to define the entropy in statistical mechanics in so many different ways.

Let the smallest possible value of the energy for given values of the external parameters be $\varepsilon_{\min}$ which in general will be a function of the a_i. Since in general it will always be possible to choose all p_i equal to zero, independent of the values of the q_i, so that $\mathscr{T}_{\min} = 0$, we have

$$\varepsilon_{\min}(a_i) = U_{\min}(a_i). \tag{5.0501}$$

We now introduce a quantity Ω by the equation

$$\Omega = \int_{\varepsilon_{\min}(a_i)}^{\varepsilon} d\Omega. \tag{5.0502}$$

From this equation we see, firstly, that Ω is the extension in phase between the two energy surfaces corresponding to $\varepsilon_{\min}$ and ε and, secondly, that Ω depends both on ε and on the a_i, the latter through $\varepsilon_{\min}$, that is,

$$\Omega = \Omega(\varepsilon, a_i). \tag{5.0503}$$

We also introduce a function ϕ defined by the equation

$$e^{\phi} = \frac{\partial \Omega}{\partial \varepsilon}, \tag{5.0504}$$

or

$$\phi(\varepsilon, a_i) = \ln \frac{\partial \Omega}{\partial \varepsilon}. \tag{5.0505}$$

We shall need presently the value of e^{ϕ} for $\varepsilon = \varepsilon_{\min}$. We shall assume for the sake of simplicity that the kinetic energy $\mathcal{T}$ depends on the p_i only. We now introduce the following two elementary volumes

$$d\Omega_p = \prod_{i=1}^{sN} dp_i, \qquad d\Omega_q = \prod_{i=1}^{sN} dq_i, \tag{5.0506}$$

which in the simplified case considered here are, apart from a multiplying constant, what Gibbs calls respectively an element of *extension in velocity* and an element of *extension in configuration*. (For a discussion of the general case, see Problem 3 of Chapter 5.)

Furthermore, we write

$$\Omega_p = \int_0^{\mathcal{T}} d\Omega_p, \qquad \Omega_q = \int_{U_{\min}}^{U} d\Omega_q, \tag{5.0507}$$

and we introduce two quantities ϕ_p and ϕ_q by the equations

$$e^{\phi_p} = \frac{\partial \Omega_p}{\partial \mathcal{T}}, \qquad e^{\phi_q} = \frac{\partial \Omega_q}{\partial U} \tag{5.0508}$$

From Eqs. (5.0502) and (5.0507) it follows that

$$\Omega = \iint d\Omega_p \, d\Omega_q = \int_{U_{\min}}^{U=\varepsilon} \Omega_p \, d\Omega_q, \tag{5.0509}$$

where Ω_p depends on the q_i through the fact that the upper limit $\mathcal{T}$ in the integral in Eq. (5.0507) must satisfy the relation

$$\mathcal{T} + U = \varepsilon. \tag{5.0510}$$

Taking the derivative with respect to ε in Eq. (5.0509) and using the fact that

$$\frac{\partial \Omega_p}{d\varepsilon} = \frac{\partial \Omega_p}{\partial \mathcal{T}}$$

we get

$$e^{\phi} = \frac{\partial \Omega}{\partial \varepsilon} = \int_{U_{\min}}^{U=\varepsilon} e^{\phi_p} \, d\Omega_q + \Omega_p e^{\phi_q} \Big|_{U=\varepsilon}, \tag{5.0511}$$

and hence by virtue of Eq. (5.0501) and the fact that $\Omega_p = 0$ for $U = \varepsilon$,

$$e^{\phi}|_{\varepsilon=\varepsilon_{\min}} = 0. \tag{5.0512}$$

If a phase function G is a function of the energy only, we have for its average value

$$\bar{G} = \int_{\Gamma} G(\varepsilon)\rho(\varepsilon)\, d\Omega, \tag{5.0513}$$

or

$$\bar{G} = \int_{\varepsilon_{\min}}^{\infty} G(\varepsilon)\rho(\varepsilon)e^{\phi}\, d\varepsilon. \tag{5.0514}$$

We must draw attention here to the different ways by which the right-hand sides of Eqs. (5.0513) and (5.0514) depend on the a_i. In the first equation, $\bar{G}$ depends on the a_i through ε, but in the second equation $\bar{G}$ depends on the a_i through ϕ and the lower limit of integration, while ε is an integration variable.

From the normalization equation we get in this way

$$e^{-\beta\psi} = \int e^{-\beta\varepsilon+\phi}\, d\varepsilon. \tag{5.0515}$$

In the preceding section we saw that the fluctuations in energy in a macrocanonical ensemble are in general very small. This means that practically all members of the ensemble will possess energies that are practically equal to $\bar{\varepsilon}$, and so Eq. (5.0515) can be written in the form

$$e^{-\beta\psi} \approx [e^{-\beta\varepsilon+\phi}]_{\varepsilon=\bar{\varepsilon}} \cdot \Delta, \tag{5.0516}$$

where Δ is the "half-width" of the energy fluctuations which is of the order of magnitude $\bar{\varepsilon}/\sqrt{N} \sim \sqrt{N}$ (compare the discussion at the end of the preceding section),

$$\Delta^2 = C\,\overline{(\varepsilon - \bar{\varepsilon})^2}. \tag{5.0517}$$

In order to derive Eq. (5.0516) we first note that, since the energy fluctuations are small, $\bar{\varepsilon}$ will be the energy possessed by the largest number of systems in the ensemble. Since the fraction of systems, $f(\varepsilon)d\varepsilon$, with energies between ε and $\varepsilon + d\varepsilon$ is given by the equation

$$f(\varepsilon)d\varepsilon = e^{\beta(\psi-\varepsilon)+\phi}\, d\varepsilon, \tag{5.0518}$$

as follows from the definitions of ρ and ϕ, $\bar{\varepsilon}$ will also be the energy for which $e^{-\beta\varepsilon+\phi}$ will be maximum, or

$$\beta = \left.\frac{\partial\phi}{\partial\varepsilon}\right|_{\varepsilon=\bar{\varepsilon}}. \tag{5.0519}$$

Developing $\phi - \beta\varepsilon$ in a Taylor series, we get

$$\phi - \beta\varepsilon = \phi|_{\varepsilon=\bar{\varepsilon}} - \beta\bar{\varepsilon} + \frac{1}{2}\frac{\partial^2\phi}{\partial\varepsilon^2}\bigg|_{\varepsilon=\bar{\varepsilon}}(\varepsilon - \bar{\varepsilon})^2 + \cdots, \tag{5.0520}$$

where we used Eq. (5.0519). From Eq. (5.520) it follows that to a first approximation

$$e^{-\beta\varepsilon+\phi} = [e^{-\beta\varepsilon+\phi}]_{\varepsilon=\bar{\varepsilon}} \cdot \exp\left[\frac{1}{2}\frac{\partial^2\phi}{\partial\varepsilon^2}\bigg|_{\varepsilon=\bar{\varepsilon}}(\varepsilon - \bar{\varepsilon})^2\right]. \tag{5.0521}$$

Substituting Eq. (5.0521) into Eq. (5.0515) and integrating, we get approximately (by integrating from 0 to ∞ instead of from $\varepsilon_{\min}$ to ∞)

$$e^{-\beta\psi} = [e^{-\beta\varepsilon+\phi}]_{\varepsilon=\bar{\varepsilon}} \cdot \left(-\frac{2}{\pi}\frac{\partial^2\phi}{\partial\varepsilon^2}\bigg|_{\varepsilon=\bar{\varepsilon}}\right)^{-1/2} \tag{5.0522}$$

On the other hand, substituting Eq. (5.0521) into the expression for $\overline{(\varepsilon - \bar{\varepsilon})^2}$, we get

$$\overline{(\varepsilon - \bar{\varepsilon})^2} = \frac{\int e^{\beta(\psi-\varepsilon)+\phi}(\varepsilon - \bar{\varepsilon})^2\, d\varepsilon}{\int e^{\beta(\psi-\varepsilon)+\phi}\, d\varepsilon} \cong \frac{\int (\varepsilon - \bar{\varepsilon})^2 \exp\left[\frac{1}{2}\frac{\partial^2\phi}{\partial\varepsilon^2}\Big|_{\varepsilon=\bar{\varepsilon}}(\varepsilon - \bar{\varepsilon})^2\right] d\varepsilon}{\int \exp\left[\frac{1}{2}\frac{\partial^2\phi}{\partial\varepsilon^2}\Big|_{\varepsilon=\bar{\varepsilon}}(\varepsilon - \bar{\varepsilon})^2\right] d\varepsilon}$$

$$\cong \left(-\frac{\partial^2\phi}{\partial\varepsilon^2}\bigg|_{\varepsilon=\bar{\varepsilon}}\right)^{-1}. \tag{5.0523}$$

Combining Eqs. (5.0522) and (5.0523), we obtain Eq. (5.0516).

We may remark here that Eq. (5.0521) expresses that the distribution in energy is to a first approximation a Gaussian distribution.

From Eq. (5.0516) it follows that

$$\beta(\bar{\varepsilon} - \psi) = \phi(\bar{\varepsilon}, a_i) + \ln \Delta. \tag{5.0524}$$

Since $\ln \Delta$ is in general of the order of magnitude $\ln N$ while $\bar{\varepsilon}$ and ψ are of the order of magnitude N, we may neglect the last term on the right-hand side of Eq. (5.0524). Using Eq. (5.0309) for the entropy we have, apart possibly from an additive constant,

$$\frac{S}{k} = \phi(\bar{\varepsilon}, a_i) = \ln \frac{\partial\Omega}{\partial\varepsilon}\bigg|_{\varepsilon=\bar{\varepsilon}}. \tag{5.0525}$$

From the well-known relation between entropy and temperature (compare Eq. (5.0308)).

$$\frac{1}{T} = \frac{\partial S}{\partial\varepsilon}\bigg|_{\varepsilon=\bar{\varepsilon}}, \tag{5.0526}$$

we get, using Eq. (5.0525),

$$\beta = \frac{1}{kT} = \left.\frac{\partial S/k}{\partial \varepsilon}\right|_{\varepsilon=\bar{\varepsilon}} = \left.\frac{\partial \phi}{\partial \varepsilon}\right|_{\varepsilon=\bar{\varepsilon}}, \tag{5.0527}$$

which we derived a moment ago by slightly different arguments (Eq. (5.0519)).

If we calculate the average value of $\partial\phi/\partial\varepsilon$ over the ensemble, we get

$$\overline{\frac{\partial \phi}{\partial \varepsilon}} = \int \frac{\partial \phi}{\partial \varepsilon} e^{\beta(\psi-\varepsilon)+\phi}\, d\varepsilon = \left[e^{\beta(\psi-\varepsilon)+\phi}\right]_{\varepsilon_{\min}}^{\infty} + \beta \int e^{\beta(\psi-\varepsilon)+\phi}\, d\varepsilon,$$

or

$$\overline{\frac{\partial \phi}{\partial \varepsilon}} = \beta, \tag{5.0528}$$

where we have used Eq. (5.0512) to show that the integrated part is zero and where we have also used Eq. (5.0515).

Comparing Eqs. (5.0527) and (5.0528), we see that the average value of $\partial\phi/\partial\varepsilon$ is the same as its value when the energy has its most probable value. This property is equally true for other quantities and is a consequence of the fact that the dispersion in energy is extremely small.

We mentioned before that by far the largest fraction of systems in the ensemble have an energy in the neighborhood of $\bar{\varepsilon}$. From Eq. (5.0518) we see that this means that e^{ϕ} must increase steeply when ε increases from $\varepsilon_{\min}$ to $\bar{\varepsilon}$. Let us compare a macrocanonical ensemble with $\eta = \beta(\psi - \varepsilon)$ and average energy $\bar{\varepsilon}$ with an F-ensemble with density given by the equation

$$\rho_F = a e^{\beta(\psi-\bar{\varepsilon})} F(\varepsilon), \tag{5.0529}$$

where $F(\varepsilon)$ is given by Eq. (5.0210) and where a is a constant, and where we have put $\varepsilon_0 = \bar{\varepsilon}$.

If G is a phase function that depends only on the energy, we have for its average value $\bar{G}_F$, taken over the F-ensemble,

$$\bar{G}_F = \int_{\Gamma} \rho_F G(\varepsilon)\, d\Omega = \int_{\varepsilon_{\min}}^{\bar{\varepsilon}} a e^{\beta(\psi-\bar{\varepsilon})} G(\varepsilon) e^{\phi}\, d\varepsilon, \tag{5.0530}$$

or, since e^{ϕ} has a steep maximum for $\varepsilon \cong \bar{\varepsilon}$,

$$\bar{G}_F \approx a\Delta e^{\beta(\psi-\bar{\varepsilon})+\phi(\bar{\varepsilon})} G(\bar{\varepsilon}). \tag{5.0531}$$

On the other hand, the average taken over the macrocanonical ensemble is given by the equation

$$\bar{G} = \int_{\Gamma} e^{\beta(\psi-\bar{\varepsilon})} G(\varepsilon)\, d\Omega \approx e^{\beta(\psi-\varepsilon)+\phi(\bar{\varepsilon})} G(\bar{\varepsilon})\Delta. \tag{5.0532}$$

We note that in deriving Eqs. (5.0531) and (5.0532) we have used Eqs. (5.0520) and (5.0523) and have neglected numerical factors of order unity as we also did in Eq. (5.0516). From Eqs. (5.0530) and (5.0532) we see that the two ensembles will lead to about the same average values and thus be more or less equivalent, if

$$a \approx 1. \tag{5.0533}$$

The normalization of the F-ensemble gives

$$1 = \int_{\varepsilon_{\min}}^{\bar{\varepsilon}} a e^{\beta(\psi - \bar{\varepsilon})}\, d\Omega = a e^{\beta(\psi - \bar{\varepsilon})} \int_{\varepsilon_{\min}}^{\bar{\varepsilon}} d\Omega = a e^{\beta(\psi - \bar{\varepsilon})}\, \Omega(\bar{\varepsilon}). \tag{5.0534}$$

Taking the logarithm and neglecting $\ln a$ with respect to the other terms and using Eq. (5.0309) for the entropy, we get, apart from a possible additive constant,

$$\frac{S}{k} = \ln \Omega(\bar{\varepsilon}); \tag{5.0535}$$

for β we get the equation

$$\beta = \left.\frac{\partial S/k}{\partial \varepsilon}\right|_{\varepsilon=\bar{\varepsilon}} = \left.\frac{\partial \ln \Omega}{\partial \varepsilon}\right|_{\varepsilon=\bar{\varepsilon}} = \left.\frac{e^{\phi}}{\Omega}\right|_{\varepsilon=\bar{\varepsilon}}. \tag{5.0536}$$

We note from Eqs. (5.0525) and (5.0535) that, once we know $\Omega(\varepsilon)$, we can evaluate the entropy. As S is one of the thermodynamic potentials, it follows that a knowledge of $\Omega(\varepsilon)$ enables us to evaluate the thermodynamic properties of the system represented by the macrocanonical ensemble.

5.6 THE COUPLING OF TWO MACROCANONICAL ENSEMBLES

In Section 5.2 we mentioned that one way to couple two stationary ensembles in such a way that the resulting ensemble is again a stationary ensemble is to couple two macrocanonical ensembles with the same value of β. We shall now justify that statement. The discussion in this section will once more follow closely Gibbs' considerations.

Let us consider two ensembles. The first one may have a density ρ_1, index of probability η_1, energy ε_1, and if it is macrocanonical, modulus $\Theta_1 = \beta_1^{-1}$, while the corresponding quantities in the second ensemble will have indices 2. Let us couple these two ensembles at $t = t'$. This coupling consists of forming a new ensemble with systems obtained by combining one system from the first ensemble with one system from the second ensemble. The coupling is supposed to be a physical one so that energy can be exchanged between the two systems that are combined to form a system of the new ensemble. The density, index of probability, and energy of the new ensemble will be denoted by ρ_{12}, η_{12}, and ε_{12}. After a while, at $t = t''$, the two systems are separated and we again have two separate ensembles. At t' we have clearly

$$\rho_{12}' = \rho_1' \cdot \rho_2', \tag{5.0601}$$

and we must show that, if

$$\rho_1 = e^{\beta_1(\psi_1 - \varepsilon_1)}, \qquad \rho_2 = e^{\beta_2(\psi_2 - \varepsilon_2)}, \tag{5.0602}$$

the equation

$$\rho_{12}' = \rho_{12}'' \tag{5.0603}$$

will be satisfied only if $\beta_1 = \beta_2$. Henceforth, as in the Eqs. (5.0601) and (5.0603), primes and double primes will indicate values at $t = t'$ and $t = t''$, respectively.

That Eq. (5.0603) is satisfied if $\beta_1 = \beta_2 = \beta_{12}$ follows easily from Eqs. (5.0601), (5.0602), and the equation

$$\varepsilon_{12} = \varepsilon_1 + \varepsilon_2, \tag{5.0604}$$

whence we have

$$\rho_{12} = e^{\beta_1(\psi_1 - \varepsilon_1)} \cdot e^{\beta_2(\psi_2 - \varepsilon_2)} = e^{\beta_{12}(\psi_{12} - \varepsilon_{12})} \qquad (\psi_{12} = \psi_1 + \psi_2), \tag{5.0605}$$

which is a stationary macrocanonical distribution.

Let us now consider what would happen if we coupled two macrocanonical ensembles with different moduli. We see immediately that the resulting ensemble is not a stationary one, since the density is not a function of ε_{12}, but contains ε_1 and ε_2 separately. We shall prove that, if

$$\beta_1 < \beta_2, \tag{5.0606}$$

a system from the first ensemble will on the average have lost energy to the system from the second ensemble, with which it was combined between t' and t''. Since β is inversely proportional to the temperature, this corresponds to the thermodynamical fact that energy will pass from a system of higher temperature to a system of lower temperature.

We must remark here that, in order that energy can be transported from a system of the first ensemble to a system from the second ensemble, it is necessary that there exist an interaction energy ε_{int} so that instead of Eq. (5.0604) we really have

$$\varepsilon_{12} = \varepsilon_1 + \varepsilon_2 + \varepsilon_{\text{int}}. \tag{5.0607}$$

However, we shall assume, firstly, that $\varepsilon_{\text{int}} \ll \varepsilon_{12}$, and, secondly, that the coupling takes place in such a way that the systems are brought together and pulled apart so gradually that both at t' and at t'' Eq. (5.0604) is satisfied, that is,

$$\varepsilon_{12}' = \varepsilon_1' + \varepsilon_2' \qquad \text{and} \qquad \varepsilon_{12}'' = \varepsilon_1'' + \varepsilon_2''. \tag{5.0608}$$

Before proving that, on the average, energy from the first ensemble will go over to the second ensemble if Eq. (5.0606) is satisfied, we shall prove a few necessary lemmas.

Consider an ensemble which is such that all of its systems consist of two parts, denoted by indices "a" and "b" while an index "ab" indicates the whole system. We then have

$$\int_{\Gamma_{ab}} \rho_{ab}\, d\Omega_{ab} = \int_{\Gamma_{ab}} e^{\eta_{ab}}\, d\Omega_{ab} = 1, \tag{5.0609}$$

where

$$d\Omega_{ab} = d\Omega_a\, d\Omega_b. \tag{5.0610}$$

Introducing

$$\rho_a = e^{\eta_a} = \int_{\Gamma_b} e^{\eta_{ab}}\, d\Omega_b, \tag{5.0611}$$

$$\rho_b = e^{\eta_b} = \int_{\Gamma_a} e^{\eta_{ab}}\, d\Omega_a, \tag{5.0612}$$

we see that from Eq. (5.0609) it follows that

$$\int_{\Gamma_a} e^{\eta_a}\, d\Omega_a = \int_{\Gamma_b} e^{\eta_b}\, d\Omega_b = 1. \tag{5.0613}$$

We also see that ρ_a and ρ_b evidently denote the densities of the subensembles made up out of Part a and Part b.

Lemma I. In the case of an ensemble such as the one just dicussed we have

$$\bar{\eta}_{ab} \geqslant \bar{\eta}_a + \bar{\eta}_b, \tag{5.0614}$$

where the averages are taken over the corresponding ensembles or subensembles.

Proof. First let us note that the expression $u \cdot e^u + 1 - e^u$ is positive for $u \neq 0$ and is zero for $u = 0$.

Secondly, let us evaluate $\bar{\eta}_{ab} - \bar{\eta}_a - \bar{\eta}_b$. Using Eqs. (5.0611) and (5.0612), we have

$$\begin{aligned}\bar{\eta}_{ab} - \bar{\eta}_a - \bar{\eta}_b &= \int_{\Gamma_{ab}} \eta_{ab} e^{\eta_{ab}}\, d\Omega_{ab} - \int_{\Gamma_b} \eta_a e^{\eta_a}\, d\Omega_a - \int_{\Gamma_b} \eta_b e^{\eta_b}\, d\Omega_b \\ &= \int_{\Gamma_{ab}} (\eta_{ab} - \eta_a - \eta_b) e^{\eta_{ab}}\, d\Omega_{ab}.\end{aligned} \tag{5.0615}$$

From Eq. (5.0613) we have

$$\int_{\Gamma_a} e^{\eta_a}\, d\Omega_a \cdot \int_{\Gamma_b} e^{\eta_b}\, d\Omega_b = \int_{\Gamma_{ab}} e^{\eta_a + \eta_b}\, d\Omega_{ab} = 1, \tag{5.0616}$$

and using Eqs. (5.0615), (5.0616), and (5.0609), we get finally

$$\bar{\eta}_{ab} - \bar{\eta}_a - \bar{\eta}_b = \int_{\Gamma_{ab}} (xe^x + 1 - e^x)e^{\eta_a + \eta_b}\, d\Omega_{ab} \geqslant 0,$$

$$(x = \eta_{ab} - \eta_a - \eta_b) \tag{5.0617}$$

by virtue of the fact that the integrand is nonnegative. This proves our lemma.

Lemma II. Consider two ensembles with indices of probability given by the equations

$$\eta_1 = \beta(\psi - \varepsilon), \qquad \eta_2 = \beta(\psi - \varepsilon) + \Delta\eta, \tag{5.0618}$$

that is, one macrocanonical ensemble and one arbitrary ensemble. We then have

$$\overline{(\eta + \beta\varepsilon)_1} \leqslant \overline{(\eta + \beta\varepsilon)_2}. \tag{5.0619}$$

Proof. We have from Eq. (5.0618)

$$\overline{(\eta + \beta\varepsilon)_2} - \overline{(\eta + \beta\varepsilon)_1} = \int_{\Gamma} [(\beta\psi + \Delta\eta)e^{\eta_1 + \Delta\eta} - \beta\psi e^{\eta_1}]\, d\Omega \tag{5.0620}$$

Moreover, from the normalization condition we have

$$\int_{\Gamma} e^{\eta_1 + \Delta\eta}\, d\Omega = \int_{\Gamma} e^{\eta_1}\, d\Omega, \tag{5.0621}$$

and combining Eqs. (5.0620) and (5.0621), we find the right-hand side of Eq. (5.0620) is equal to

$$\int_{\Gamma} \Delta\eta e^{\eta_1 + \Delta\eta}\, d\Omega = \int_{\Gamma} [\Delta\eta e^{\Delta\eta} + 1 - e^{\Delta\eta}]e^{\eta_1}\, d\Omega \geqslant 0, \tag{0.0622}$$

from which our second lemma follows.

We can now consider the joining together and separating of the two ensembles. From Eq. (5.0601) it follows that

$$\eta_{12}' = \eta_1' + \eta_2', \tag{5.0623}$$

and thus

$$\bar{\eta}_{12}' = \bar{\eta}_1' + \bar{\eta}_2', \tag{5.0624}$$

where, in going over from Eq. (5.0623) to Eq. (5.0624), we have used the fact that ρ_1 and ρ_2 are normalized.)*

* From Eq. (5.0623) one has

$$\int_{\Gamma_{12}} \eta_{12}'\rho_{12}\, d\Omega_{12} = \int_{\Gamma_{12}} \eta_1'\rho_{12}\, d\Omega_{12} + \int_{\Gamma_{12}} \eta_2'\rho_{12}\, d\Omega_{12}$$

$$= \int_{\Gamma_1} \eta_1'\rho_1\, d\Omega_1 \cdot \int_{\Gamma_2} \rho_2\, d\Omega_2 + \int_{\Gamma_1} \rho_1\, d\Omega_1 \int_{\Gamma_2} \eta_2'\rho_2\, d\Omega_2.$$

At t'' we have (see Eq. (5.0614))

$$\bar{\eta}_{12}'' \geqslant \bar{\eta}_1'' + \bar{\eta}_2''. \tag{5.0625}$$

We saw in Section 5.3 that for a macrocanonical ensemble $-\bar{\eta}$, apart from possible additive and multiplying constants, corresponded to the entropy. We can also easily see that, in general, $\bar{\eta}$ should correspond to a generalization of Boltzmann's H, as defined by Eq. (2.301), as

$$\bar{\eta} = \int_\Gamma \eta e^{\eta} \, d\Omega = \int_\Gamma \rho \ln \rho \, d\Omega, \tag{5.0626}$$

since ρ can be considered to be a generalization of the distribution function f.

From the second law of thermodynamics, or from a generalized H-theorem (see Section 5.14), we should then expect that

$$\bar{\eta}_{12}'' \leqslant \bar{\eta}_{12}'. \tag{5.0627}$$

Since at t' the two ensembles were macrocanonical, while this is no longer necessarily so at t'', the conditions under which Lemma II is valid are satisfied and we have from inequality (5.0699)

$$\bar{\eta}_1' + \beta_1 \bar{\varepsilon}_1' \leqslant \bar{\eta}_1'' + \beta_1 \bar{\varepsilon}_1'' \tag{5.0628}$$

and

$$\bar{\eta}_2' + \beta_2 \bar{\varepsilon}_2' \leqslant \bar{\eta}_2'' + \beta_2 \bar{\varepsilon}_2'' \tag{5.0629}$$

Combining Eqs. (5.0624), (5.0625), and (5.0627) to (5.0699), we get

$$\beta_1(\bar{\varepsilon}_1'' - \bar{\varepsilon}_1') + \beta_2(\bar{\varepsilon}_2'' - \bar{\varepsilon}_2') \geqslant 0. \tag{5.0630}$$

From Eqs. (5.0608) and the principle of conservation of energy, we have

$$\bar{\varepsilon}_1'' - \bar{\varepsilon}_1' + \bar{\varepsilon}_2'' - \bar{\varepsilon}_2' = 0, \tag{5.0631}$$

and hence, combining Eqs. (5.0630) and (5.0631),

$$(\bar{\varepsilon}_1'' - \bar{\varepsilon}_1')(\beta_1 - \beta_2) \geqslant 0. \tag{5.0632}$$

From Eqs. (5.0606) and (5.0632) we then have

$$\bar{\varepsilon}_1' \geqslant \bar{\varepsilon}_1'',$$

or, the system with the smaller β will, on the average, have lost energy to the other system.

We have just seen that, if two macrocanonical ensembles are physically coupled, they will form a stationary ensemble if the values of their β are equal, and that the one with the smaller β will on the average lose energy to the other one if the values of their β are not equal.

If we couple two microcanonical ensembles, the resulting ensemble will also be a microcanonical ensemble; hence one might ask whether there is any reason to prefer the macrocanonical to the microcanonical ensembles as we have done. The reason is that while in both cases the resulting ensemble is again a stationary one, separation at t'' reproduces the situation before t' in the case of macrocanonical ensembles, but not in the case of microcanonical ones. This last point we can see as follows. Consider two systems, A and B, which together form one system of the combined ensemble. If ε_1 is the energy of the first and ε_2 that of the second microcanonical ensemble, System A had an energy ε_1 and System B an energy ε_2 before t'. During the interval t', t'' the two systems were coupled and could exchange energy. When they are separated at t'' there is no reason to assume that all systems A still have the same energy, and hence they will no longer form a microcanonical ensemble.

The case is different, however, when two macrocanonical ensembles are coupled. Since, according to our assumptions, the interaction between the systems of the two coupled ensembles is negligibly small, we can at any time (also between t' and t'') calculate the average values of the powers of ε_1. We then find that they correspond at any time to the average values in a macrocanonical ensemble with given β so that, indeed, after t'' we will again have two macrocanonical ensembles.

The average value of $\varepsilon_1{}^n$ can be calculated as follows. Let $p_1', \cdots p_{s'N'}'$, $q_1', \cdots, q_{s'N'}'$ be the coordinates pertaining to the first ensemble and $p_1'', \cdots$, $p_{s''N''}'', q_1'', \cdots, q_{s''N''}''$ those pertaining to the second ensemble. We have then

$$\varepsilon_1 = \varepsilon_1(p',q'), \quad \varepsilon_2 = \varepsilon_2(p'',q''), \quad \varepsilon_{\text{tot}} = \varepsilon_1(p',q') + \varepsilon_2(p'',q'')$$

and

$$d\Omega_{\text{tot}} = d\Omega' \cdot d\Omega'',$$

where the index "tot" indicates the combined ensemble.

The average value of $\varepsilon_1{}^n$ in the combined ensemble then satisfies the equations

$$\begin{aligned}\langle \varepsilon_1{}^n \rangle_{\text{av tot}} &= \int_{\Gamma_{\text{tot}}} \varepsilon_1{}^n(p',q')\rho_{\text{tot}}\, d\Omega_{\text{tot}} \\ &= \int_{\Gamma'} \varepsilon_1{}^n(p',q')\rho'\, d\Omega' \cdot \int_{\Gamma''} \rho''\, d\Omega'' \\ &= \int_{\Gamma'} \varepsilon_1{}^n(p',q')\rho'\, d\Omega' = \langle \varepsilon_1{}^n \rangle_{\text{av 1}}.\end{aligned}$$

5.7 MICROCANONICAL ENSEMBLES

In this section we shall consider microcanonical ensembles. We saw in Section 5.1 and know from theoretical mechanics that the energy of a system is one of its integrals. We might therefore be tempted to prefer microcanonical to macrocanonical ensembles, especially as long as our considerations are classical so that in principle we could hope to determine the energy of the system under observation as accurately as we wished.

A different reason why microcanonical ensembles are often used is that one can show that — under conditions to be discussed briefly in Section 5.14 — the time average of a phase function is equal to the average of that phase function taken over a suitable microcanonical ensemble. We shall derive this result as follows. Consider a system of a given energy ε, the representative point of which moves over the appropriate energy surface. The time average $\tilde{G}$ of a phase function G is defined by the equation

$$\tilde{G} = \lim_{T \to \infty} \frac{1}{2T} \int_{-T}^{+T} G(p,q)\, dt, \tag{5.0701}$$

where p,q denote the values of the coordinates of the representative point at the time t. Consider now the orbit of the system on the energy surface (see Fig. 5.2). Let P_0 be the position of the representative point at $t = t_0$, P_1 at

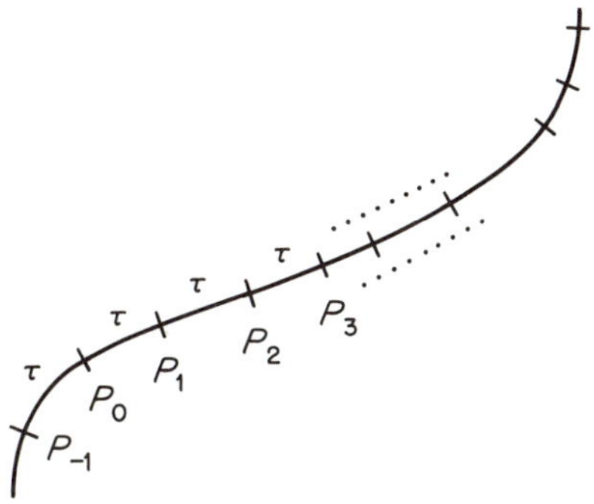

FIG. 5.2 The path of a representative point on the energy surface.

$t_0 + \tau$, P_2 at $t_0 + 2\tau, \cdots$, while P_{-1}, $P_{-2}, \cdots$ were the positions at $t_0 - \tau$, $t_0 - 2\tau, \cdots$, in general, P_i at $t_0 + i\tau$ $(i = 0, \pm 1, \pm 2, \cdots)$. The points P_i form a stationary ensemble: any collection of points in Γ-space forms an ensemble and since the P_i at a later time, say $t_0 + n\tau$, can be obtained from the P_i at t_0 by putting

$$P_i(t = t_0 + n\tau) = P_{i-n}(t = t_0), \tag{5.0702}$$

the ensemble is stationary.

Let us now consider the average $\langle G \rangle_{av}$ of a phase function $G(p,q)$ over the P_i,

$$\langle G \rangle_{av} = \frac{\cdots + G_{-2} + G_{-1} + G_0 + G_1 + G_2 + \cdots}{\cdots + 1 + 1 + 1 + 1 + 1 + \cdots}, \qquad (5.0703)$$

where G_i is the value of $G(p,q)$ at P_i. If we now take the limit as $\tau \to 0$, the right-hand side of Eq. (5.0703) will become the time average $\tilde{G}$. On the other hand, we see that the average on the right-hand side of Eq. (5.0703) resembles strongly an average taken over the whole of the energy surface, that is, an average over a microcanonical ensemble. If we accept that identification for a moment, and if we denote here and henceforth the average over a microcanonical ensemble by a double bar, we have

$$\bar{\bar{G}} = \tilde{G}. \qquad (5.0704)$$

From the fact that it can be proved (see Section 5.14) that Eq. (5.0704) is valid for most physical systems and from the fact that, while one measures time averages, ensemble averages can nearly always be much more easily computed, it follows that it might be advantageous to consider in slightly more detail the microcanonical ensemble. We shall then see that averages over a microcanonical ensemble will be the same as averages taken over a suitably chosen macrocanonical ensemble. We see thus that we can for actual calculations practically always confine ourselves to macrocanonical ensembles, which is rather gratifying, since they are more amenable to mathematical treatment and since there are also other arguments that point to macrocanonical ensembles, as we have seen and shall see again in Section 5.13.

We now introduce the microcanonical ensemble by Eq. (5.0205). Let us consider the identity

$$\int_\Gamma \delta(\varepsilon - \varepsilon_0)\, d\Omega = \int_{\varepsilon_{\min}(a_i)}^{\infty} \delta(\varepsilon - \varepsilon_0)\, e^{\phi(\varepsilon, a_i)}\, d\varepsilon. \qquad (5.0705)$$

If we take the derivative with respect to a_j both of the left-hand side and of the right-hand side, the result is

$$\left[\bar{\bar{A}}_j(\varepsilon)\delta(\varepsilon - \varepsilon_0)\, e^{\phi}\right]_{\varepsilon_{\min}}^{\infty} + \int_{\varepsilon_{\min}}^{\infty} \left(\frac{\partial \bar{\bar{A}}_j}{\partial \varepsilon} + \bar{\bar{A}}_j \frac{\partial \phi}{\partial \varepsilon}\right)\delta(\varepsilon - \varepsilon_0)\, e^{\phi}\, d\varepsilon$$

$$= \int_{\varepsilon_{\min}}^{\infty} \frac{\partial \phi}{\partial a_j}\, \delta(\varepsilon - \varepsilon_0)\, e^{\phi}\, d\varepsilon - \left[\delta(\varepsilon - \varepsilon_0)\, e^{\phi} \frac{\partial \varepsilon}{\partial a_j}\right]_{\varepsilon_{\min}} \qquad (5.0706)$$

Equation (5.0706) is derived as follows. First of all, we note that while the left-hand side of Eq. (5.0705) depends on a_j because ε depends on it, the right-hand side depends on a_j through ϕ and $\varepsilon_{\min}$. Finally, we used the identity

$$\int_{\Gamma} G\,\delta(\varepsilon-\varepsilon_0)\,d\Omega = \int_{\varepsilon_{\min}}^{\infty} \overline{\overline{G}}(\varepsilon)\,\delta(\varepsilon-\varepsilon_0)e^{\phi}\,d\varepsilon, \tag{5.0707}$$

where $\overline{\overline{G}}(\varepsilon)$ is the average of G over a microcanonical ensemble with energy ε.

As the integrated parts clearly vanish, we get from Eq. (5.0706)

$$e^{\phi}\frac{\partial \overline{\overline{A}}_j}{\partial\varepsilon} + \overline{\overline{A}}_j\, e^{\phi}\frac{\partial\phi}{\partial\varepsilon} = e^{\phi}\frac{\partial\phi}{\partial a_j}\left(= \frac{\partial^2\Omega}{\partial a_j\,\partial\varepsilon}\right). \tag{5.0708}$$

This equation can be integrated and the result is*

$$\overline{\overline{A}}_j\, e^{\phi} = \frac{\partial\Omega}{\partial a_j}, \tag{5.0709}$$

or

$$\overline{\overline{A}}_j = e^{-\phi}\frac{\partial\Omega}{\partial a_j}. \tag{5.0710}$$

The variation of Ω is now given by the equation

$$\delta\Omega = e^{\phi}\left(\delta\varepsilon + \sum_j \overline{\overline{A}}_j\,\delta a_j\right). \tag{5.0711}$$

This last equation can be written in the form

$$\delta \ln \Omega = \frac{\delta\varepsilon + \sum_j \overline{\overline{A}}_j\,\delta a_j}{e^{-\phi}\Omega}, \tag{5.0712}$$

and comparison with the thermodynamical equation for the variation of the entropy shows that we can take $\ln \Omega$ as being proportional to the entropy of the system and $e^{-\phi}\Omega$ as a measure for the temperature. Since that gives us the same as expressions (5.0535) and (5.0536), which were derived for macrocanonical ensembles, we have found the result that the averages over a microcanonical ensemble with energy ε_0 are the same as those taken over a macrocanonical ensemble for which $\bar{\varepsilon} = \varepsilon_0$.† We shall consider in somewhat more detail the relationship between micro- and macrocanonical ensembles in Section 5.13.

* There should be an integration const $K_j(a_i)$ on the right-hand side of Eq. (5.0709). One can, however, prove by writing down the variation of Ω (compare Eq. (5.0711)), considering such variations of the a_i and of ε that $\delta\varepsilon_{\min} = 0$, and using Eq. (5.0512) that all the K_j are equal to zero.

† It must be borne in mind that, for instance, in Eq. (5.0710) all expressions on the right-hand side must be taken for $\varepsilon = \varepsilon_0$.

5.8 APPLICATION: THE PERFECT GAS

We shall briefly discuss here the simple case of a perfect gas in order to illustrate the theory developed in the preceding sections.

The discussion was started in Section 5.3, where an expression for $e^{\beta\psi}$, or rather for the partition function Z_Γ, was derived. We had there (Eqs. (5.0325) and (5.0310))

$$e^{-\beta\psi} = V^N\left(\frac{2\pi m}{\beta}\right)^{3N/2}. \tag{5.0801}$$

From this equation we can calculate, for instance, the fluctuations in the energy, using Eq. (5.0407), which gives

$$\overline{(\varepsilon - \bar{\varepsilon})^2} = -\frac{\partial^2(\beta\psi)}{\partial\beta^2} = \frac{3N}{2\beta^2}, \tag{5.0802}$$

and combining this with the result for the average energy

$$\bar{\varepsilon} = \frac{3N}{2\beta}, \tag{5.0803}$$

we have

$$\frac{\overline{(\varepsilon - \bar{\varepsilon})^2}}{\bar{\varepsilon}^2} = \frac{2}{3N}, \tag{5.0804}$$

showing that the dispersion is indeed very small as soon as N is large.

Let us now consider the fluctuations in the pressure. We mentioned in Section 5.4 that in this case we cannot use formula (5.0409) as long as we use ideal walls, that is, as long as we use the function $F(x_i,y_i,z_i)$ of Eq. (5.0325) for the potential energy function. We must now consider the time average of the pressure and its fluctuations. We can do this in two steps. The first step consists in reminding ourselves of the fact that, for a system of many degrees of freedom, we have found in Section 5.7 that we obtain the same results whether we consider a macro- or a microcanonical ensemble. The second step then follows immediately, since we also saw that time averages and averages taken over the corresponding microcanonical ensemble are the same. We can now use Eq. (5.0710) for an average in a microcanonical ensemble; thus we have

$$\tilde{P} = \bar{\bar{P}} = e^{-\phi}\frac{\partial\Omega}{\partial V}, \tag{5.0805}$$

and since

$$e^{\phi} = \frac{\partial\Omega}{\partial\varepsilon}, \tag{5.0806}$$

we have

$$\tilde{P} = \frac{\dfrac{\partial \Omega}{\partial V}}{\dfrac{\partial \Omega}{\partial \varepsilon}}, \tag{5.0807}$$

and where all quantities are evaluated for $\varepsilon = \bar{\varepsilon}$, since we are dealing with a microcanonical ensemble.

In the case of a perfect gas, Ω can easily be computed from the equations

$$\Omega(\varepsilon, V) = \int_{\varepsilon=0}^{\varepsilon} d\Omega = \int_{\varepsilon=0}^{\varepsilon} \prod_i d^3\mathbf{p}_i \prod_i d^3\mathbf{r}_i, \tag{5.0808}$$

$$\varepsilon = \sum_i \left(\frac{\mathbf{p}_i^{\,2}}{2m} + F(x_i, y_i, z_i) \right). \tag{5.0809}$$

The integration over the x_i, y_i, and z_i gives a factor V^N and the integral over the $\mathbf{p}_i$ is the formula for the volume of a $3N$-dimensional sphere with radius $\sqrt{2m\varepsilon}$. The formula for this volume is different for even or odd N, but we are only interested in the dependence of Ω on ε and V and can therefore write

$$\Omega(\varepsilon, V) = K\varepsilon^{3N/2} V^N, \tag{5.0810}$$

where K is a constant.

From Eqs. (5.0807) and (5.0810) we get the familiar equation (see Eq. (4.715))

$$\tilde{P} = \frac{2}{3} \frac{\bar{\varepsilon}}{V}. \tag{5.0811}$$

From Eqs. (5.0811) and (5.0804) we now get for the fluctuations in the time average of the pressure

$$\frac{\overline{(\tilde{P} - \bar{\tilde{P}})^2}}{\bar{\tilde{P}}^2} = \frac{\overline{(\varepsilon - \bar{\varepsilon})^2}}{\bar{\varepsilon}^2} = \frac{2}{3N}, \tag{5.0812}$$

which proves that, in this case at least, the fluctuations are negligible for large N (see also the discussion in Section 8.2).

5.9 GRAND ENSEMBLES

In the last chapter of his famous monograph Gibbs considers systems in which chemical reactions can take place. This means that the number of particles of a given kind is no longer constant, and we must thus abandon the condition of a constant total number of particles, just as we abandoned the condition of a constant total energy when we went from microcanonical ensembles to macrocanonical ensembles. Gibbs called ensembles with a constant number of particles *petit ensembles* and those without a restriction on the total number of particles *grand ensembles.* The ensembles discussed so far in this chapter were all petit ensembles. Apart from systems in which chemical reactions can take place, the need for grand ensembles also arises when we are dealing with systems at very high temperatures in which pair creation and annihilation processes may take place. In Section 5.13 we shall discuss other, more fundamental, reasons for introducing grand ensembles.

Just as we imagined the temperature of a system to be fixed through the interaction between the system and a "temperature bath," we can now imagine an interaction between the system and a "particle bath" or "particle bank" with which it can exchange particles, just as it could exchange energy with the "temperature bath." To retain as close an analogy as possible with the term "temperature bath," such a particle bath should be called "activity bath," "fugacity bath," or "chemical potential bath."

We shall start out to develop the theory of the grand ensembles in a rather general way and assume that our system contains r different kinds of particles. The number of particles of the ith kind will be denoted by ν_i, and the density of our representative grand ensemble will now not only be a function of the appropriate p's and q's, but also of the ν_i,

$$\rho = \rho(\nu_1,\nu_2,\cdots, \nu_r;p,q), \tag{5.0901}$$

where, of course, the number of degrees of freedom, that is, the number of the p's and q's, depends on the actual values of the ν_i. If each particle of the ith kind possesses s_i degrees of freedom, the total number of degrees of freedom s of a system characterized by ν_1, $\nu_2,\cdots$, ν_r will be given by the equation

$$s = \sum_i \nu_i s_i. \tag{5.0902}$$

The density of a grand ensemble is defined in such a way that

$$\rho(\nu_1,\nu_2,\cdots,\nu_r;p_1,\cdots,p_s,q_1,\ldots,q_s)\, dp_1 \cdots dp_s\, dq_1 \cdots dq_s$$

is the fraction of systems in the ensemble with ν_i particles of the ith kind and with the p_i and q_i within the phase-space volume element $dp_1 \cdots dp_s\, dq_1 \cdots dq_s$. We shall write for the sake of brevity

$$d\Omega = dp_1 \cdots dp_s\, dq_1 \cdots dq_s, \tag{5.0903}$$

where we have to remember that $d\Omega$ depends on the ν_i, since s depends on the ν_i.

From the above definition of the density it follows that ρ satisfies the normalization condition

$$\sum_{\nu_1=0}^{\infty} \sum_{\nu_2=0}^{\infty} \cdots \sum_{\nu_r=0}^{\infty} \int_{\Gamma} \rho(\nu_1,\nu_2,\cdots,\nu_r;p,q)\, d\Omega = 1. \tag{5.0904}$$

Let us now discuss briefly the implications of the fact that the ν_i atoms of the ith kind are identical. Two situations differing only in the interchange of two particles of the same kind can be considered to correspond either to different phases of the system, or to the same phase. If they are considered to give us the same phase, we are dealing with *generic phases*, as Gibbs calls them, while if they are considered to correspond to different phases, we call the phase defined in that way a *specific phase*, another term due to Gibbs. It is easily seen that for given values of the ν_i each generic phase is made up out of $\Pi \nu_i!$ different specific phases.

We shall only be interested in stationary grand ensembles and, moreover, in *canonical grand ensembles* which are defined by the following specific density:

$$\rho_{\text{spec}} = \frac{1}{\nu_1!\nu_2! \cdots \nu_r!} e^{-q+\alpha_1\nu_1+\alpha_2\nu_2+\cdots+\alpha_r\nu_r-\beta\varepsilon}, \tag{5.0905}$$

where $\varepsilon = \varepsilon(\nu_i;p,q;a_k)$ is the total energy of the system. Instead of β Gibbs had a parameter Θ, the *modulus* of the ensemble, which is the reciprocal of β. The quantity q does not depend on the p's and q's, but will depend on β, the α_i, and the external parameters a_k. We shall see presently the physical meaning of q, β, and the α_i (see Eqs. (5.0907) to (5.0909)).

From Eq. (5.0905) it follows that the generic density ρ_{gen} is given by the equation

$$\rho_{\text{gen}} = \sum \rho_{\text{spec}} = e^{-q+\alpha_1\nu_1+\alpha_2\nu_2+\cdots+\alpha_r\nu_r-\beta\varepsilon}, \tag{5.0906}$$

where the summation is over all different specific phases making up one generic phase and where we have used the fact that the energies of all the specific phases corresponding to the same generic phase will be the same.

Even as a macrocanonical (petit) ensemble can be considered to be built up out of microcanonical ensembles, each canonical grand ensemble is the aggregate of a large number of macrocanonical petit ensembles, as can be seen from Eq. (5.0905) or (5.0906), since for each of the petit ensembles the ν_i are fixed and the formula reduces to the familiar formula for the density of a macrocanonical ensemble.

The quantity q, which is called by Kramers the *grand potential*, is determined by the normalization condition (5.0904), and we have

$$e^q = \sum_{\nu_1} \cdots \sum_{\nu_r} \frac{e^{\alpha_1\nu_1+\cdots+\alpha_r\nu_r}}{\nu_1! \cdots \nu_r!} \int_{\Gamma} e^{-\beta\varepsilon}\, d\Omega. \tag{5.0907}$$

Using for the integral in Eq. (5.0907) Eq. (5.0204) from the theory of the macrocanonical ensembles, we can write Eq. (5.0907) in the form

$$e^q = \sum_{\nu_1} \cdots \sum_{\nu_r} \frac{e^{\alpha_1\nu_1 + \cdots + \alpha_r\nu_r - \beta\psi}}{\nu_1! \cdots \nu_r!}. \tag{5.0908}$$

From Eqs. (5.0907) or (5.0908) we see that

$$q = q(\beta,\alpha_i,a_k). \tag{5.0909}$$

We shall denote the average over a canonical grand ensemble by pointed brackets: $\langle \cdot \cdot \rangle$. The average of a phase function $G(p,q;\nu_i)$, which now will in general also be a function of the ν_i is given by the equation

$$\langle G \rangle = \sum_{\nu_i} \int_{\text{spec}} G(p, q_j\nu_i)\rho_{\text{spec}}\, d\Omega = \sum_{\nu_i} \int_{\text{gen}} G(p, q_j\nu_i)\rho_{\text{gen}}\, d\Omega, \tag{5.0910}$$

where $\int_{\text{spec}}$ indicates the integration over all different specific phases and $\int_{\text{gen}}$ the integration over all different generic phases only.

Let us consider the variation of q as a function of the variations of β, the α_i, and the a_k. From Eqs. (5.0907), (5.0905), and (5.0910) we get, after a straightforward calculation,

$$\delta q = \sum_i \langle \nu_i \rangle \delta\alpha_i - \langle \varepsilon \rangle \delta\beta - \beta\langle \delta\varepsilon \rangle, \tag{5.0911}$$

where

$$\langle \delta\varepsilon \rangle = -\sum_k \langle A_k \rangle \delta a_k, \tag{5.0912}$$

$$A_k = -\frac{\partial\varepsilon}{\partial a_k}. \tag{5.0913}$$

We can write Eq. (5.0911) in the form

$$\delta[-q + \sum_i \langle \nu_i \rangle \alpha_i - \beta\langle \varepsilon \rangle] = \sum_i \alpha_i \delta\langle \nu_i \rangle - \beta(\delta\langle \varepsilon \rangle - \langle \delta\varepsilon \rangle), \tag{5.0914}$$

and by the same arguments as those used in Sections 2.7, 4.4, or 5.3 we find that

$$\beta^{-1} = kT, \tag{5.0915}$$

$$\frac{S}{k} = q - \sum_i \alpha_i \langle \nu_i \rangle + \beta\langle \varepsilon \rangle, \tag{5.0916}$$

where T is the temperature and S the entropy.*

* It is perhaps well to remind ourselves that when we talk glibly about "the entropy," we really mean "the entropy of the system that is represented by the canonical grand ensemble under discussion," and similarly about other quantities.

Comparing Eqs. (5.0916) and (5.0906), we see that, apart from a multiplying constant, the entropy is equal to the average of the logarithm of the generic density,

$$S = -k\langle \ln \rho_{\text{gen}} \rangle.$$

We can compare this result with the relationship that we found in Chapter 1 between Boltzmann's H and the entropy. There we saw that H, which is the average of the logarithm of the density function, and $-S/k$ were the same.

From Eq. (5.0916) it follows that the Helmholtz free energy F expressed in terms of q, β, the α_i, and the $\langle \nu_i \rangle$ is given by the equation

$$F = \langle \varepsilon \rangle - ST = -\frac{q}{\beta} + \sum_i \mu_i \langle \nu_i \rangle, \tag{5.0917}$$

where

$$\mu_i = \frac{\alpha_i}{\beta}. \tag{5.0918}$$

The variation of βF is given by the equation

$$\delta \beta F = \sum_i \alpha_i \delta \langle \nu_i \rangle + \langle \varepsilon \rangle \delta \beta + \beta \langle \delta \varepsilon \rangle, \tag{5.0919}$$

and we see that the μ_i are the partial (Helmholtz) free energies,

$$\mu_i = \left(\frac{\partial F}{\partial \langle \nu_i \rangle} \right)_{\beta, a_k} \tag{5.0920}$$

Let us assume for a moment that the volume V is one of the external parameters and, furthermore, that it is the only one. Introducing the thermal potential, or Gibbs free energy, G,

$$G = F + PV = -\frac{q}{\beta} + \sum_i \mu_i \langle \nu_i \rangle + PV, \tag{5.0921}$$

and considering the variation of βG,

$$\delta \beta G = \sum_i \alpha_i \delta \langle \nu_i \rangle + (\langle \varepsilon \rangle + PV) \delta \beta + \beta V \delta P, \tag{5.0922}$$

we see that the μ_i are also the partial thermal potentials or the partial (Gibbs) free energies,

$$\mu_i = \left(\frac{\partial G}{\partial \langle \nu_i \rangle} \right)_{\beta, P} \tag{5.0923}$$

It may be mentioned in passing that the α_i are related to the absolute *activities*, z_i, which are sometimes called the *fugacities** or the active number density† and are also sometimes denoted by λ_i, by the equation

$$z_i = e^{\alpha i}. \tag{5.0924}$$

If the system under consideration is homogeneous in all kinds of particles, G will be a homogeneous function of the first degree in the $\langle \nu_i \rangle$, and we have

$$G = \sum_i \langle \nu_i \rangle \frac{\partial G}{\partial \langle \nu_i \rangle} = \sum_i \mu_i \langle \nu_i \rangle. \tag{5.0925}$$

Combining Eqs. (5.0921) and (5.0925), we get in this case‡

$$q = \frac{PV}{kT}. \tag{5.0926}$$

It will have been noticed that the grand potential resembles in many ways the q-potential discussed in Chapter 4. Indeed, Eq. (5.0926) is the same as Eq. (4.426), while Eqs. (5.0911) and (5.0916) have as their counterparts Eqs. (4.421) and (4.420). In order to see the analogy between Eqs. (5.0916) and (4.420) more clearly, we write Eq. (5.0916) in the form

$$q = \beta\left(ST - \sum_i \mu_i \langle \nu_i \rangle - \langle \varepsilon \rangle\right). \tag{5.0927}$$

From Eq. (5.0911) it follows that the grand potential can be used to derive various averages, as follows (compare Eqs. (4.418) to (4.420)):

$$\langle \nu_i \rangle = \frac{\partial q}{\partial \alpha_i}, \tag{5.0928}$$

$$\langle \varepsilon \rangle = -\frac{\partial q}{\partial \beta}, \tag{5.0929}$$

$$\langle A_k \rangle = \frac{\partial q}{\beta \partial a_k}. \tag{5.0930}$$

In the special case where $a_k \equiv V$ we have from Eq. (5.0930)

$$P = kT \frac{\partial q}{\partial V}, \tag{5.0931}$$

which will give us the equation of state, once q has been computed. Of

* J. E. and M. G. Mayer, *Statistical Mechanics*, New York, 1940.

† J. O. Hirschfelder, C. F. Curtiss, and R. B. Bird, *Molecular Theory of Gases and Liquids*, New York and London, 1954.

‡ The grand potential q is related to the quantity Ω introduced by Gibbs by the equation $\Omega = -q/\beta$.

course, in the case of a homogeneous system, the equation of state follows directly from Eq. (5.0926), which is then the same as Eq. (5.0931), since in this particular case q will be directly proportional to V.

5.10 FLUCTUATIONS IN A CANONICAL GRAND ENSEMBLE

In order to show that canonical grand ensembles can be used to describe systems in statistical equilibrium, we must consider the fluctuations of the various quantities around their average values and show that these in general will be small, at least if the total number of particles involved is large. Let us immediately remark here that sometimes subsystems of a larger system will be considered, and then the fluctuations may not be small but quite substantial, for instance, when we consider density fluctuations in a gas at a temperature near the critical temperature. In that case, however, these fluctuations are real; thus it is a virtue and not a drawback of the canonical grand ensembles that they contain the possibility of fluctuations.

The calculation of the fluctuations in the ν_i, the A_k, and ε is completely analogous to the calculation of the fluctuations in Section 5.4, and we shall therefore only indicate the way in which these fluctuations can be calculated and give the results.

Taking the derivative of the normalizing condition with respect to α_i, a_k, or β, respectively, we obtain Eqs. (5.0928) to (5.0930). Once more taking the derivative with respect to the same variables, we get, respectively,

$$\langle(\nu_i - \langle\nu_i\rangle)^2\rangle = \frac{\partial^2 q}{\partial \alpha_i{}^2}\left(= \frac{\partial\langle\nu_i\rangle}{\partial\alpha_i}\right), \tag{5.1001}$$

$$\langle(A_k - \langle A_k\rangle)^2\rangle = \frac{1}{\beta^2}\frac{\partial^2 q}{\partial a_k{}^2} + \frac{1}{\beta}\left\langle\frac{\partial^2 \varepsilon}{\partial a_k{}^2}\right\rangle. \tag{5.1002}$$

$$\langle(\varepsilon - \langle\varepsilon\rangle)^2\rangle = \frac{\partial^2 q}{\partial\beta^2}\left(= -\frac{\partial\langle\varepsilon\rangle}{\partial\beta}\right). \tag{5.1003}$$

To derive Eqs. (5.1001) to (5.1003), we write the normalizing condition in the form

$$\sum_{\nu_i}\int_\Gamma\left[\prod_i \frac{e^{\alpha_i\nu_i}}{\nu_i!}\right] e^{-q-\beta\varepsilon}\, d\Omega = 1.$$

Taking the derivative with respect to, for example, a_k, we find that

$$\sum_{\nu_i}\int_\Gamma\left[\prod_i \frac{e^{\alpha_i\nu_i}}{\nu_i!}\right] e^{-q-\beta\varepsilon}\left\{-\frac{\partial q}{\partial a_k} - \beta\frac{\partial\varepsilon}{\partial a_k}\right\} d\Omega = 0,$$

and using Eqs. (5.0910), (5.0905), and (5.0913), that Eq. (5.0930) follows.

Taking once more the derivative with respect to, for example, a_k, we get

$$\sum_{\nu_i}\int_{\Gamma}\left[\prod_i \frac{e^{\alpha_i \nu_i}}{\nu_i!}\right] e^{-q-\beta\varepsilon}\left\{\left(-\frac{\partial q}{\partial a_k}-\beta\frac{\partial \varepsilon}{\partial a_k}\right)^2-\frac{\partial^2 q}{\partial {a_k}^2}-\beta\frac{\partial^2 \varepsilon}{\partial {a_k}^2}\right\} d\Omega=0,$$

from which Eq. (5.1002) follows. Equations (5.1001) and (5.1003) can be derived in a similar manner.

In general q will be proportional to the total number of particles, and

$$\frac{\langle(\varepsilon-\langle\varepsilon\rangle)^2\rangle}{\langle\varepsilon\rangle^2} \quad \text{and} \quad \frac{\langle(A_k-\langle A_k\rangle)^2\rangle}{\langle A_k\rangle^2}$$

will be of the order of magnitude of $1/N$, where $N=\sum_i\langle\nu_i\rangle$, while

$$\frac{\langle(\nu_i-\langle\nu_i\rangle)^2\rangle}{\langle\nu_i\rangle^2}$$

will be of the order of magnitude of $\langle\nu_i\rangle^{-1}$. As in Section 5.4 we can by similar means evaluate the correlations between various quantities.

In Section 5.12 we shall apply some of the results we have just obtained to the case of a perfect gas.

5.11 THE COUPLING OF TWO CANONICAL GRAND ENSEMBLES

In Section 5.6 we discussed in some detail the coupling of two petit ensembles. Since the situation is much the same in the case of grand ensembles, we shall discuss only those points where new features are appearing. Consider two canonical grand ensembles, A and B, and assume for the sake of simplicity that in each of the two systems only one kind of particle is present. Their specific densities are

$$\rho_{\text{spec}}{}^{\text{A}}=\frac{1}{\nu_{\text{A}}!}e^{q_{\text{A}}+\alpha_{\text{A}}\nu_{\text{A}}-\beta_{\text{A}}\varepsilon_{\text{A}}},\ \rho_{\text{spec}}{}^{\text{B}}=\frac{1}{\nu_{\text{B}}!}e^{q_{\text{B}}+\alpha_{\text{B}}\nu_{\text{B}}-\beta_{\text{B}}\varepsilon_{\text{B}}}, \qquad (5.1101)$$

and their generic densities are

$$\rho_{\text{gen}}{}^{\text{A}}=e^{q_{\text{A}}+\alpha_{\text{A}}\nu_{\text{A}}-\beta_{\text{A}}\varepsilon_{\text{A}}},\ \rho_{\text{gen}}{}^{\text{B}}=e^{q_{\text{B}}+\alpha_{\text{B}}\nu_{\text{B}}-\beta_{\text{B}}\varepsilon_{\text{B}}}. \qquad (5.1102)$$

If the two ensembles are physically coupled to form an ensemble AB and if we may neglect the interaction energy, we see that the specific density of the ensemble AB is given by the equation

$$\rho_{\text{spec}}{}^{\text{AB}}=\rho_{\text{spec}}{}^{\text{A}}\rho_{\text{spec}}{}^{\text{B}}=\frac{1}{\nu_{\text{A}}!\nu_{\text{B}}!}e^{q_{\text{A}}+\alpha_{\text{A}}\nu_{\text{A}}-\beta_{\text{A}}\varepsilon_{\text{A}}}e^{q_{\text{B}}+\alpha_{\text{B}}\nu_{\text{B}}-\beta_{\text{B}}\varepsilon_{\text{B}}}, \qquad (5.1103)$$

and its generic density by the equation

$$\rho_{\text{gen}}{}^{\text{AB}} = \rho_{\text{gen}}{}^{\text{A}} \rho_{\text{gen}}{}^{\text{B}} = e^{q_{\text{A}} + \alpha_{\text{A}} \nu_{\text{A}} - \beta_{\text{A}} \varepsilon_{\text{A}}} e^{q_{\text{B}} + \alpha_{\text{B}} \nu_{\text{B}} - \beta_{\text{B}} \varepsilon_{\text{B}}}. \tag{5.1104}$$

We see from Eqs. (5.1103) and (5.1104) that as long as the particles in System A are different from those in System B, we have for the ensemble AB the correct relation between specific and generic densities. Moreover, if $\beta_{\text{A}} = \beta_{\text{B}}$, the ensemble AB will again be a canonical grand ensemble. However, if the systems A and B contain the same kind of particles, we see that the ensemble AB is only canonically distributed as far as the generic phases are concerned—provided $\beta_{\text{A}} = \beta_{\text{B}}$ and $\alpha_{\text{A}} = \alpha_{\text{B}}$, the two familiar conditions of equality of temperature and thermal potential—but not as far as specific phases are concerned. We see here the importance of the generic phases which was clearly recognized by Gibbs. We can also understand now why $\langle \ln \rho_{\text{gen}} \rangle$ and not $\langle \ln \rho_{\text{spec}} \rangle$ might play the role of Boltzmann's H. Once this point of the pre-eminence of the generic phases is accepted, we must start from Eq. (5.1104) and if the particles in the systems A and B are the same, we must write for the specific density, instead of Eq. (5.1103),

$$\rho_{\text{spec}}{}^{\text{AB}} = \frac{1}{\nu_{\text{AB}}!} \rho_{\text{gen}}{}^{\text{AB}} = \frac{1}{\nu_{\text{AB}}!} e^{q_{\text{AB}} + \alpha_{\text{AB}} \nu_{\text{AB}} - \beta_{\text{AB}} \varepsilon_{\text{AB}}}, \tag{5.1105}$$

where*

$$\left.\begin{aligned} \nu_{\text{AB}} = \nu_{\text{A}} + \nu_{\text{B}}, \quad q_{\text{AB}} = q_{\text{A}} + q_{\text{B}}, \quad \varepsilon_{\text{AB}} = \varepsilon_{\text{A}} + \varepsilon_{\text{B}}, \\ \alpha_{\text{AB}} = \alpha_{\text{A}} = \alpha_{\text{B}}, \quad \beta_{\text{AB}} = \beta_{\text{A}} = \beta_{\text{B}}, \end{aligned}\right\} \tag{5.1106}$$

The choice of

$$S = -k \langle \ln \rho_{\text{gen}} \rangle \tag{5.1107}$$

entails the introduction of the factor $1/N!$ in Eq. (5.0311) as can be seen as follows. From Eq. (5.1107) and the equation $F = \langle \varepsilon \rangle - ST$ we get

$$e^{-\beta F} = e^{q - \sum \alpha_i \langle \nu_i \rangle}. \tag{5.1108}$$

If now the $\langle \nu_i \rangle$ are sufficiently large so that their dispersion is negligible, the canonical grand ensemble will practically be a macrocanonical ensemble and we can substitute in all equations $\nu_i \cong \langle \nu_i \rangle$. This corresponds to the equivalence of the macro- and microcanonical ensembles which we discussed in the preceding chapter. In that way we get from Eq. (5.1108) by using Eq. (5.0907)

$$e^{-\beta F} = \frac{1}{\prod_i \langle \nu_i \rangle!} \left[\int_\Gamma e^{-\beta \varepsilon} \right]_{\nu_i = \langle \nu_i \rangle}, \tag{5.1109}$$

* Note that the indices "A," "B," and "AB" denote different ensembles here and not, as in Eq. (5.1115), different particles taking part in a reaction.

which reduces to Eq. (5.0311) in the case of only one kind of particle—apart from the factor h^{sN}, which we should have introduced throughout in order to get formulas that are the limiting cases of the quantum-mechanical formulas to be derived in the next chapter.

We may remark here that, although we see from this discussion the plausibility of the introduction of the factor $1/N!$ in Z_Γ, the only decisive argument is the one given by considering quantum statistics and considering the classical limit of the quantum-mechanical expressions. Since these lead to the factor $1/N!$, we can be sure of their correctness instead of just feeling rather safe in including them.

In Section 5.6 we discussed at some length the coupling of two petit ensembles. We noted then that the only way in which we could at time t' couple two ensembles physically and at time t'' decouple them again in such a way that the situation after t'' is the same as that before t' was to couple two macrocanonical ensembles with the same value of β. In proving that and also in proving that, if we couple two macrocanonical ensembles with different values of β, energy will flow from systems in the ensemble with the lower value of β to those in the ensemble with the larger value of β, we used certain lemmas as well as the consequences of a generalized H-theorem.

We can now proceed in a similar manner. If we introduce the index of probability η by Eq. (5.0121), we see that Eq. (5.1107) can be written in the form

$$S = -k\langle \eta_{\text{gen}} \rangle, \tag{5.1110}$$

which is the counterpart of Eq. (5.0309) for the case of petit ensembles. From the generalized H-theorem we now get instead of Eq. (5.0627)

$$\langle \eta_{\text{gen}} \rangle_{12}'' \leqslant \langle \eta_{\text{gen}} \rangle_{12}', \tag{5.1111}$$

where the index "12" again indicates a combined ensemble. From Eq. (5.1111) we now get by calculations completely similar to those in Section 5.6 instead of Eq. (5.0630) the inequality

$$\beta_1(\langle \varepsilon \rangle_1'' - \langle \varepsilon \rangle_1') + \beta_2(\langle \varepsilon \rangle_2'' - \langle \varepsilon \rangle_2') \\ + \sum_i [\alpha_{i1}(\langle \nu_i \rangle_1' - \langle \nu_i \rangle_1'') + \alpha_{i2}(\langle \nu_i \rangle_2' - \langle \nu_i \rangle_2'')] \geqslant 0, \tag{5.1112}$$

where the equal sign holds, if the two ensembles before the coupling were canonical with $\beta_1 = \beta_2$ and $\alpha_{i1} = \alpha_{i2}$ (all i).

If we consider for a moment ensembles with only one kind of particle, and if we consider the coupling of two canonical grand ensembles for which $\beta_1 = \beta_2$, but $\alpha_1 \neq \alpha_2$, it follows from Eq. (5.1112) and the conservation of particle number, that

$$\langle \nu \rangle_1' < \langle \nu \rangle_1'', \qquad \text{if } \alpha_1 < \alpha_2, \tag{5.1113}$$

that is, the system with the larger value of α will, on the average have lost particles to the other system.

Let us now consider the case in which we have three kinds of particles in the systems in the grand ensemble: A, B, and AB, and let us assume that the following reaction can take place:

$$\mathrm{A} + \mathrm{B} \rightleftarrows \mathrm{AB}. \tag{5.1114}$$

Equation (5.1112) can be the case where we couple two ensembles with $\beta_1 = \beta_2$ be written in the form

$$(\alpha_{\mathrm{A}} + \alpha_{\mathrm{B}} - \alpha_{\mathrm{AB}})(\langle \nu_{\mathrm{A}} \rangle' - \langle \nu_{\mathrm{A}} \rangle'') \geq 0, \tag{5.1115}$$

where we have used the relations

$$\langle \nu_{\mathrm{A}} \rangle' - \langle \nu_{\mathrm{A}} \rangle'' = \langle \nu_{\mathrm{B}} \rangle' - \langle \nu_{\mathrm{B}} \rangle'' = \langle \nu_{\mathrm{AB}} \rangle'' - \langle \nu_{\mathrm{AB}} \rangle', \tag{5.1116}$$

following from Eq. (5.1114).

Just as from Eq. (5.0630) we find the condition for equilibrium $\beta_1 = \beta_2$, so we find from Eq. (5.1115) the well-known relation

$$\alpha_{\mathrm{A}} + \alpha_{\mathrm{B}} = \alpha_{\mathrm{AB}}, \qquad \text{or} \qquad \mu_{\mathrm{A}} + \mu_{\mathrm{B}} = \mu_{\mathrm{AB}}, \tag{5.1117}$$

as an equilibrium condition.

5.12 APPLICATION OF THEORY OF CLASSICAL GRAND ENSEMBLES TO A PERFECT GAS

Although the results that we shall obtain are not new and were derived previously in Chapters 1, 2, and 4 and Section 8 of the present chapter, we shall once more illustrate the theory by considering the case of a perfect gas enclosed in a volume V. The energy of the system is given by the equation

$$\varepsilon = \sum_{i=1}^{\nu} \left(\frac{\mathbf{p}_i^2}{2m} + F_i \right), \tag{5.1201}$$

where

$$F_i \equiv F(x_i, y_i, z_i) = 0, \qquad \text{if } x_i, y_i, z_i \text{ lies inside } V,$$

and

$$F(x_i, y_i, z_i) = \infty, \qquad \text{if } x_i, y_i, z_i \text{ lies outside } V.$$

From Eq. (5.0907) we get for the grand potential the equation

$$e^q = \sum_{\nu=0}^{\infty} \frac{e^{\alpha\nu}}{\nu!} \int_\Gamma \exp\left[-\beta \sum_i \left(\frac{\mathbf{p}_i^2}{2m} + F_i \right) \right] d\Omega. \tag{5.1202}$$

Integration over the x_i, y_i, and z_i gives a factor V^ν, and we get

$$e^q = \sum_\nu \frac{e^{\alpha\nu} V^\nu}{\nu!} \left[\int_{-\infty}^{+\infty} e^{-\beta u^2/2m}\, du \right]^{3\nu},$$

or

$$e^q = \sum_\nu \frac{1}{\nu!}\left[e^\alpha V(2\pi mkT)^{3/2}\right]^\nu,$$

which leads to the equation (compare Eq. (4.610))

$$q = e^\alpha V(2\pi mkT)^{3/2}. \tag{5.1203}$$

Using Eq. (5.0928), we get for $\langle \nu \rangle$

$$\langle \nu \rangle = \frac{\partial q}{\partial \alpha} = q, \tag{5.1204}$$

and combining this with Eq. (5.0926), we have

$$\langle \nu \rangle = \frac{pV}{kT}, \qquad \text{or} \qquad pV = \langle \nu \rangle kT, \tag{5.1205}$$

the perfect gas law.

We find from Eq. (5.101) for the fluctuations in the number of particles

$$\langle (\nu - \langle \nu \rangle)^2 \rangle = \frac{\partial^2 q}{\partial \alpha^2} = \langle \nu \rangle, \tag{5.1206}$$

which shows that, indeed,

$$\frac{\langle (\nu - \langle \nu \rangle)^2 \rangle}{\langle \nu \rangle^2} = \frac{1}{\langle \nu \rangle}. \tag{5.1207}$$

The average energy and its fluctuations satisfy the equations

$$\langle \varepsilon \rangle = -\frac{\partial q}{\partial \beta} = \frac{3q}{2\beta} = \frac{3}{2}\langle \nu \rangle kT\left(= \frac{3}{2}PV\right), \tag{5.1208}$$

$$\langle (\varepsilon - \langle \varepsilon \rangle)^2 \rangle = \frac{\partial^2 q}{\partial \beta^2} = \frac{15\langle \nu \rangle}{4\beta^2}, \tag{5.1209}$$

$$\frac{\langle (\varepsilon - \langle \varepsilon \rangle)^2 \rangle}{\langle \varepsilon \rangle^2} = \frac{5}{3\langle \nu \rangle}. \tag{5.1210}$$

The free energy and the entropy of the system are given by the equations

$$F = -qkT + \alpha\langle \nu \rangle = \langle \nu \rangle kT[\ln P - \tfrac{5}{2}\ln T - \tfrac{3}{2}\ln M + \text{const}], \tag{5.1211}$$

and

$$S = qk - k\alpha\langle \nu \rangle - \frac{\langle \varepsilon \rangle}{T} = \langle \nu \rangle k\left[\frac{5}{2}\ln T - \ln P + \frac{3}{2}\ln M + \text{const}\right], \tag{5.1212}$$

where M is the molecular weight of the gas and where we have used for α

the equation

$$e^{\alpha} = \langle v \rangle V^{-1}(2\pi mkT)^{-3/2}, \tag{5.1213}$$

which follows from Eqs. (5.1203) and (5.1204).

We shall conclude this section by discussing the famous paradox that bears Gibbs' name, which is briefly mentioned in the last paragraph of his monograph as well as discussed in some detail in his famous papers on the equilibrium of heterogeneous substances.* To discuss this paradox we consider two systems, one consisting of particles of the kind A in a volume V_A at a temperature T, the other of particles of the kind B in a volume V_B at the same temperature T. We now join these two systems together and remove any walls dividing them, so that both systems now have a volume $V_A + V_B$ at their disposal. We then ask what the value of the entropy of the combined system will be, assuming throughout that we can treat both the separate systems and the combined system as perfect gases. From Eq. (5.1107) and calculations similar to those leading to Eq. (5.1212) we find—provided the particles of kind A are different from those of kind B—that

$$S_{AB} = S_A + S_B + \langle v_A \rangle k \ln \frac{V_A + V_B}{V_A} + \langle v_B \rangle k \ln \frac{V_A + V_B}{V_B}, \tag{5.1214}$$

where S_A and S_B are given by Eq. (5.1212) with the appropriate subscripts. We see that the entropy is increased by the removal of any dividing walls. This is as it was to be expected, since the removal of a dividing wall is an irreversible process and the mixing entropy, as the extra term is called, is a consequence of the diffusion of the two gases into one another.

If, however, the particles of kind A and of kind B were identical, and if to simplify the calculations we assume that $V_A = V_B (= V)$ and $\langle v_A \rangle = \langle v_B \rangle (= N)$ —we are dealing with a system of $2N$ particles in a volume $2V$ and we find now, as should be the case,

$$S_{AB} = S_A + S_B. \tag{5.1215}$$

Equation (5.1215) is, of course, not the limit of Eq. (5.1214), but differs from it in the limit by a term $2Nk \ln 2$. The paradoxical result is that somehow or other the particles must have sensed that they were identical.

It is at this point that we can once again demonstrate the reason why the generic density rather than the specific density occurs in the definition of the entropy. If we had used instead of Eq. (5.1107) the equation

$$S' = -k\langle \ln \rho_{\text{spec}} \rangle, \tag{5.1216}$$

and evaluated the value of S_{AB}' for the case where the two kinds of particles

* See *The Collected works of J. Willard Gibbs*, Vol. I, New Haven, 1948, p. 165.

were identical, we would have found instead of Eq. (5.1215) the incorrect equation

$$S_{AB}' = S_A' + S_B' + 2kN \ln 2. \tag{5.1217}$$

This shows that we must, indeed, use the generic density.

5.13 THE RELATIONSHIP BETWEEN ENSEMBLES AND ACTUALLY OBSERVED SYSTEMS

In this and the next section we shall be concerned with some of the fundamental problems of statistical mechanics referring to the foundations of the subject. We cannot enter very deeply here into a discussion of these problems for which we refer to the literature mentioned in the bibliographical notes. (A more extensive discussion will be given in Vol. II of this textbook.)

One could argue that the fact that statistical mechanics has been able to predict accurately and successfully the behavior of physical systems under equilibrium and near-equilibrium conditions should be a sufficient justification for the methods used. However, a great deal of time and effort has been spent in attempts to show that the statistical formalism follows in a straightforward manner from classical or quantum mechanics. The two following questions are the main ones with which we shall be concerned here:

(A) "Why is it possible to describe the behavior of almost all physical systems by considering only equilibrium situations?" with the corollary (A1) "How can one define an equilibrium situation?"

(B) "Why is it possible to describe the behavior of an actual physical system by considering a large number of identical systems (in ensemble theory) and identifying the average behavior of this group of systems with the physical system in which we are interested?" with the corollary (B1) "How can we construct an ensemble such that it will "represent" an actual given physical system?"

Question A had occupied Boltzmann in his "Gastheorie" who showed by his H-theorem (see Sections 1.5 and 2.3) that any nonequilibrium situation would develop in the direction of an equilibrium situation. The answer to Question A1 was in that case that the equilibrium situation is the most probable situation, compatible with a few restricting conditions. In its original, unrestricted form the H-theorem proved that any system will tend toward equilibrium, if equilibrium does not originally exist. This would mean that, provided one waited a sufficiently long time, one would find the system in an equilibrium situation and, moreover, this equilibrium would persist forever after. From this follows that equilibrium would be the rule and nonequilibrium the exception, so that Question A is answered.

We shall not consider here the various problems concerning the limitations of the unrestricted H-theorem and the necessity—stressed especially by the Ehrenfests—of the statistical nature of the H-theorem in kinetic theory.

Apart from introducing the H-theorem to explain (A), Boltzmann also tried to prove that the average behavior of a system is the same as its equilibrium behavior. This statement is meant in the following way: The time average of any phase function taken over a sufficiently long period should be equal to the value of the phase function at equilibrium. To prove this, one has to prove Eq. (5.0704); this is done by the so-called ergodic theory, which we shall discuss in the next section.

We now come to Questions B and B1. One might ask what reason there is to introduce ensembles at all since in actual cases one is always dealing with one system. Indeed, Gibbs introduced ensembles in order to use them for statistical considerations rather than to illustrate the behavior of physical systems, even though he did not adhere strictly to his original intentions as set out in the introduction to his monograph. Einstein considered rather more carefully the possibility that the average behavior of a system in a macrocanonical ensemble would represent the actual behavior of a system in thermodynamical equilibrium. However, the Ehrenfests, Ornstein* and Uhlenbeck† considered ensemble theory still mainly as a mathematical trick to lighten the calculation of average values of phase functions with a given total energy and given total number of particles, much in the same way as our discussion of Eq. (5.0705) in the preceding section. Uhlenbeck therefore considers Gibbs' method and the Darwin-Fowler method (see Section 4.5) as more or less equivalent. However, Tolman in his monograph succeeded in showing the importance of ensemble theory from a physical point of view. We shall give some of his arguments in the present section, although referring the reader to his monograph for a thorough discussion of this point.

We have mentioned before that there are very severe limitations to the possibilities of obtaining accurate knowledge about the initial condition, say, of the physical system that interests us. One mole of a monatomic gas contains 6.10^{23} atoms, and instead of obtaining values for the 36.10^{23} coordinates and momenta which we need to know in order to characterize the situation completely, we can only obtain a few relations between them, for instance, by measuring the pressure, the linear momentum, and the angular momentum of the system. Moreover, we certainly will not know the total number of particles in the system: Avogadro's number is only known to a few parts in 10^5. Having obtained this very scant information, we now want to make predictions about the future behavior of the system, starting from this information. Clearly the only way to do so is by using probability theory and statistical methods. If we could have obtained the values of all the coordinates and momenta at one moment—and if we could have solved the equations of motion—the behavior of the system would be completely determined, as we would then be dealing with a problem of classical mechanics with a sufficient number of boundary conditions to determine the orbit of the representative point of the system in phase space. However, as we cannot obtain these values, we are reduced to making statements about the most probable behavior of the system. In order to determine this most probable behavior, we compare various systems possessing the same value of those quantities that have been measured, but

* L. S. Ornstein, *Toepassing der Statistische Mechanica van Gibbs op Molekulair-theoretische Vraagstukken*, Thesis, Leiden, 1908.

† G. E. Uhlenbeck, *Over Statistische Methoden in de Theorie der Quanta*, Thesis, Leiden, 1927.

which otherwise may differ widely. This means that we construct an ensemble to represent our system. The question remains as to how we can construct such a representative ensemble. This introduces the problem of a priori probabilities. In constructing a classical ensemble, we use the assumption of equal a priori probabilities for equal volumes in Γ-space. We shall briefly discuss the plausibility or justification of this choice at the end of this section.

To see what kind of ensemble will best represent a system, we shall state some results, but only prove some of them (proofs of the other statements are asked for as problems at the end of this chapter). We find that

(1) If a system is completely isolated so that it can exchange neither energy nor particles with its surroundings, it will be represented by a microcanonical ensemble;

(2) If a system can exchange energy, but cannot exchange particles with its surroundings, it will be represented by a macrocanonical ensemble;

(3) If a system can exchange both energy and particles with its surroundings, it will be represented by a canonical grand ensemble.

We shall only prove Statement 3 here. Consider the system together with its surroundings and let this large system be isolated so that the total number of particles (we shall assume for the sake of simplicity that only one kind of particle is present) and the total energy of the large system are constant. Let $\rho(\rho')$, $\varepsilon(\varepsilon')$, and $\nu(\nu')$ denote, respectively the (generic) ensemble density, the energy, and the number of particles in the system (surroundings). We then have the following equations to be satisfied by ρ and ρ':

$$\sum \int_{\Gamma} \rho \, d\Omega = 1, \tag{5.1301}$$

$$\sum \int_{\Gamma'} \rho' \, d\Omega' = 1, \tag{5.1302}$$

$$\sum \int_{\Gamma_{\text{tot}}} (\varepsilon + \varepsilon')\rho\rho' \, d\Omega \, d\Omega' = \text{const}, \tag{5.1303}$$

$$\sum (\nu + \nu') \int_{\Gamma_{\text{tot}}} \rho\rho' \, d\Omega \, d\Omega' = \text{const}. \tag{5.1304}$$

Furthermore we assume that an equilibrium situation has been reached. This means, as we shall see in the next section, that the average value of the index of probability is a minimum for the large system, or

$$\langle \eta_{\text{gen}} \rangle = \sum \int_{\Gamma_{\text{tot}}} (\rho\rho') \ln (\rho\rho') \, d\Omega \, d\Omega' = \text{minimum}. \tag{5.1305}$$

We can replace Eqs. (5.1303) and (5.1304) by the equations

$$\sum \int_{\Gamma} \varepsilon\rho \, d\Omega + \sum \int_{\Gamma'} \varepsilon'\rho' \, d\Omega' = \text{const}, \tag{5.1306}$$

$$\sum \nu \int_{\Gamma} \rho \, d\Omega + \sum \nu' \int_{\Gamma'} \rho' \, d\Omega' = \text{const}, \tag{5.1307}$$

where we have used Eqs. (5.1301) and (5.1302). From Eqs. (5.1301), (5.1302), and (5.1305) to Eq. (5.1307) we now find in the usual way, by the method of Lagrangian multipliers, that the equilibrium expression for ρ will be

$$\rho = e^{-q+\alpha\nu-\beta\varepsilon}, \tag{5.1308}$$

which is the same as Eq. (5.0906). This concludes our proof.

We should mention here that we have tacitly introduced our assumption about equal a priori probabilities for equal volumes in Γ-space by not introducing any weight function into Eqs. (5.1301) to (5.1304).

We note that from our analysis it follows that any part of an isolated system will be represented by a canonical ensemble. The analogy between Eq. (5.1308) and the expression for the Maxwell distribution is not accidental, as we are there considering an example of a subsystem (one atom) in an isolated system (the gas).

Before considering the question of a priori probabilities, let us briefly discuss the idea that energy and temperature are complementary quantities. The term "complementarity" is used here in a different manner from its use when it enters in the discussion of the Heisenberg relations. While in quantum mechanics we cannot define two canonically conjugate quantities simultaneously with arbitrary accuracy, the point here is that we cannot reduce the fluctuations in both temperature and energy arbitrarily. We discussed one side of this question, namely, that *in a system in temperature equilibrium energy fluctuations occur*. On the other hand, if we consider an isolated system, say, a volume of helium gas, its total energy is fixed. Is it possible to speak about the temperature of such a system? In order to do this, we must envisage a temperature measurement, for instance, by considering a small part of the gas and measuring its mean kinetic energy per atom.* Since we have just seen that such a subsystem will behave as the average member of a macroconical ensemble, we shall find fluctuations in the value of tne mean kinetic energy per atom, and hence in the temperature. *In an isolated system we must expect temperature fluctuations.*

Let us now consider the problem of a priori probabilities. Let us restrict ourselves for the sake of simplicity to petit ensembles. Consider two regions Ω_0 and Ω_0' of Γ-space. Let those two regions at t_0 be filled with representative points of an ensemble, and let $p(\Omega_0)$ and $p(\Omega_0')$ be the a priori probabilities for the two regions. At $t_1, t_2, \cdots$ the representative points, which at t_0 filled up Ω_0 and Ω_0', will fill up regions Ω_1 and Ω_1', Ω_2 and Ω_2', $\cdots$ with a priori probabilities $p(\Omega_1)$ and $p(\Omega_1')$, $p(\Omega_2)$ and $p(\Omega_2')$, $\cdots$. One can clearly only tolerate such a choice of a priori probabilities that

$$\frac{p(\Omega_0)}{p(\Omega_0')} = \frac{p(\Omega_1)}{p(\Omega_1')} = \frac{p(\Omega_2)}{p(\Omega_2')} = \cdots, \tag{5.1309}$$

and these relations are satisfied, if we take

$$p(\Omega) = W(\Omega), \qquad p(\Omega') = W(\Omega'), \tag{5.1310}$$

* We cannot introduce an external thermometer, because this would destroy the isolation of the system.

where $W(\Omega)$ or $W(\Omega')$ is the volume of the region Ω or Ω' in Γ-space. We remark here in passing that Eq. (5.1310) will only satisfy Eq. (5.1309) if we are dealing with a phase space made up of canonically conjugated p's and q's (compare the remark in Section 2.3).

Instead of Eq. (5.1310) one could choose the general form.

$$p(\Omega) = \int_{\Omega} F(\varepsilon,\varphi_i)\, d\Omega, \tag{5.1311}$$

where the integration extends over the region Ω and where $F(\varepsilon,\varphi_i)$ is a once and for all chosen function of the energy ε and the other constants of motion φ_i of Eqs. (5.0104).

The reason for choosing $F(\varepsilon,\varphi_i)$ equal to 1, which means choosing Eq. (5.1310) instead of the more general form (5.1311), was originally historical and was based upon the assumption that ergodic systems (see next section) existed. Since we are nowadays convinced that most mechanical systems are quasi-ergodic (again see Section 5.14), which means essentially that the various possible values of φ_i all occur with practically equal frequency in each part of phase space, we can immediately restrict ourselves to what the Ehrenfests call the ergodic choice,

$$p(\Omega) = \int_{\Omega} F(\varepsilon)\, d\Omega. \tag{5.1312}$$

However, since the density ρ of the representative ensembles of equilibrium systems are functions of the energy only, the choice of Eq. (5.1312) instead of Eq. (5.1310) would correspond to a density

$$\rho'(\varepsilon) = \rho(\varepsilon)/F(\varepsilon) \tag{5.1313}$$

instead of the $\rho(\varepsilon)$ given, for instance, by Eq. (5.1308). The $F(\varepsilon)$ would occur in all integrals; but the final results would be the same, as the $F(\varepsilon)$ in the denominator of ρ' would counteract the same function, which would now occur as a weight factor in all integrals.

5.14 ERGODIC THEORY AND THE *H*-THEOREM IN ENSEMBLE THEORY

In the preceeding section, we based the proof that the canonical grand ensemble is an equilibrium ensemble upon Eq. (5.1305). To derive that equation we need a generalized *H*-theorem. Before discussing this, we want to dwell briefly on Eq. (5.0704), stating the equivalence of time- and ensemble averages.

Boltzmann expected the existence of so-called *ergodic* systems.* An ergodic system is one such that the orbit in Γ-space will go through every point of its energy surface. Once the existence of ergodic systems is accepted, one can easily see the advantage of considering time averages, since the proof of Eq.

* From ἔργον = work (used here as "energy"; compare the erg as unit of energy) and ὁδός = path; the orbit passes through every point of the energy surface. This name was introduced by Boltzmann (*J. f. Math.*, **100,** 201, 1887).

(5.0704) is now complete. One can, however, easily see that ergodic systems do not exist, as was pointed out by Rosenthal and Plancherel. Plancherel's argument was that the points of an orbit form a set of measure zero on the energy surface the measure of which itself is not zero.*

We are thus left to find a different proof of Eq. (5.0704). This is done in ergodic theory, which is in fact a branch of pure mathematics rather than one of physics. We shall therefore only in the briefest possible way sketch the main points of ergodic theory. First of all, we mention that Birkhoff has proved that (1) for bounded mechanical motion and for "reasonable" phase functions G the time average $\tilde{G}$ almost always exists.† and (2) for metrically transitive systems this time average is equal to the (microcanonical) ensemble average. A mechanical system is metrically transitive, if the energy surface cannot be divided into two finite regions such that orbits starting from points in one of the regions always remain in that region. Whether or not the systems considered in physics are metrically transitive—as many people accept without proof—is still an open question.

We now come to the *H*-theorem in ensemble theory. First of all, we shall prove the following statements — or rather we shall prove the last of them, leaving as problems the proof of the other two.

1. If our ensemble is such that all systems have the same number of particles N (we restrict the discussion for the sake of simplicity to systems containing only one kind of particles) and that the energy ε of all systems lies in the interval $E \leqslant \varepsilon \leqslant E + \delta E$, the average value $\langle \eta \rangle$ of the generic index of probability, given by the equation

$$\langle \eta \rangle = \sum_{\nu} \int_{\Gamma} \rho_{\text{gen}} \ln \rho_{\text{gen}} \, d\Omega, \tag{5.1401}$$

will be minimum, if ρ (we drop here and henceforth the index "gen") is given by the equation

$$\left.\begin{aligned} \rho &= \text{const} \times \delta(\nu - N), \qquad && E \leqslant \varepsilon \leqslant E + \delta E; \\ \rho &= 0 \qquad\qquad\qquad\quad , && \varepsilon < E,\; E + \delta E < \varepsilon, \end{aligned}\right\} \tag{5.1402}$$

where $\delta(x)$ is the Dirac δ-function.

2. If our ensemble is such that all systems have the same number of particles, but only the average energy is given, that is, ρ must satisfy the condition‡

$$\sum \int_{\Gamma} \varepsilon \rho \, d\Omega = E, \tag{5.1403}$$

* The measure of a point set on a surface is defined as follows (see, for example, P. R. Halmos, *Measure Theory*, New York, 1950). Let $f(\mathrm{P})$ be a function that is 1, if P belongs to the point set, and which is 0 otherwise. The Lebesgue integral $\int f \, d\sigma$, which extends over the whole of the surface, is then called the Lebesgue measure of the point set on the surface.

† The expression "almost always", must be interpreted as meaning "in by far the most cases" and can only be made into a more rigorous statement by using the language of measure theory.

‡ The summation over ν is here trivial as only one term of the sum contributes.

then $\langle\eta\rangle$ will be minimum, if ρ satisfies the equation

$$\rho = \delta(\nu - N)e^{\beta(\psi-\varepsilon)}. \tag{5.1404}$$

3. If only the average number of particles and the average energy of the systems in the ensemble are given so that ρ must satisfy Eq. (5.1403) and the condition

$$\sum \nu \int_{\Gamma} \rho \, d\Omega = N, \tag{5.1405}$$

$\langle\eta\rangle$ will be a minimum, if ρ satisfies the equation

$$\rho = e^{q+\alpha\nu-\beta\varepsilon}. \tag{5.1406}$$

The proof of the minimum properties of $\langle\eta\rangle$ rests mainly on the fact that

$$xe^x - e^x + 1 \geqslant 0, \tag{5.1407}$$

for all x, the equal sign holding only when $x = 0$.

We now prove Statement (3) by considering two densities, ρ_1 and ρ_2, where ρ_1 is given by Eq. (5.1406), while ρ_2 is given by

$$\rho_2 = \rho_1 e^{\Delta\eta}, \tag{5.1408}$$

where $\Delta\eta$ is an unspecified function of the p's, the q's and of ν. Both ρ_1 and ρ_2 satisfy Eqs. (5.1403) and (5.1405), as well as the normalization condition

$$\sum \int_{\Gamma} \rho \, d\Omega = 1. \tag{5.1409}$$

We now consider the equation

$$\langle\eta\rangle_2 - \langle\eta\rangle_1 = \sum \int_{\Gamma} (\rho_2 \ln \rho_2 - \rho_1 \ln \rho_1) \, d\Omega. \tag{5.1410}$$

Adding to the right-hand side of Eq. (5.1410) the expression

$$\sum \int_{\Gamma} [q - \alpha\nu + \beta\varepsilon - 1](\rho_2 - \rho_1) \, d\Omega,$$

which vanishes because of the normalization condition and of Eqs. (5.1403) and (5.1405), we find after some straightforward transformations

$$\langle\eta\rangle_2 - \langle\eta\rangle_1 = \sum \int_{\Gamma} \rho_1[\Delta\eta \, e^{\Delta\eta} - e^{\Delta\eta} + 1] \, d\Omega \geqslant 0, \tag{5.1411}$$

which concludes our proof.

Comparing expression (5.1401) with the definition (2.301) of Boltzmann's H, we might hope to be able to prove that $d\langle\eta\rangle/dt$ is nonpositive, and thus that there is always a tendency to establish equilibrium ensembles. However, if we compare the values of $\langle\eta\rangle$ at times t' and t'', we have

$$\langle\eta\rangle'' = \sum \int_\Gamma \rho'' \ln \rho'' \, d\Omega'' = \sum \int_\Gamma \rho' \ln \rho' J \, d\Omega'$$

$$= \sum \int_\Gamma \rho' \ln \rho' \, d\Omega' = \langle\eta\rangle', \tag{5.1412}$$

which shows that in fact $d\langle\eta\rangle/dt = 0$. In Eq. (5.1412) ρ' is a shorthand notation for $\rho(p',q';t')$ and similarly for ρ'', where the relation between p'', q'' and p', q' is such that a representative point p', q' at time t' will have moved to p'', q'' by t'', while J is the Jacobian of the transformation from p'', q'' to p', q' which by virtue of Liouville's theorem is equal to 1.

Although we have just seen that $\langle\eta\rangle$ is constant, there is still in a certain sense an approach to a stationary state. Let us first illustrate this by an example introduced by Gibbs. Consider a container with a liquid, say water, in which is put some coloring material; let us assume that this coloring material is non-diffusible and consists of colloidal particles. It is a well-known empirical fact that, if we start from a state where the coloring material is unevenly distributed, practically any kind of stirring will produce a situation where the color distribution is, as far as our eye can see, uniform. That means that stirring will prodice an "equilibrium" state. However, if we look at the system very closely, we will still find that, in microscopic volumes, part of the space is occupied by the water and part occupied by the colloidal particles. Although the coarse distribution is uniform, the finer distribution is still uneven.

From this example it follows that it might be advantageous to introduce apart from the *fine-grained density* ρ *a coarse-grained density* P (read "capital ρ") defined as follows: Divide Γ-space into finite, but small, cells Ω_i and let P_i be the average of ρ taken over Ω_i, or

$$\mathrm{P}_i = \frac{1}{W^\nu(\Omega_i)} \int_{\Omega_i} \rho \, d\Omega, \tag{5.1413}$$

where $W^\nu(\Omega_i)$ is the volume of the cell Ω_i corresponding to a system with ν particles.*

From the normalization of ρ we then get

$$\sum_\nu \sum_i \mathrm{P}_i W^\nu(\Omega_i) = 1. \tag{5.1414}$$

We can now introduce the coarse-grained density $\mathrm{P}(p,q)$ by putting it constant in each cell Ω_i and equal to P_i. Equation (5.1414) can then be written in the form

$$\sum_\nu \int_\Gamma \mathrm{P}(p,q) \, d\Omega = 1. \tag{5.1415}$$

* We should really also have lumped together a range of ν-values, but since this would not have added anything new to the discussion we have left this complication for the reader to carry out.

Instead of $\langle\eta\rangle$ we can now introduce a function $\langle \mathrm{H}\rangle$ by the equation*

$$\langle \mathrm{H}\rangle = \sum_i \mathrm{P}_i \ln \mathrm{P}_i W^{\nu}(\Omega_i). \tag{5.1416}$$

Using the coarse-grained density P, Eq. (5.1416) takes the form

$$\langle \mathrm{H}\rangle = \sum_{\nu} \int_{\Gamma} \mathrm{P} \ln \mathrm{P}\, d\Omega \tag{5.1417}$$

Since $\ln \mathrm{P}$ is constant over each cell Ω_i and $\int_{\Omega_i} \mathrm{P}\, d\Omega = \mathrm{P}_i W^{\nu}(\Omega_i) = \int_{\Omega_i} \rho\, d\Omega$, we can also write Eq. (5.1417) in the form

$$\langle \mathrm{H}\rangle = \sum_{\nu} \int_{\Gamma} \rho \ln \mathrm{P}\, d\Omega, \tag{5.1418}$$

and we see that

$$\langle \mathrm{H}\rangle = \langle \ln \mathrm{P}\rangle. \tag{5.1419}$$

The conclusions which we reached at the beginning of this section about $\langle\eta\rangle$ will also hold for $\langle \mathrm{H}\rangle$, and one can, indeed, easily show, for instance, that in an ensemble of which the constituent systems have all energies within the range E, $E+\delta E$, $\langle \mathrm{H}\rangle$ will be minimum if P is constant in the region between the two corresponding energy surfaces. And when there is a slight interaction between the different members of the ensemble, $\langle \mathrm{H}\rangle$ will be minimum if P corresponds to a coarse-grained macrocanonical distribution, and so on.

We can now consider the change of $\langle \mathrm{H}\rangle$ with time. Let us assume that we have made some observations about a physical system at $t = t'$. Since these observations will never give us the maximum possible information, we can construct an ensemble, the average properties of which at t' correspond to the observed properties of the system under observation at t'. Because of experimental limitations we shall at most be able to give $\rho(p,q)$ changing from one cell Ω_i to another, if we have chosen the size of the cells Ω_i in accordance with the experimental limitations, as we shall assume that we have done. In accordance with the principle of equal a priori probabilities for equal volumes in Γ-space, the fine-grained density will be chosen constant in each Ω_i and we have thus at t'

$$\rho' = \mathrm{P}', \tag{5.1420}$$

where the prime indicates values at t'. We have thus for $\langle \mathrm{H}\rangle'$

$$\langle \mathrm{H}\rangle' = \sum_{\nu} \int_{\Gamma} \mathrm{P}' \ln \mathrm{P}'\, d\Omega = \sum_{\nu} \int_{\Gamma} \rho' \ln \rho'\, d\Omega.$$

If the situation at t' already corresponds to an equilibrium situation—either microcanonical, macrocanonical, or grand canonical, according to

* The symbol $\langle \mathrm{H}\rangle$ can be read either as the ensemble average of Boltzmann's H or rather as the average of a function "capital η".

whether or not interactions between the members of the ensemble are excluded from our considerations—the situation will be the same at a later time, since it is a stationary distribution. Let us therefore assume that ρ' is not a stationary distribution. It can then easily be seen that at a later moment, t'',* Eq. (5.1420) will no longer hold and we have

$$\rho'' \neq \mathrm{P}''. \tag{5.1421}$$

This inequality is due to the fact that, although ρ will stay constant in extensions in phase of unchanging magnitude Ω_i, the shape of these extensions will change and at a later moment each of the Ω_i will in general be covered by points which at t' belonged to many different cells.

We have now for $\langle \mathrm{H} \rangle''$

$$\langle \mathrm{H} \rangle'' = \sum_{\nu} \int_{\Gamma} \mathrm{P}'' \ln \mathrm{P}'' \, d\Omega, \tag{5.1422}$$

and we can no longer replace P'' by ρ''. Comparing $\langle \mathrm{H} \rangle'$ and $\langle \mathrm{H} \rangle''$, we have after some manipulations

$$\langle \mathrm{H} \rangle' - \langle \mathrm{H} \rangle'' = \sum_{\nu} \int_{\Gamma} \mathrm{P}''[\Delta e^{\Delta} + 1 - e^{\Delta}] \, d\Omega > 0, \tag{5.1423}$$

where Δ is defined by the equation

$$\rho'' = \mathrm{P}'' e^{\Delta}. \tag{5.1424}$$

Equation (5.1423) is derived as follows:

$$\begin{aligned}
\langle \mathrm{H} \rangle' - \langle \mathrm{H} \rangle'' &= \sum_{\nu} \int_{\Gamma} (\rho' \ln \rho' - \mathrm{P}'' \ln \mathrm{P}'') \, d\Omega \\
&= \sum_{\nu} \int_{\Gamma} (\rho'' \ln \rho'' - \rho'' \ln \mathrm{P}'') \, d\Omega \\
&= \sum_{\nu} \int_{\Gamma} [\rho'' \ln \rho'' - \rho'' \ln \mathrm{P}'' + \mathrm{P}'' - \rho''] \, d\Omega.
\end{aligned}$$

The second equality follows from Eqs. (5.1412), (5.1417), and (5.1418), and the last equality from Eqs. (5.1409) and (5.1415).

We see now that $\langle \mathrm{H} \rangle''$ will be less than $\langle \mathrm{H} \rangle'$ by virtue of the fact that ρ'' and P'' are no longer everywhere equal. Comparing the present situation with the case of the coloring material in a liquid, we may expect that, as time goes on, ρ and P will differ more and more and that $\langle \mathrm{H} \rangle$ will continually decrease until a distribution has been reached for which either

$$\left.\begin{aligned}
&\mathrm{P} = \text{const} \times \delta(\nu - N), && E \leqslant \varepsilon \leqslant E + \delta E, \\
&\mathrm{P} = 0, && \varepsilon < E,\ E + \delta E < \varepsilon,
\end{aligned}\right\} \tag{5.1425}$$

* It would be more general to say at *another* moment, leaving open the question whether t'' is later or earlier than t'. However, since we have set up our ensemble to represent a system that was observed at t', we shall be interested only in predictions that can actually be verified, that is, in times for which $t'' > t'$.

or

$$\mathrm{P} = A\delta(\nu - N)e^{-\beta\varepsilon}, \tag{5.1426}$$

or

$$\mathrm{P} = Be^{\alpha\nu - \beta\varepsilon}, \tag{5.1427}$$

depending on the situation which we are studying.

Since $\langle \mathrm{H} \rangle$ is minimum when the distributions (5.1425) to (5.1427) are reached, it will no longer decrease. If these distributions are reached we would, on observation, in general come to the conclusion that the state of the physical system would best be represented by a micro- or a macrocanonical ensemble. We have thus found a justification for using these ensembles to describe systems in thermodynamical equilibrium.

We saw that the decrease of $\langle \mathrm{H} \rangle$ resulted from the fact that we came from a state where $\rho = \mathrm{P}$ to a state where ρ was no longer everywhere equal to P. The decrease of $\langle \mathrm{H} \rangle$ thus corresponds to a decrease of the specific information about our physical system which we could obtain. Since $\langle \mathrm{H} \rangle$ can be regarded to be a "coarse-grained" $\langle \eta \rangle$, and since $\langle \eta \rangle$ was seen to be equal to the negative entropy (apart from a constant factor, see Eq. (5.0309)), we see how *increase of lack of knowledge* corresponds to an increase in entropy. We shall discuss this aspect of entropy in Section 6.4.

PROBLEMS

1. Use a canonical ensemble to derive a recursion formula for the following quantity:

$$\frac{\overline{(\mathscr{T} - \overline{\mathscr{T}})^n}}{\overline{\mathscr{T}}^n}, \quad n = 0, 1, 2, \cdots,$$

where $\mathscr{T}$ is the kinetic energy (see Eq. (5.0102)), and hence evaluate this quantity for the special cases when $n = 2$, 3, or 4.

2. Evaluate the following correlations:

$$\langle (A_k - \langle A_k \rangle)(A_l - \langle A_l \rangle) \rangle \quad \text{and} \quad \langle (\varepsilon - \langle \varepsilon \rangle)(A_k - \langle A_k \rangle) \rangle.$$

3. If Δ denotes the Jacobian

$$\frac{\partial(\dot{q}_1, \dot{q}_2, \cdots, \dot{q}_{sN})}{\partial(p_1, p_2, \cdots, p_{sN})},$$

and Δ' its reciprocal, we can, with Gibbs, introduce elements of extension in configuration,

$$\sqrt{\Delta'}\, dq_1 \cdots dq_{sN},$$

and elements of extension in velocity,

$$\sqrt{\Delta}\, dp_1 \cdots dp_{sN},$$

which can easily be seen to reduce—apart from a constant factor—to the expressions (5.0506) for the case considered in Section 5.5. Show that these two elements of extension are invariants.

4. Evaluate e^{ϕ_p} of Eq. (5.0508).

5. If we consider the band structure of a semiconductor, say, we find that the band widths are functions of the lattice constants. As the lattice constants will be functions of the temperature, we are dealing here with the at first sight paradoxical concept of temperature-dependent energy levels.* The reason for this difficulty is, of course, that we are treating the motion of the lattice as being given, before we evaluate the energy levels available to the electrons; that is, instead of considering the complete system, we consider only a subsystem, while we make some assumptions about the influence of the remainder of the system.

   Consider a system, for which the energy levels can be written in the form

$$\varepsilon = \varepsilon(p',q') + \varepsilon(p,q),$$

   where p, q stands for all coordinates that we want to ignore in our considerations and p', q' for those in which we are interested. For instance, the $\varepsilon(p',q')$ might be the energy of the electrons and the $\varepsilon(p,q)$ the energy of the lattice. In many calculations we would be tempted to use as a partition function the expression

$$Z_\Gamma' = \int e^{-\beta\varepsilon(p',q')}\, dp'\, dq'. \tag{A}$$

   Prove that a proper averaging over the ignored degrees of freedom (p,q) leads to the conclusion that we should replace in Eq. (A) $\varepsilon(p',q')$ by the Helmholtz free energy $F(p',q')$ related to $\varepsilon(p',q')$ through the Gibbs-Helmholtz equation

$$\varepsilon(p',q') = \frac{\partial F}{\partial \beta}.$$

6. Evaluate $\Omega(\varepsilon)$ for a system of N one-dimensional harmonic oscillators of frequency Ω, and hence find the average energy of this system as function of the temperature.

7. Prove Eq. (5.1112).

8. Use the perfect gas law and Eq. (5.1212) to prove Eq. (5.1214).

* See G. S. Rushbrooke, *Trans. Faraday Soc.*, **36**, 1055, 1940.

9. Prove statements (1) and (2) from Section 5.13.
10. Use arguments similar to the ones leading to Eq. (5.1308) to prove Eq. (5.1117).
11. Prove Statements (1) and (2) from Section 5.14.
12. Consider the following model of a vapor.* Assume the vapor to consist of noninteracting droplets, and assume that each drop can be represented by a volume v_l in which l atoms move independently in a smoothed-out negative potential $-\chi_l$ $(l = 1, 2, \cdots)$, where χ_l and v_l vary with l as follows:

$$\chi_l = \frac{l-1}{l}\chi, \quad v_l = v_1 l^{(l-2)/(l-1)}.$$

Express the free energy of a drop in terms of the various parameters of the problem, and after that find the equation of state. By an analysis similar to the one of the Einstein condenstaion in Section 4.7, show that this model shows condensation. Evaluate the average number of droplets containing l atoms.

BIBLIOGRAPHICAL NOTES

This chapter is largely based on the following references:

1. J. W. Gibbs, *Elementary Principles in Statistical Mechanics* (Vol. II of his Collected Works), New Haven, 1948.
2. R. C. Tolman, *The Principles of Statistical Mechanics*, Oxford, 1938.
3. P. and T. Ehrenfest, *Enzyklopädie der mathematischen Wissenschaften*, Vol. IV, Pt. 32, Leipzig-Berlin, 1911; *The Conceptual Foundations of the Statistical Approach in Mechanics*, Ithaca, N. Y., 1959.

See also

4. *A Commentary on the Scientific Works of J. W. Gibbs*, edited by A. Haas, New Haven, 1936; especially the articles R, S, T, and U by A. Haas and P. S. Epstein.

Independently of Gibbs, Einstein introduced ensemble theory at about the same time:

5. A. Einstein, *Ann. Physik*, **9,** 417, 1902.
6. A. Einstein, *Ann. Physik*, **11,** 170, 1903.

It should also be mentioned here that Boltzmann had made use of ensembles, especially, of course, in connection with the ergodic theorem. See, for instance,

7. L. Boltzmann, *Wien. Ber.*, **63,** 679, 1871.

* H. Wergeland, *Trans. Norwegian Acad. Sci., Oslo, Math.-Phys. Section*, **1943,** No. 11.

Section 5.1. Liouville's theorem was announced in 1838:

8. J. Liouville, J. de Math., **3,** 348, 1838.

Section 5.2 For a discussion of the term "macrocanonical" see

9. V. F. Lenzen, *Am. J. Phys.*, **22,** 638, 1954.
10. D. ter Haar, *Am. J. Phys.*, **22,** 638, 1954.

Section 5.9. See References 1 and 2 and also

11. H. A. Kramers, *Proc. Kon. Ned. Akad. Wet.* (*Amsterdam*), **41,** 10, 1938.
12. R. H. Fowler, *Proc. Cambridge Phil. Soc.*, **34,** 382, 1938.
13. E. A. Guggenheim, *J. Chem. Phys.*, **7,** 103, 1939.
14. R. C. Tolman, *Phys. Rev.*, **57,** 1160, 1940.
15. R. Becker, *Z. Physik. Chem.*, **196,** 181, 1950.

Section 5.12. For a discussion of a simple model related to the Gibbs paradox, see

16. M. J. Klein, *Am. J. Phys.* **26,** 80, 1958; Ned. Tijds. Natuurk., **25,** 73, 1959.

Section 5.13 and 5.14. See References 2, 3, and 14 and also

17. D. ter Haar, *Elements of Statistical Mechanics*, 1st ed., New York, 1954, App. I.
18. D. ter Haar, *Rev. Mod. Phys.*, **27,** 289, 1955.
19. G. E. Uhlenbeck and G. W. Ford, *Lectures in Statistical Mechanics*, Providence, R.I., Chap. 1.
20. R. Jancel, *Les Fondements de la Mécanique Statistique Classique et Quantique*, Paris, 1963 (English translation, Oxford, in course of publication).
21. I. E. Farquhar, *Ergodic Theory in Statistical Mechanics*, New York, 1964.

We refer to References 3, 18, 20, and 21 for extensive bibliographies.

For a simple, not completely rigorous, proof of the equivalence of micro-canonical and time averages, see

22. H. Wergeland, *Acta Chem. Scand.*, **12,** 1117, 1958.

6

The Ensembles in Quantum Statistics

6.1 THE DENSITY MATRIX

It is, of course, well known that modern theoretical physics is to a large extent dominated by quantum mechanics. Nevertheless, there are large fields of physics in which one can use classical theory because the quantum-mechanical corrections are far too small to enter into the discussion. Large parts of statistical mechanics belong to classical physics. On the other hand, statistical mechanics is one of the branches of physics in which the successes of quantum mechanics have been most noteworthy. We saw as instances of the successful intervention of quantum mechanics the solution of the paradox of the specific heats. Other instances are the theory of metals and many branches of modern astrophysics. The discussion of quantum statistics was started in Chapters 3 and 4, where we discussed, firstly, the quantum-mechanical analogy to the Maxwell-Boltzmann distribution and, secondly, the implications of the exclusion principle. However, we still must discuss quantum-mechanical ensemble theory, and this will be done in the present chapter.

In discussing quantum-mechanical ensemble theory one must be extremely careful to make a clear distinction between the statistical aspects inherent in quantum mechanics and the statistical aspects introduced by the ensembles. The statistical nature of quantum mechanics is connected with the fact that quantum mechanics gives us a wave function to describe a system and that from this wave function one then constructs a probability density, a probability current, and probable (or average) values of physical quantities. Only in exceptional cases will it be possible to ascribe to a quantity a completely defined value. If this is the case, it results in a complete lack of definition of the *complementary* quantity. In general, both of the two complementary quantities will be defined only within certain limits, and these limits cannot be chosen independently but must satisfy the so-called Heisenberg relations. We do not wish to discuss this point in any more detail, since it really belongs to a discussion of the principles of quantum mechanics. In the following we shall assume that the reader is familiar with this aspect of quantum mechanics and will use it whenever necessary.

We now wish to consider the statistical aspects of the ensembles in quantum statistics. We consider thereto a large number of identical systems. Let $\hat{\mathscr{H}}$ be the energy operator (or Hamiltonian) and let ψ^k be the normalized wave function describing the kth system. This wave function will then satisfy the Schrödinger equation

$$\hat{\mathscr{H}}\psi^k = i\hbar\dot{\psi}^k. \tag{6.101}$$

It is useful to introduce a complete set of orthonormal functions φ_n into which the wave functions or any other function can be developed.*

Expanding ψ^k in terms of the φ_n, we have

$$\psi^k = \sum_n a_n{}^k \varphi_n, \tag{6.102}$$

where

$$a_n{}^k = \int \varphi_n{}^* \psi^k \, d\tau. \tag{6.103}$$

where a star indicates the conjugate complex and where $d\tau$ is a volume element of coordinate space.

The coefficients $a_n{}^k$ can be used to describe the kth system instead of ψ^k.

* In order not to complicate our equations we shall usually not explicitly introduce degeneracy, spin, or continuous spectra. It is not difficult to extend all formulas to the most general case, and we leave that to the reader.

The two representations are equivalent and the transformed Schrödinger equation that the a_n^k must satisfy is

$$i\hbar\dot{a}_n^k = \sum_m H_{nm} a_m^k, \tag{6.104}$$

where the H_{nm} are the matrix elements of the Hamiltonian,

$$H_{nm} = \int \varphi_n^\star \hat{\mathscr{H}} \varphi_m \, d\tau. \tag{6.105}$$

The physical significance of the a_n^k is that they are probability amplitudes, $|a_n^k|^2$ being the probability that the kth system is characterized by the function φ_n. From the normalization of the ψ^k and the fact that the φ_n form a complete orthonormal set, we have

$$\sum_n |a_n^k|^2 = 1. \tag{6.106}$$

In the classical ensemble theory we discussed a density function $\rho(p,q)$. Now we introduce a density operator which we shall define by its matrix elements. The *density matrix* is defined through its elements by

$$\rho_{mn} = \frac{1}{\mathscr{N}} \sum_{k=1}^{\mathscr{N}} a_m^k a_n^{k\star}, \tag{6.107}$$

where $\mathscr{N}$ is the number of systems in the ensemble.

From Eq. (6.104) we can find the equation that ρ_{mn} must satisfy. This equation is

$$i\hbar\dot{\rho}_{mn} = \sum_l (H_{ml}\rho_{ln} - \rho_{ml}H_{ln}). \tag{6.108}$$

In deriving Eq. (6.108) we used the fact that $\mathscr{H}$ is a Hermitian operator, that is, that

$$H_{kl} = H_{lk}^\star. \tag{6.109}$$

Defining the commutator $[\hat{A},\hat{B}]_-$ of two operators $\hat{A}$ and $\hat{B}$ by the equation

$$[\hat{A},\hat{B}]_- = \hat{A}\hat{B} - \hat{B}\hat{A}, \tag{6.110}$$

or, in matrix notation,

$$[\hat{A},\hat{B}]_{-,kl} = \sum_m [A_{km}B_{ml} - B_{km}A_{ml}], \tag{6.111}$$

we can write Eq. (6.108) in matrix form, as follows,

$$i\hbar\dot{\hat{\rho}} = [\hat{\mathscr{H}},\hat{\rho}]_-. \tag{6.112}$$

This equation is the quantum-mechanical analogy of Liouville's theorem (Eqs. (5.0107) or (5.0115)) of the classical theory. Of course, it is well known that the commutators are the quantum-mechanical counterparts of the Poisson brackets of classical mechanics.

The average value $\bar{G}$ of a physical quantity is defined by the equation

$$\bar{G} = \frac{1}{\mathscr{N}} \sum_{k=1}^{\mathscr{N}} \int \psi^{k*} \hat{G} \psi^k \, d\tau, \tag{6.113}$$

where $\bar{G}$ is the quantum-mechanical operator corresponding to the classical quantity G. Equation (6.113) shows that $\bar{G}$ is twice an average. First we take the quantum-mechanical average of G in a system described by the wave function ψ^k, and, secondly, we take the average over the ensemble.

Introducing the $a_n{}^k$ instead of the ψ^k, we can write Eq. (6.113) in the form

$$\bar{G} = \frac{1}{\mathscr{N}} \sum_{k=1}^{\mathscr{N}} \sum_{m,n} a_m{}^k a_n{}^{k*} G_{nm}, \tag{6.114}$$

or, introducing the density matrix by means of Eq. (6.107),

$$\bar{G} = \sum_{m,n} \rho_{mn} G_{nm},$$

or

$$\bar{G} = \sum_m (\rho G)_{mm} = \operatorname{Tr} \hat{\rho} \hat{G}, \tag{6.115}$$

where Tr indicates the trace, that is, the sum of diagonal elements.

A special case of this equation is the normalization condition which we get by putting $G \equiv 1$,

$$\operatorname{Tr} \hat{\rho} = 1. \tag{6.116}$$

This equation can also be proven directly from the definition of $\hat{\rho}$ and Eq. (6.106),

$$\operatorname{Tr} \hat{\rho} = \sum_m \rho_{mm} = \frac{1}{\mathscr{N}} \sum_k \sum_m a_m{}^k a_m{}^{k*} = \frac{1}{\mathscr{N}} \sum_{k=1}^{\mathscr{N}} 1 = 1.$$

We shall restrict our discussion once again mainly to stationary ensembles. In that case $\hat{\rho}$ and the energy operator $\mathscr{H}$ must commute. Moreover, we shall assume that $\hat{\rho}$ is a function of the energy only,

$$\hat{\rho} = f(\mathscr{H}), \tag{6.117}$$

and even that we are dealing with a canonical ensemble:

$$\hat{\rho} = e^{\beta(\psi - \mathscr{H})}. \tag{6.118}$$

This last equation has a meaning only if we understand it as being an abbreviated notation for

$$\hat{\rho} = e^{\beta\psi} \sum_{n=0}^{\infty} \frac{1}{n!} [-\beta\mathscr{H}]^n. \tag{6.119}$$

Before discussing the statistical implications of our formalism, we shall consider some of the mathematical aspects of the density matrix. First of all, we note that it follows from the definition (6.107) that $\hat{\rho}$ is a Hermitian matrix:

$$\rho_{mn} = \rho_{nm}^{\star}, \tag{6.120}$$

which is normalized according to Eq. (6.116). It follows from this that (1) the diagonal elements of $\hat{\rho}$ are real, and (2) they must satisfy the relations

$$\sum_n \rho_{nn} = 1, \qquad \text{and thus } 0 \leqslant \rho_{nn} \leqslant 1. \tag{6.121}$$

From the definition (6.107) it is clear that ρ_{nn} is the (normalized) probability that φ_n is realized in the ensemble.

Let us now consider a change from the φ-representation to another representation, say, the χ-representation. Let the unitary transformation matrix S_{np} characterize this transformation, so that

$$\chi_p = \sum_n \varphi_n S_{np}, \tag{6.122}$$

$$S_{np}^{\star} = (S^{-1})_{pn}, \tag{6.123}$$

$$\sum_n S_{np}^{\star} S_{nq} = \delta_{pq}. \tag{6.124}$$

We now rewrite Eq. (6.102) as follows

$$\psi^k = \sum_n a_n{}^k \varphi_n = \sum_p b_p{}^k \chi_p, \tag{6.125}$$

and Eqs. (6.122) and (6.125) lead to

$$a_n{}^k = \sum_q S_{nq} b_q{}^k, \tag{6.126}$$

or

$$b_q{}^k = \sum_n a_n{}^k S_{nq}. \tag{6.127}$$

If the transformed density matrix is denoted by $\hat{\rho}'$, so that

$$\rho_{pq}{}' = \frac{1}{\mathcal{N}} \sum_{k=1}^{\mathcal{N}} b_p{}^k b_q{}^{k\star}, \tag{6.128}$$

we find from Eqs. (6.127) and (6.107)

$$\rho_{pq}{}' = \sum_{m,n} (S^{-1})_{pm} \rho_{mn} S_{mq}, \tag{6.129}$$

or

$$\hat{\rho}' = \hat{S}^{-1} \hat{\rho} \hat{S}, \tag{6.130}$$

which is the normal equation for the transformation of an operator. We can thus treat $\hat{\rho}$ in all respects as an ordinary quantum-mechanical operator.

Let us now assume that we have made a transformation to a representation in which $\hat{\rho}$ is diagonal. We then have

$$\rho_{mn} = \rho_m \delta_{mn}. \tag{6.131}$$

Consider now the average value of $\hat{\rho}$. Using Eqs. (6.121) and (6.131), we have

$$\bar{\rho} = \text{Tr}\,\hat{\rho}^2 = \sum_m {\rho_m}^2 \leqslant \left(\sum_m \rho_m\right)^2 = (\text{Tr}\,\hat{\rho})^2 = 1, \tag{6.132}$$

and hence, in any representation

$$\text{Tr}\,\hat{\rho}^2 \leqslant 1, \qquad \text{or,} \qquad \sum_{n,m} \rho_{nm}\rho_{mn} = \sum_{n,m} |\rho_{mn}|^2 \leqslant 1, \tag{6.133}$$

which imposes a severe limitation upon all elements of the density matrix.

In many cases we are interested in the change in time of the properties of the system. As long as we are working in the Schrödinger representation, that is, describe the system by means of a time-dependent wave function and time-independent operators, the equation of motion of $\hat{\rho}$ is given by Eq. (6.112), while we get from Eqs. (6.115), (6.112), and the cyclic property of the trace:

$$\begin{aligned} i\hbar\dot{\bar{G}} &= i\hbar\,\text{Tr}\,\dot{\hat{\rho}}\hat{G} = \text{Tr}[\mathscr{H},\hat{\rho}]_-\hat{G} = \text{Tr}(\mathscr{H}\hat{\rho}\hat{G} - \hat{\rho}\mathscr{H}\hat{G}) = \text{Tr}\,\hat{\rho}(\hat{G}\mathscr{H} - \mathscr{H}\hat{G}) \\ &= \overline{[\hat{G},\mathscr{H}]_-}. \end{aligned} \tag{6.134}$$

Note that the signs in Eqs. (6.112) and (6.134) are different. In fact, in the Heisenberg representation, where we work with time-dependent operators and time-independent state functions, the operators satisfy the equation of motion

$$i\hbar\dot{\hat{G}} = [\hat{G},\mathscr{H}]_-, \tag{6.135}$$

corresponding to the sign of Eq. (6.134).

6.2 PURE CASE AND MIXED CASE ("REINER FALL" AND "GEMISCH")

Let us briefly consider once again the reasons for introducing ensembles. As in the classical case, the necessity to introduce representative ensembles arises from the fact that our knowledge about the physical system under consideration is practically always far removed from the maximum attainable knowledge. The situation is now, however, slightly complicated because of the added statistical aspects coming from quantum mechanics. In an ideal (and practically always idealized) case we may know that our system is in a characteristic state of one of the quantum-mechanical operators. Let us assume for the sake of simplicity that this operator is the energy operator, so that we know that the system is in a stationary state.* In that case we are dealing with a "Reiner Fall," as von Neumann calls it. We shall call it a *pure case*. All the systems in the representative ensemble will possess the same wave function, namely, that corresponding to the characteristic state in question, and of the two averaging processes that are prescribed by Eq. (6.113) only one remains, namely, that of taking the quantum-mechanical average. In fact, we are in a situation where ensemble theory does not need to be used, at any rate, as long as we are not concerned with the question of how to measure the properties of the system.† This situation has no classical analogy because we are still in a position where we must use statistical considerations, even though our knowledge is maximal.‡

In general, however, the situation will not be so favorable and then we must construct a representative ensemble. Each system in the ensemble will then have a wave function which is a superposition, or mixture, of the various characteristic functions of, for instance, the energy operator, and the averages both over ψ^k and over the ensemble must be taken in order to evaluate $\bar{G}$. We are now dealing with a "Gemisch" or *mixed case*. As in the classical case, we must make certain assumptions about a priori probabilities in order to be able to construct the representative ensemble. The assumption that is made is the following one. We assume equal a priori probabilities and random a priori phases for the various nondegenerate states. Once the representative ensemble has been set up, we are in the same position as in the classical case and can then make predictions about the most probable behavior of our system by considering the average behavior of a system in the ensemble. An important point is again the smallness of the fluctuations.

At this point we want to discuss briefly two different approaches to the density matrix: the operational approach§ and the quantum-mechanical approach.¶

* It is easy to extend the discussion to include the cases where the system is in a characteristic state of two or more commuting operators.

† For a discussion of the theory of measurement and its connection with density matrix theory, see D. ter Haar, *Repts. Progr. Phys.*, **24,** 304, 1961, Section 10; and especially the treatment by London and Bauer in their monograph *La Théorie de l'Observation en Mécanique Quantique*, Paris, 1939 (English translation to be published by Princeton University Press).

‡ We have tacitly assumed that the stationary state in question is nondegenerate or, if it is a degenerate state, that it is at the same time a characteristic state of other operators (for example, the operators corresponding to the square and the z-component of the angular momentum) in such a way that there is only one wave function that can describe the state of the system.

To some extent we might say that ordinary mechanics is the classical analogy to a pure case, as was pointed out to me by H. Margenau.

§ U. Fano, *Rev. Mod. Phys.*, **29,** 74, 1957.

¶ See, for instance, R. McWeeny, *Rev. Mod. Phys.*, **32,** 335, 1960.

In the operational approach, one recognizes, on the one hand, the lack of sufficiently detailed information—which is at the basis of the introduction of representative ensembles, as we discussed in Section 5.13—and, on the other hand, one omits from the start a detailed description of those degrees of freedom in which he is not interested. A typical example is the discussion of polarization experiments. If one studies the polarization of an electron beam, the only quantity of interest is the polarization vector. For a spin-$\frac{1}{2}$ particle the polarization vector is completely determined by three numbers, and the density matrix to describe the polarization will therefore be a two-by-two matrix, which by virtue of its Hermiteicity and normalization has exactly three independent components. Fano (*loc. cit*) has shown how one can, indeed, define a suitable density matrix to describe polarization experiments and also how one can find the equation of motion for the polarization vector from the equation of motion of the density matrix (see Problem 12 in Chapter 6).

In the quantum-mechanical approach, one is also led to averaging processes that lead one to the so-called *reduced density matrices*. To fix our ideas, let us consider an atom in its ground state. Even though we know the Hamiltonian of the atom, we are not able to find the exact ground state wave function. If we wish to evaluate the ground state energy, we must have recourse to approximation methods. As, however, we usually assume that the Hamiltonian will be a sum of operators involving either one or two electrons, it is tempting to introduce suitable averages of the many-body wave function—or rather the product of this wave function and its adjoint as this product occurs in the integrals defining the energy—over all particles but two. Such averages, which determine the energy of the atom, were called by Husimi reduced density matrices. One can show that, in fact, these reduced density matrices are the density matrix, as we have introduced it, for two electrons in the coordinate representation (compare Problem 4 at the end of this chapter).

Let us now derive the necessary and sufficient condition that the density matrix describe a pure case. If we work in a representation in which the density matrix is diagonal, so that Eq. (6.131) holds, we have from Eq. (6.107)

$\rho_m = 1$

$$\rho_m = \frac{1}{\mathcal{N}} \sum_{k=1}^{\mathcal{N}} |a_m{}^k|^2, \tag{6.201}$$

and we see that ρ_m is the probability of finding an arbitrary system from the ensemble in the state characterized by φ_m. Since all $|a_m{}^k|^2$ are positive and at most equal to one,

$$0 \leqslant \rho_m \leqslant 1, \tag{6.202}$$

which also follows from Eqs. (6.121) and (6.131). The last equal sign holds only if for all k

$$|a_m{}^k|^2 = 1, \tag{6.203}$$

which entails that all systems are in the state characterized by φ_m as wave function. Hence, $\rho_m = 1$ is a sufficient condition for a pure case. If $\rho_m = 1$,

$$\rho_n{}^2 = \rho_n \tag{6.204}$$

for any n, or, in matrix notation,

$$\hat{\rho}^2 = \hat{\rho}. \tag{6.205}$$

Since from Eq. (6.202) it follows that all characteristic values of $\hat{\rho}$ are positive and smaller than or at most equal to one, Eq. (6.205) can be satisfied only if one characteristic value is equal to one and all the others are equal to zero. Hence Eq. (6.205) is a different way of expressing a sufficient condition for a pure case.

It is also possible to show that Eq. (6.205) is a necessary condition for a pure case. If we are dealing with a pure case, all $a_m{}^k$ with a given m are the same, and we have

$$\rho_{mn} = \frac{1}{\mathcal{N}} \sum_{k=1}^{\mathcal{N}} a_m{}^k a_n{}^{k\star} = a_m a_n{}^\star. \tag{6.206}$$

Using Eq. (6.106) we can then show that

$$[\rho^2]_{mn} = \sum_l \rho_{ml}\rho_{ln} = \sum_l a_m a_l{}^\star a_l a_n{}^\star = a_m a_n{}^\star = \rho_{mn}, \tag{6.207}$$

whence follows Eq. (6.205) as a necessary condition.

6.3 MACROCANONICAL PETIT ENSEMBLES IN QUANTUM STATISTICS

In order that we shall be able fully to apply quantum statistics, it is necessary to develop first the formalism of quantum-mechanical ensemble theory to the same extent as we developed that of classical ensemble theory in the preceding two chapters. In principle everything is contained in the discussion of Section 6.1, but we shall enlarge slightly on that discussion here. Let us first consider petit ensembles, and more specifically, macrocanonical petit ensembles with a density matrix

$$\hat{\rho} = e^{\beta(\psi - \hat{\mathscr{H}})}, \tag{6.301}$$

where $\hat{\rho}$ and $\hat{\mathscr{H}}$ are operators, ψ and β are c-numbers, that is, quantities that commute with any operator, and where the right-hand side of Eq.(6.301) is understood in the way discussed at the end of Section 6.1. From the normalizing equation (6.116) we get for ψ the equation

$$e^{-\beta\psi} = \mathrm{Tr}\, e^{-\beta\hat{\mathscr{H}}}, \tag{6.302}$$

where we have used the fact that ψ and β are c-numbers. Equation (6.302) is the quantum-mechanical equivalent of Eq. (5.0302). Once again we shall only indicate the proof that ψ is the (Helmholtz) free energy and that, apart from a multiplying constant, β is equal to the reciprocal of the temperature. We must bear in mind that $\hat{\mathscr{H}}$ will depend on the external parameters a_k and that the generalized forces are now operators given by the equations

$$\hat{A}_k = -\frac{\partial \hat{\mathscr{H}}}{\partial a_k}. \tag{6.303}$$

We can write Eq. (6.302) in the form

$$\mathrm{Tr}\, \hat{\rho} = e^{\beta\psi}\, \mathrm{Tr}\, e^{-\beta\hat{\mathscr{H}}} = 1, \tag{6.304}$$

and, considering two situations with slightly different values of β and the a_k, we get

$$\delta\beta\psi = \delta\beta\, \mathrm{Tr}\, \hat{\rho}\hat{\mathscr{H}} - \beta \sum_k \delta a_k\, \mathrm{Tr}\, \hat{\rho}\hat{A}_k \tag{6.305}$$

or, using Eq. (6.115) for average values, we find

$$\delta\beta\psi = \overline{\mathscr{H}}\delta\beta - \beta \sum_k \bar{A}_k \delta a_k. \tag{6.306}$$

If we write

$$\overline{\delta A} = \sum_k \bar{A}_k \delta a_k, \tag{6.307}$$

Eq. (6.306) can be written in the form

$$\delta\beta\psi = \overline{\mathscr{H}}\delta\beta - \beta\overline{\delta A}. \tag{6.308}$$

This equation is completely analogous to Eq. (5.0305), and it follows therefore that ψ can be considered to be the free energy, that

$$\beta^{-1} - kT, \tag{6.309}$$

and that, apart from a possible additive constant, the entropy S is given by the equation

$$S = k\beta(\overline{\mathscr{H}} - \psi). \tag{6.310}$$

Comparing Eqs. (6.302) and (5.0311), we see that $\mathrm{Tr}\, e^{-\beta\hat{\mathscr{H}}}$ now takes the place of the partition function of classical theory.

Before we go over to the quantum-mechanical grand ensembles we may apply the formalism of the quantum-mechanical macrocanonical ensemble to the case of a system of independent particles, for the moment neglecting exclusion effects.

The Hamiltonian of a system consisting of N independent particles is the sum of the Hamiltonians of the particles,

$$\mathscr{H} = \sum_{i=1}^{N} \mathscr{H}_i, \tag{6.311}$$

where $\mathscr{H}_i$ is the energy operator of the ith particle.

In order to simplify our calculations we shall assume that the φ_n are the characteristic functions of the Hamiltonian. This brings the energy matrix on diagonal form,

$$H_{kl} = E_k \delta_{kl}, \tag{6.312}$$

and the density matrix is hence also diagonal,

$$\rho_{kl} = \delta_{kl} e^{\beta(\psi - E_k)}, \tag{6.313}$$

so that we get from Eq. (7.302)

$$e^{-\beta\psi} = \sum_k \gamma_k e^{-\beta E_k}, \tag{6.314}$$

where γ_k is the multiplicity of the energy level E_k. We see that Eq. (6.314) resembles closely Eq. (3.401) with one important difference, namely, that in Eq. (3.401) the ε_n were the energy levels of single molecules, whereas now the E_k are the energy levels of the whole system. This corresponds to the transition from μ-space to Γ-space. Since the particles in the system are independent, we have (compare Eq. (6.311))

$$E_k = \sum_{i=1}^{N} \varepsilon_i. \tag{6.315}$$

In Eq. (6.315) the summation is over all particles in the system. We can replace this by a summation over all energy levels, by taking together all particles with the same energy. If there are n_1 particles with energy ε_1, n_2 with energy ε_2, $\cdots$, we can clearly write Eq. (6.315) in the form

$$E_k = \sum_{\Sigma n_s = N} n_s \varepsilon_s, \tag{6.316}$$

where all ε_s are now different.

The degeneracy γ_k arises from two factors. The first one is the fact that the levels ε_s may be degenerate, say, with a degeneracy g_s. The second factor comes from the degeneracy which derives from the fact that N particles can be divided into groups containing, respectively, $n_1, n_2, \cdots, n_s, \cdots$ in several ways. We obtain

$$\gamma_k = \frac{N!}{\prod_s n_s!} \prod_s g_s^{n_s}, \tag{6.317}$$

and substituting Eqs. (6.316) and (6.317) into Eq. (6.314), we get

$$e^{-\beta\psi} = \sum_{n_s=N} N!\left(\prod_s \frac{g_s^{n_s}}{n_s!}\right)e^{-\beta\Sigma n_s\varepsilon_s},$$

or, using Eq. (1.704),

$$e^{-\beta\psi} = \left[\sum_s g_s e^{-\beta\varepsilon_s}\right]^N. \tag{6.318}$$

The expression inside the square brackets is the partition function for one particle. In Section 4.6 we calculated this partition function for a system of spin-less point particles under the assumption that the energy levels were lying so densely that the sum could be replaced by an integral. There is no point in writing down the result here, since we see immediately from Eq. (6.318) that for a system of N independent particles the free energy is equal to N times the free energy of one particle, as would be expected. We are thus back to the situation encountered in Chapter 3 and no new features will be found.

As in the case of the classical macrocanonical ensembles it is possible to calculate the fluctuations in energy and generalized forces. The formulas, which are derived by the same methods as before, are entirely the same as those found in Section 5.4, the only difference being in the actual value of ψ. We refer therefore to the discussion of that section.

6.4 CANONICAL GRAND ENSEMBLES IN QUANTUM STATISTICS

We come now to one of the most powerful tools of modern statistical mechanics, the quantum-mechanical canonical grand ensembles. In Vol. II we shall have ample opportunity to apply this tool.

In the preceding chapter we discussed the classical grand ensembles and saw that in many cases these must be preferred to the petit ensembles. It is thus desirable to develop their quantum-mechanical counterparts. Since in quantum mechanics the indistinguishability of identical particles finds its ready expression in the exclusion principle, we shall now have no reason to distinguish between specific and generic densities. This makes the quantum-mechanical ensemble theory simpler than the classical one. Moreover, due to the statistical aspects of quantum mechanics the physical interpretation of the quantum-mechanical ensemble theory will be much more natural.

We shall assume that our system is confined to a finite volume so that the energy levels will form a discrete spectrum. The energy levels will, of course, depend on the ν_i, the number of particles of the various kinds that may be present in the system. We shall write for the energy levels $E(\nu_i;n)$ where n numbers the levels.

The density matrix of a stationary ensemble must now commute not only with $\hat{\mathscr{H}}$ but also with the number operators $\hat{\nu}_i$. These $\hat{\nu}_i$ are the well-known Jordan-Klein and Jordan-Wigner matrices from quantum mechanics.* Their characteristic values are the integers ν_i.

We shall discuss only canonical grand ensembles. Their density operator is given by the equation

$$\hat{\rho} = e^{-q+\sum_i \alpha_i \hat{\nu}_i - \beta \hat{\mathscr{H}}}, \tag{6.401}$$

where q, β, and the α_i are c-numbers. The grand potential q is determined by the normalization condition, and we have

$$e^q = \mathrm{Tr}\, e^{\sum_i \alpha_i \hat{\nu}_i - \beta \hat{\mathscr{H}}}, \tag{6.402}$$

where in Eqs. (6.401) and (6.402) the exponentials are defined by the power series.

If we choose for the φ_n the characteristic functions of both $\hat{\mathscr{H}}$ and all the $\hat{\nu}$, Eq. (6.402) becomes

$$e^q = \sum_{\nu_i, n} e^{\sum \alpha_i \nu_i - \beta E(\nu_i, n)}, \tag{6.403}$$

which is the quantum-mechanical equivalent of Eq. (5.0907). Since $\hat{\mathscr{H}}$, and thus also the $E(\nu_i;\, n)$, depend on the external parameters, we see that q is a function of β and the α_i and a_k, as in the classical case,

$$q = q(\beta, \alpha_i, a_k). \tag{6.404}$$

The average $\langle G \rangle$ of a quantity G over the canonical grand ensemble is given by Eq. (6.113) which can be put in the form

$$\langle G \rangle = \sum_{\nu_i} \sum_{n} G(\nu_i;n)\rho(\nu_i;n), \tag{6.405}$$

where $\rho(\nu_i;n)$ is the diagonal element of the density matrix which, due to our choice of the φ_n, is on diagonal form, and where $G(\nu_i;n)$ is the quantum-mechanical expectation value of G for the state characterized by the ν_i and by n,

$$G(\nu_i;n) = \int \psi^*(\nu_i;n)\hat{G}\psi(\nu_i;n)\, d\tau. \tag{6.406}$$

By using Eq. (6.401) for $\hat{\rho}$, Eq. (6.405) can be written in the form

$$\langle G \rangle = e^{-q} \sum_{\nu_i} e^{\sum \alpha_i \nu_i} \sum_{n} G(\nu_i;n) e^{-\beta E(\nu_i;n)}. \tag{6.407}$$

In order to see the physical meaning of β and the α_i we again consider two situations with slightly different values of β, α_i, and a_k. We shall give

* H. A. Kramers, *Quantum Mechanics*, Amsterdam, 1957, Section 72.

only the relevant results without their derivations, since the calculations are exactly the same as those given on numerous previous occasions. We have for the variation of q the equation

$$\delta q = \sum_i \langle \nu_i \rangle \delta\alpha_i - \langle \varepsilon \rangle \delta\beta + \beta \sum_k \langle A_k \rangle \delta a_k. \tag{6.408}$$

The entropy S and (Helmholtz) free energy F are given by the equations

$$\frac{S}{k} = q - \sum_i \alpha_i \langle \nu_i \rangle + \beta \langle \varepsilon \rangle, \tag{6.409}$$

$$F = \langle \varepsilon \rangle - ST = -\frac{q}{\beta} + \frac{\sum_i \alpha_i \langle \nu_i \rangle}{\beta}. \tag{6.410}$$

Introducing

$$\mu_i = \frac{\alpha_i}{\beta}, \tag{6.411}$$

we can show that the μ_i are the partial free energies or the partial thermal potentials,

$$\mu_i = \left(\frac{\partial F}{\partial \langle \nu_i \rangle} \right)_{\beta, a_k} = \left(\frac{\partial G}{\partial \langle \nu_i \rangle} \right)_{\beta, P}, \tag{6.412}$$

where $G = F + PV$ is the thermal potential, or Gibbs free energy.

The temperature T is related to β by the familiar equation

$$\beta^{-1} = kT, \tag{6.413}$$

and in the case of a homogeneous system we have

$$q = \frac{PV}{kT}. \tag{6.414}$$

The grand potential gives us all thermodynamic quantities and we have, for instance, from Eq. (6.408)

$$\langle \nu_i \rangle = \frac{\partial q}{\partial \alpha_i}, \tag{6.415}$$

$$\langle \varepsilon \rangle = -\frac{\partial q}{\partial \beta}, \tag{6.416}$$

$$\langle A_k \rangle = \frac{1}{\beta} \frac{\partial q}{\partial a_k}, \tag{6.417}$$

$$P = \frac{1}{\beta} \frac{\partial q}{\partial V}. \tag{6.418}$$

One can calculate the fluctuations similarly. The easiest way is to use the normalization condition and take the derivative with respect to two of the quantities β, α_i, or a_k. In this way we get, for instance,

$$\langle(\nu_i - \langle\nu_i\rangle)^2\rangle = \frac{\partial^2 q}{\partial {\alpha_i}^2}, \qquad \langle(\varepsilon - \langle\varepsilon\rangle)^2\rangle = \frac{\partial^2 q}{\partial \beta^2}, \cdots \tag{6.419}$$

In general the dispersions will be small if the $\langle\nu_i\rangle$ are large, and in such cases we are dealing with ensembles with practically completely determined values of E and the ν_i.

Before applying the formulas of this section to a perfect gas and investigating the consequences of the exclusion principle, we shall briefly discuss a few aspects of quantum-mechanical ensemble theory. Both points that we shall examine also enter classical ensemble theory, be it sometimes in a slightly different way, and we have discussed them before in connection with the classical theory.

The first point is the question of fluctuations. In a phenomenological thermodynamical theory all quantities such as temperature and energy, thermal potentials and numbers of particles of a given kind, volume and pressure, and so on, have well-defined values. However, if a canonical grand ensemble is considered, only the first of each of these pairs has a sharply defined value, the second quantity in each pair fluctuating around an average value. In some cases these fluctuations enter because our very inadequate knowledge about the physical system forces us to use a representative ensemble and we cannot expect to obtain completely defined predictions from very incomplete data. Sometimes, however, the fluctuations may not only present the uncertainty of our predictions, but also correspond to a physical phenomenon. If we consider, for instance, our system embedded in a large thermostat which at the same time can exchange particles with the system, the fluctuations in energy and numbers of particles will occur and under favorable circumstances even be measurable.* In most cases these fluctuations will be extremely small and actually too small to be observable. This is important, since we can then safely apply the theory to thermodynamical systems.

It might be feared that some fluctuations, for instance, in the pressure or in other forces might be large (compare similar considerations in Section 5.4). There are two reasons why these fluctuations are not important. First of all, the large fluctuations often arise only due to modifications introduced to simplify calculations, such as, for instance, idealized walls and rigid molecules. Secondly, any measurement of a physical quantity will require a finite time interval so that we are not really concerned with $\langle G\rangle$, but with

* It is possible to consider our system to be one of an ensemble, the other members of the ensemble forming the temperature and particle bath. We do not wish to stress this point of view, since it does not fully acknowledge the importance of representative ensembles.

the time average of $\langle G\rangle$. From quantum mechanics it follows, however, that fluctuations in a time average are extremely small and decrease rapidly with increasing length of the period over which the average is taken.*

It is possible to take a point of view complementary to the one taken when one considers canonical grand ensembles. One then considers systems with a constant energy and constant numbers of particles. Such systems are represented by microcanonical petit ensembles which it is possible to consider in quantum mechanics, though we shall not do so. If the temperature or the thermal potentials are measured, they will show fluctuations of the same order of magnitude as the fluctuations in energy and particle numbers in a canonical grand ensemble. As long as the number of particles in the system is very large, all fluctuations will be negligible and the results obtained by using microcanonical petit ensembles are the same as those obtained by using canonical grand ensembles.

The second point that we wish to discuss at this juncture is the entropy. Our considerations will be similar to those of Section 5.5. Let dZ be the total number of energy levels for which the energy lies between ε and $\varepsilon + d\varepsilon$ and for which the ν_i have values between ν_i and $\nu_i + d\nu_i$. If we write

$$dZ = W(\nu_i;\varepsilon)d\varepsilon \prod_i d\nu_i, \tag{6.420}$$

the normalization equation can be written in the form

$$\int e^{-q+\sum_i \alpha_i \nu_i - \beta\varepsilon}\, W(\nu_i;\varepsilon)\, d\varepsilon \prod_i d\nu_i = 1, \tag{6.421}$$

where, if necessary, the integral on the left-hand side must be treated as a Stieltjes integral.

Consider now a situation where the $\langle \nu_i\rangle$ are so large that the fluctuations are negligible. In that case the values of ε and ν_i for which the integrand in Eq. (6.421) is maximum will be practically equal to $\langle\varepsilon\rangle$ and $\langle \nu_i\rangle$ and we can write Eq. (6.421) in the form

$$\left[e^{-q+\sum_i \alpha_i \nu_i - \beta\varepsilon}\, W(\nu_i;\varepsilon)\right]_{\substack{\nu_i=\langle\nu_i\rangle\\ \varepsilon=\langle\varepsilon\rangle}} \Delta\varepsilon \prod_i \Delta\nu_i = 1, \tag{6.422}$$

where $\Delta\varepsilon$ and $\Delta\nu_i$ are the half-widths of the fluctuations, that is,

$$(\Delta\varepsilon)^2 = C\langle(\varepsilon - \langle\varepsilon\rangle)^2\rangle, \qquad (\Delta\nu_i)^2 = C_i\langle(\nu_i - \langle\nu_i\rangle)^2\rangle, \tag{6.423}$$

where C and the C_i are constants of the order of magnitude of unity, and where we have approximately

$$\Delta\varepsilon \sim \frac{\langle\varepsilon\rangle}{\sqrt{\sum\langle\nu_i\rangle}}, \qquad \Delta\nu_i \sim \sqrt{\langle\nu_i\rangle}. \tag{6.424}$$

* Compare H. A. Kramers, *Quantum Mechanics*, Amsterdam, 1957, Section 52.

Taking the logarithm of both sides of Eq. (6.422), we have

$$q - \sum_i \langle \nu_i \rangle \alpha_i + \beta \langle \varepsilon \rangle = \ln\left[W(\langle \nu_i \rangle ; \langle \varepsilon \rangle) \Delta\varepsilon \prod_i \Delta \nu_i \right]. \qquad (6.425)$$

The left-hand side of this equation is equal to S/k, as can be seen from Eq. (6.409), and we have thus

$$S = k \ln\left[W(\langle \nu_i \rangle ; \langle \varepsilon \rangle) \Delta\varepsilon \prod_i \Delta \nu_i \right]. \qquad (6.426)$$

From the general relation between entropy and probability it follows that $W(\nu_i;\varepsilon)$ can be considered to be the probability of finding certain values of ε and ν_i, or that $W(\nu_i,\varepsilon)\Delta\varepsilon \,\Pi\, \Delta\nu_i$ is the probability of finding ε and ν_i within certain limits. This corresponds closely to the definition of W. Indeed, $W(\nu_i;\varepsilon)\Delta\varepsilon \,\Pi\, \Delta\nu_i$ was defined as being the number of energy states corresponding to ε and ν_i within given limits, and this together with the fact that earlier we mentioned that all stationary states should have equal a priori probabilities shows that, indeed, $W(\nu_i;\varepsilon)\Delta\varepsilon \,\Pi\, \Delta\nu_i$ can be considered to be the probability for the energy and the numbers of particles lying inside given intervals. We can express this slightly differently by stating that the entropy measures our *lack of knowledge* or *lack of detailed information*, since $W(\nu_i;\varepsilon)\Delta\varepsilon \,\Pi\, \Delta\nu_i$ gives us the number of possibilities that are left, once we know that ε and ν_i are confined to certain given intervals. A similar situation is present in the classical case (compare Eq. (5.0524); $(\partial\Omega/\partial\varepsilon)\Delta$ gives us a measure for the volume in Γ-space in which the representative point can be found once the energy is stated to lie within a given interval).

The relationship between entropy and lack of information has led many authors, notably Shannon, to introduce "entropy" as a measure for the information transmitted by cables and so on, and in this way entropy has figured largely in recent discussions in information theory. It must be stressed here that the entropy introduced in information theory is *not* a thermodynamical quantity and that the use of the same term is rather misleading. It was probably introduced because of a rather loose use of the term "information."

In this connection we may briefly discuss *Maxwell's demon*. In 1871 Maxwell introduced his famous demon, "a being whose faculties are so sharpened that he can follow every molecule in its course, and would be able to do what is at present impossible to us. . . . Let us suppose that a vessel is divided into two portions A and B by a division in which there is a small hole, and that a being who can see the individual molecules opens and closes this hole, so as to allow only the swifter molecules to pass from A to B, and only the slower ones to pass from B to A. He will, thus, without expenditure of work raise the temperature of B and lower that of A, in contradiction to the second law of thermodynamics."

Maxwell's demon has been widely discussed and various authors have set out to show that various attempts to circumvent the second law by using the demon are bound to fail.* Although their discussions differ in some respects, they have a few points in common. The first point is the observation that one should take the demon to be part of the total system, and then one must consider the total entropy of the original system and the demon. The second point which was most clearly developed for the first time by Szilard† is that the demon, in order to be able to operate the trap-door through which the molecules pass, must receive information. Its own entropy increases therefore; it is now the question whether the increase of the demon's entropy is smaller or larger than the decrease of the entropy of the gas. Both Szilard and Brillouin‡ consider possible arrangements and show that in those cases the net change of entropy is positive. Szilard analyzes the problem very thoroughly and shows that one can describe a generalized Maxwell's demon as follows. By some means an operation on a system is determined by the result of a measurement on the system that immediately precedes the operation. In Maxwell's original scheme the operation was the opening of the trap-door and the measurement was the determination of the velocity of an approaching molecule. The result of the operation will be a decrease of entropy, but the preceding measurement will be accompanied by an increase in entropy, and once again one must consider the balance.

Wiener takes a simpler point of view.§ He considers the situation, where the demon acts, as a metastable state and writes: "In the long run, the Maxwell demon is itself subject to a random motion corresponding to the temperature of its environment and it receives a large number of small impressions until it falls into 'a certain vertigo' and is incapable of clear perceptions. In fact, it ceases to act as a Maxwell demon." This point of view is probably too simplified and we prefer that of Szilard's and refer the reader to his paper for a more extensive discussion.

In Section 5.5 we saw that there were several ways of defining the entropy, which is also the case here. One can prove that, because of the very steep increase of $W(v_i;\varepsilon)$ with both ε and v_i, the entropy can also be defined by the equation

$$S = k \ln V, \tag{6.427}$$

where

$$V = \int_{\varepsilon_{\min}}^{\langle\varepsilon\rangle} \int_{v_i=0}^{\langle v_i\rangle} \cdots \int W(v_i;\varepsilon)\, d\varepsilon \prod_i dv_i. \tag{6.428}$$

Equation (6.427) corresponds to Eq. (5.0535) of the classical theory.

* Let us briefly point out that these proofs of the general validity of the second law are different from the proofs of the general validity of the Heisenberg relations in quantum mechanics. In both cases one sets up an idealized experiment which should be suited to get around the theoretical limitations, but while in quantum mechanics one proves that it is *never* possible to break the limitations, in the thermodynamical case one can prove only that *on the average* it is not possible to violate the second law.

† L. Szilard, *Z. Physik*, **53,** 840, 1929.

‡ L. Brillouin, *J. Appl. Phys*., **22,** 334, 1951.

§ N. Wiener, *Cybernetics*, New York, 1948, p. 72; see also Brillouin, *loc. cit*., Section I.

We have included the factor $\Delta\varepsilon\Pi\Delta v_i$ in Eq. (6.426) in order that the expression under the logarithm would be dimensionless. Actually, Eq. (6.427) is probably an easier one to use for calculations, since otherwise $\Delta\varepsilon$ and the Δv_i must be computed first. In Chapter 5 we were not concerned with the actual value of the entropy and therefore did not pay any attention to the fact that neither $\partial\Omega/\partial\varepsilon$ nor Ω is dimensionless. We could have corrected this by writing instead of Eqs. (5.0525) and (5.0535)

$$S = k \ln\left[\frac{\partial\Omega}{\partial\varepsilon} \cdot \Delta \cdot h^{-3sN}\right] \tag{6.429}$$

and

$$S = k \ln(\Omega \cdot h^{-3sN}), \tag{6.430}$$

where we can see now even more clearly the correspondence between the classical and the quantum-mechanical case. In Eq. (6.427), for instance, V is the total number of states with $\varepsilon < \langle\varepsilon\rangle$ and $v_i < \langle v_i \rangle$, while in equation (6.430), which relates to a petit ensemble, Ω is the total volume in Γ-space with $\varepsilon < \bar{\varepsilon}$. The factor h^{-3sN} arises from the fact that each nondegenerate state corresponds to a volume h^{3sN} in Γ-space, and we see that, indeed, the two Eqs. (6.427) and (6.430) will give the same result in the limiting case of very high temperatures or very low densities, when all quantum-mechanical effects are negligible.

6.5 THE *H*-THEOREM IN QUANTUM STATISTICS

In Sections 5.13 and 5.14 we discussed the question of representative ensembles and the approach to equilibrium from the point of view of classical ensemble theory. We shall now consider similarly these kinds of problems from the point of view of quantum-mechanical ensemble theory. It is convenient for our discussion to introduce once again the index of probability which now is an operator defined by the equation

$$\hat{\rho} = e^{\hat{\eta}}. \tag{6.501}$$

From Eq. (6.501) we have once again for equilibrium ensembles the relation

$$S = -k\langle\eta\rangle \tag{6.502}$$

(compare Eqs. (6.310) and (6.409)), which is the quantum-mechanical counterpart of Eq. (5.1107) with the difference that now $\hat{\eta}$ is an operator. As in the relevant discussion in Chapter 5 we shall consider grand ensembles only, leaving the reader to make the necessary adjustments needed for the discussion of petit ensembles, even though when suitable we give results pertaining to petit ensembles.

We remind ourselves that the discussion in Chapter 5, which showed that the microcanonical, the macrocanonical, or the canonical grand ensemble was the equilibrium ensemble, depending on the conditions, fell into three parts. First of all, we showed that $\langle\eta\rangle$ was a minimum for the above-mentioned ensembles. Secondly, we showed that $d\langle\eta\rangle/dt \leqslant 0$, and thirdly, we showed that from the condition that $\langle\eta\rangle$ be a minimum and that the other conditions hold, the canonical ensemble densities follow as the representative ensemble densities, proving that the representative ensembles are, indeed, the equilibrium ones. We leave the last part to the reader as a problem, and shall only discuss briefly here the proofs that $\langle\eta\rangle$ is a minimum for the canonical density matrices and that $d\langle\eta\rangle/dt$ is nonpositive. As in Chapter 5, we shall give the proof only for the case where both the average energy and the average number of particles are given, while we shall also restrict our discussion again to the case where there is only one kind of particle present. The conditions that the average energy, E, and the average number of particles, N, are given are expressed by the equations

$$\operatorname{Tr} \hat{\rho}\hat{\mathscr{H}} = E, \tag{6.503}$$

$$\operatorname{Tr} \hat{\rho}\hat{\nu} = N. \tag{6.504}$$

We must prove now that $\langle\eta\rangle$ is a minimum, if $\hat{\rho}$ satisfies the equation

$$\hat{\rho} = e^{-q+\alpha\hat{\nu}-\beta\hat{\mathscr{H}}}. \tag{6.505}$$

To prove this, we consider two ensembles with density matrices $\hat{\rho}_1$ and $\hat{\rho}_2$, with $\hat{\rho}_1$ given by Eq. (6.505) and $\hat{\rho}_2$ by the equation

$$\hat{\rho}_2 = \hat{\rho}_1\, e^{\widehat{\Delta\eta}}. \tag{6.506}$$

We now consider the quantity $\langle\eta\rangle_2 - \langle\eta\rangle_1$ which is given by the equation

$$\langle\eta\rangle_2 - \langle\eta\rangle_1 = \operatorname{Tr}(\hat{\eta}_2 e^{\hat{\eta}_2} - \hat{\eta}_1 e^{\hat{\eta}_1}). \tag{6.507}$$

The right-hand side of Eq. (6.507) can be transformed as follows:

$$\begin{aligned}\operatorname{Tr}(\hat{\eta}_2 e^{\hat{\eta}_2} - \hat{\eta}_1 e^{\hat{\eta}_1}) &= \operatorname{Tr}[(\hat{\eta}_2 - \hat{\eta}_1)e^{\hat{\eta}_2}] + \operatorname{Tr}[\hat{\eta}_1(\hat{\rho}_2 - \hat{\rho}_1)] \\ &= \operatorname{Tr}[(\hat{\eta}_2 - \hat{\eta}_1)e^{\hat{\eta}_2}] + \operatorname{Tr}[(-q + \alpha\hat{\nu} - \beta\hat{\mathscr{H}})(\hat{\rho}_2 - \hat{\rho}_1)] \\ &- \operatorname{Tr}[(\hat{\eta}_2 - \hat{\eta}_1)e^{\hat{\eta}_2}],\end{aligned} \tag{6.508}$$

where we have used the fact that both $\hat{\rho}_1$ and $\hat{\rho}_2$ satisfy Eqs. (6.503), (6.504), and the normalization condition (6.116). It thus follows that

$$\langle\eta\rangle_2 - \langle\eta\rangle_1 = \operatorname{Tr}[(\hat{\eta}_2 - \hat{\eta}_1)e^{\hat{\eta}_2}]. \tag{6.509}$$

We now must prove that the right-hand side of Eq. (6.509) is positive, except when $\widehat{\Delta\eta} = 0$. We first of all note that for any Hermitean operator $\hat{A}$ we have* (see Problem 26 at the end of the chapter).

$$[e^{\hat{A}}]_{kk} \geqslant e^{A_{kk}}, \tag{6.510}$$

the equal sign holding only when $\hat{A}$ is on diagonal form.

We now choose a representation in which $\hat{\eta}_2$ is diagonal; then, using the

* R. Peierls, *Phys. Rev.*, **54,** 918, 1938.

normalization condition for $\hat{\rho}_1$ and $\hat{\rho}_2$, Eq. (6.510), and Eq. (5.1407) we find

$$\begin{aligned}\mathrm{Tr}\,[(\hat{\eta}_2-\hat{\eta}_1)e^{\hat{\eta}_2} &= \mathrm{Tr}\,[(\hat{\eta}_2-\hat{\eta}_1)e^{\hat{\eta}_2}-e^{\hat{\eta}_2}+e^{\hat{\eta}_1}]\\ &=\sum_k [(\eta_{2k}-\eta_{1kk})e^{\eta_{2k}}-e^{\eta_{2k}}+(e^{\eta_1})_{kk}]\\ &\geqslant \sum_k [(\eta_{2k}-\eta_{1kk})e^{\eta_{2k}}-e^{\eta_{2k}}+e^{\eta_{1kk}}]\\ &=\sum_k e^{\eta_{1kk}}[(\eta_{2k}-\eta_{1kk}-1)e^{\eta_{2k}-\eta_{1kk}}+1]\geqslant 0, \end{aligned} \tag{6.511}$$

which concludes our proof.

As all averages are invariant under unitary transformations and as we can transform from one instant, t', to another, t'', by a unitary transformation, it follows that, as in the classical case, $\langle\eta\rangle' = \langle\eta\rangle''$.

Now we again introduce a coarse-grained density, or rather a coarse-grained density matrix $\hat{\mathrm{P}}$ (read again "capital rho"). We choose a representation in which $\hat{\mathscr{H}}$ and $\hat{\nu}$ are diagonal. Furthermore, we shall divide the stationary states into groups. Let Z_i be the number of levels in the ith group.* We now define $\hat{\mathrm{P}}$ by its matrix elements in the chosen representation, as follows:

$$\mathrm{P}_{kl} = \frac{\delta_{kl}}{Z_i}\sum_j \rho_{jj}, \tag{6.512}$$

where the energy level E_k belongs to the ith group and where the summation is over all the states in the ith group. We see that $Z_i\mathrm{P}_{kk}$ is the probability of finding a system of the ensemble in a state belonging to the ith group. Equation (6.512) defines the coarse-grained density matrix in a particular representation; its matrix elements in any other representation follow by the usual transformation rules.

From Eq. (6.512) and the normalization of $\hat{\rho}$ it follows that $\hat{\mathrm{P}}$ is also normalized,

$$\mathrm{Tr}\,\hat{\mathrm{P}} = 1. \tag{6.513}$$

We now define a quantity $\hat{\mathrm{H}}$ (read: "capital eta") by the equation

$$\hat{\mathrm{H}} = \ln \hat{\mathrm{P}}, \tag{6.514}$$

and we have

$$\langle \mathrm{H}\rangle = \mathrm{Tr}\,\hat{\rho}\ln\hat{\mathrm{P}} = \sum_k \rho_{kk}\ln \mathrm{P}_{kk} = \sum_k \mathrm{P}_{kk}\ln \mathrm{P}_{kk} = \mathrm{Tr}\,\hat{\mathrm{P}}\ln\hat{\mathrm{P}}, \tag{6.515}$$

where we have used Eq. (6.512) and where we have worked throughout in the representation in which Eq. (6.512) was written.

If, as in the classical case, we assume the coarse graining to be in accordance

* We note the difference between the grouping here and that done in Chapter 4. Now we group energy levels of the whole system; then we grouped energy levels of individual particles.

with experimental limitations, which means that we assume that by means of our available observational methods we can discriminate between the different groups of energy levels, but not within them, we can at a time t' determine through our observations the coarse-grained density matrix $\hat{\mathrm{P}}$. The fine-grained density matrix $\hat{\rho}$ can then only be chosen to be given by the equation (we use here the assumption of equal a priori probabilities for non-degenerate states; compare the discussion of Eq. (5.1420)):

$$\hat{\rho}' = \hat{\mathrm{P}}', \tag{6.516}$$

while we have for $\langle \mathrm{H} \rangle'$:

$$\langle \mathrm{H} \rangle' = \mathrm{Tr}\, \hat{\mathrm{P}}' \ln \hat{\mathrm{P}}' = \mathrm{Tr}\, \hat{\rho}' \ln \hat{\rho}'. \tag{6.517}$$

If the situation at t' did not correspond to an equilibrium situation so that $\hat{\rho}'$ (or $\hat{\mathrm{P}}'$) did not correspond to a canonical density matrix, we cannot expect Eq. (6.516) to hold at a later time t'', but we would have

$$\hat{\rho}'' \neq \hat{\mathrm{P}}'', \tag{6.518}$$

from which follows

$$\langle \mathrm{H} \rangle'' < \langle \mathrm{H} \rangle'. \tag{6.519}$$

Equation 6.519 can be proved as follows. We first of all rewrite the expression for $\langle \mathrm{H} \rangle' - \langle \mathrm{H} \rangle''$:

$$\begin{aligned} \langle \mathrm{H}' \rangle' - \langle \mathrm{H} \rangle'' &= \mathrm{Tr}\, \hat{\mathrm{P}}' \ln \hat{\mathrm{P}}' - \mathrm{Tr}\, \hat{\mathrm{P}}'' \ln \hat{\mathrm{P}}'' \\ &= \mathrm{Tr}\, \hat{\rho}' \ln \hat{\rho}' - \mathrm{Tr}\, \hat{\rho}'' \ln \hat{\mathrm{P}}'' \\ &= \sum_k \rho_{kk}' \ln \rho_{kk}' - \sum_n \rho_{nn}'' \ln \mathrm{P}_{nn}''. \end{aligned} \tag{6.520}$$

We now add to the right-hand side of Eq. (6.520) first of all the expression

$$\sum_k \rho_{nn}'' \ln \rho_{nn}'' - \sum_k \rho_{kk}' \ln \rho_{kk}', \tag{6.521}$$

which by Klein's lemma* is never positive and only vanishes when $\hat{\rho}''$ is a diagonal matrix (see Problem 27 at the end of this chapter), and secondly the expression $\mathrm{Tr}\, \hat{\mathrm{P}}'' - \mathrm{Tr}\, \hat{\rho}''$ which vanishes as both $\hat{\rho}''$ and $\hat{\mathrm{P}}''$ are normalized. We thus get

$$\langle \mathrm{H} \rangle' - \langle \mathrm{H} \rangle'' = \sum_n (\rho_{nn}'' \ln \rho_{nn}'' - \rho_{nn}'' \ln \mathrm{P}_{nn}'' - \rho_{nn}'' + \mathrm{P}_{nn}'') \geqslant 0, \tag{6.522}$$

where the last inequality follows from Eq. (5.1407) after the substitution $x = \ln (\rho_{nn}''/\mathrm{P}_{nn}'')$ and Eq. (6.121) from which follows that $P_{nn}'' \geqslant 0$.

We want to discuss also a different way of looking at the behavior of $\langle \mathrm{H} \rangle$. First of all, we note that the probability P_i of finding a system in the ensemble in a state of the ith group is given by

$$P_i = Z_i \mathrm{P}_{kk}, \tag{6.523}$$

where P_{kk} is one of the Z_i equal diagonal elements belonging to the ith group.

* O. Klein, *Z. Physik*, **72**, 767, 1931.

We can express $\langle \mathrm{H} \rangle$ in terms of the P_i, using Eq. (6.523) as follows:

$$\langle \mathrm{H} \rangle = \sum_i \sum_k \frac{P_i}{Z_i} \ln \frac{P_i}{Z_i} = \sum_i P_i \ln \frac{P_i}{Z_i}, \tag{6.524}$$

where the summation over i is over all groups and the summation over k over the Z_i members of the ith group.

As $\sum P_i = 1$ (see Eq. (6.513)) so that $\sum dP_i/dt = 0$, we have for the time derivative of $\langle \mathrm{H} \rangle$

$$\frac{d\langle \mathrm{H} \rangle}{dt} = \sum_i (\ln P_i - \ln Z_i) \frac{dP_i}{dt}. \tag{6.525}$$

We now need an expression for the average transition probability $N_{i \to j}$ from the ith to the jth group. This is given by the equation*

$$N_{i \to j} = A_{i \to j} Z_j P_i. \tag{6.526}$$

If we accept the assumptions of equal a priori probabilities and random a priori phases for the various states, it can be shown† that, if we exclude the case where a magnetic field is present, as we shall do,

$$A_{i \to j} = A_{j \to i}. \tag{6.527}$$

Using this equation it follows from Eq. (6.526) that

$$\frac{dP_i}{dt} = \sum_j A_{i \to j} [Z_i P_j - Z_j P_i], \tag{6.528}$$

and we finally get after a few manipulations (substituting Eq. (6.528) into Eq. (6.525) and taking the arithmetic mean of the ensuing expression and the expression obtained by interchanging i and j)

$$\frac{d\langle \mathrm{H} \rangle}{dt} = \frac{1}{2} \sum_{i,j} A_{i \to j} [P_j Z_i - P_i Z_j] \ln \frac{P_i Z_j}{P_j Z_i}. \tag{6.529}$$

The right-hand side of Eq. (6.529) is negative unless $P_i Z_j = P_j Z_i$ (compare the discussion of Eq. (1.504)), as the $A_{i \to j}$ are never negative.

We note that $d\langle \mathrm{H} \rangle/dt$ vanishes only when $P_i Z_j = P_j Z_i$, that is, when

$$N_{i \to j} = N_{j \to i}, \tag{6.530}$$

and we see that at equilibrium there are as many transitions from the ith to the jth group as the other way round. This is an example of the so-called *principle of detailed balancing.*

The principle of detailed balancing states that at equilibrium the number of processes destroying situation A and creating situation B will be equal to the number of processes producing A and destroying B. This principle is valid in a

* See R. C. Tolman, *The Principles of Statistical Mechanics*, Oxford, 1938, Section 99, or D. ter Haar, *Rev. Mod. Phys.*, **27**, 289, 1955, Appendix V.

† R. C. Tolman, *ibid.*

great many cases, but not universally so. If it is not valid, one can still prove an *H*-theorem by considering cycles of transitions. The importance of the principle of detailed balancing lies in its many applications. Often one can calculate the rate at which a process goes in one direction more easily than the rate at which the reverse process takes place. The principle has also been used in nuclear physics, for instance, in determining the spin of the pion.

Equation (6.530), expressing the principle of detailed balancing, was a consequence of the vanishing of $d\langle \text{H} \rangle / dt$ and of Eq. (6.527) which is an example of *microscopic reversibility.*

6.6 THE PERFECT BOLTZMANN GAS

As in Chapters 1, 2, 4, and 5, we shall apply our formalism to the case of a perfect gas. We shall apply here the method of the canonical grand ensembles. We saw in Chapter 4 that all gases will behave like a Boltzmann gas at sufficiently high temperatures. We shall therefore first consider a perfect gas without taking into account exclusion effects, even though this is an apparent contradiction, since the exclusion principle is an integral part of quantum mechanics. This can be justified, however, by observing that the formulas obtained in the present section are, indeed, the limiting cases of the formulas obtained when the exclusion effects are duly considered. We shall not discuss the limiting process, but refer to Chapter 4, where a similar situation occurred.

In the case of a perfect gas of only one constituent we can drop the indices on the α_i and ν_i. The total energy and number of particles are given by the equations

$$\varepsilon = \sum_s n_s \varepsilon_s, \tag{6.601}$$

$$\nu = \sum_s n_s, \tag{6.602}$$

where n_s is the number of particles in the energy level ε_s.

The degeneracy γ' of a state with energy ε is given by the equation (compare Eq. (6.317))

$$\gamma' = \nu! \prod_s \frac{g_s^{\,n_s}}{n_s!}. \tag{6.603}$$

We mentioned that we are not including in our discussion the exclusion principle. However, since we are interested in obtaining formulas that will turn out to be the limiting cases of the formulas obtained when the exclusion principle is taken into account, we must realize that, because of symmetry effects, there is only one specific phase for each generic phase so that we must introduce an extra factor $1/\nu!$ into the equation for the degeneracy, as we did in Chapter 4. Instead of Eq. (6.603) we thus have

$$\gamma = \prod_s \frac{g_s^{n_s}}{n_s!}, \tag{6.604}$$

and we get from Eq. (6.403) for the grand potential the equation

$$e^q = \sum_{\nu=0}^{\infty} e^{\alpha\nu} \sum_{\Sigma n_s = \nu} \gamma e^{-\beta \Sigma n_s \varepsilon_s}, \tag{6.605}$$

or

$$e^q = \sum_{n_s=0}^{\infty} \cdots \sum \left(\prod_s \frac{g_s^{n_s}}{n_s!}\right) e^{\Sigma n_s(\alpha - \beta\varepsilon_s)},$$

$$= \prod_s \left[\sum_{n_s=0}^{\infty} \frac{g_s^{n_s}}{n_s!} e^{n_s(\alpha - \beta\varepsilon_s)}\right],$$

$$= \exp\left[\sum_s g_s e^{\alpha - \beta\varepsilon_s}\right],$$

or

$$q = \sum_s g_s e^{\alpha - \beta\varepsilon_s}. \tag{6.606}$$

Equation (6.606) looks very much the same as Eq. (4.607). There is, however, one important difference. While the energies in Eq. (4.607) were representative values for a group of levels, the energies in Eq. (6.606) are the separate levels. Moreover, for the derivation of Eq. (6.606) it is not necessary to assume that the levels are lying densely. Equation (6.606) has thus a much wider field of application.

The average number of particles $\langle\nu\rangle$ is given by Eq. (6.415) and we have

$$\langle\nu\rangle = \frac{\partial q}{\partial \alpha} = q. \tag{6.607}$$

We can use Eq. (6.419) for the fluctuations in the number of particles and thus find

$$\langle(\nu - \langle\nu\rangle)^2\rangle = \frac{\partial^2 q}{\partial \alpha^2} = q = \langle\nu\rangle, \tag{6.608}$$

or

$$\frac{\langle(\nu - \langle\nu\rangle)^2\rangle}{\langle\nu\rangle^2} = \frac{1}{\langle\nu\rangle}. \tag{6.609}$$

We can also calculate the fluctuations in the numbers of particles occupying the various energy levels. The energy values ε_s are then considered as external parameters, since we have from Eq. (6.601)

$$n_s = \frac{\partial \varepsilon}{\partial \varepsilon_s}, \tag{6.610}$$

and we see that $-\varepsilon_s$ plays the role of an external parameter. From Eqs. (6.417) and (6.606) we get thus for $\langle n_s \rangle$ the formula

$$\langle n_s \rangle = \frac{1}{\beta} \frac{\partial q}{\partial(-\varepsilon_s)} = g_s e^{\alpha - \beta \varepsilon_s}. \tag{6.611}$$

We see that we obtain the formula for the Maxwell-Boltzmann distribution and we have found here one of the most sophisticated derivations of that formula. Not only that, but we are now also in a position to compute the fluctuations around the Maxwell-Boltzmann distribution. We have for the fluctuations of n_s the equation (compare Eq. (5.1002))

$$\langle (n_s - \langle n_s \rangle)^2 \rangle = \frac{1}{\beta^2} \frac{\partial^2 q}{\partial {\varepsilon_s}^2} + \frac{1}{\beta} \left\langle \frac{\partial^2 \varepsilon}{\partial {\varepsilon_s}^2} \right\rangle. \tag{6.612}$$

Since

$$\frac{\partial^2 \varepsilon}{\partial {\varepsilon_s}^2} = 0,$$

Eq. (6.612) gives us

$$\langle (n_s - \langle n_s \rangle)^2 \rangle = \frac{1}{\beta^2} \frac{\partial^2 q}{\partial {\varepsilon_s}^2} = g_s e^{\alpha - \beta \varepsilon_s} = \langle n_s \rangle, \tag{6.613}$$

and we see that, as soon as $\langle n_s \rangle$ is large compared to one, the fluctuations in n_s will be small.

The results for the n_s and for v were independent of the kind of particles that were present in the gas. This will no longer be true when we wish to consider the average energy and its fluctuations. Let us therefore assume now that we are dealing with a perfect monatomic gas. We can then use the fact that for all temperatures the energy levels will be lying so densely that the sum in Eq. (6.606) can be replaced by an integral; then, of course, we get the same equation for q as in Chapter 4,

$$q = \left(\frac{2\pi m}{\beta h^2} \right)^{3/2} e^{\alpha} V. \tag{6.614}$$

From Eq. (6.614) we find that the perfect gas law follows in the usual way, and for the average energy we have

$$\langle \varepsilon \rangle = \frac{3\langle v \rangle}{2\beta}. \tag{6.615}$$

The fluctuations in the energy follow again from Eq. (7.419), and we have

$$\langle(\varepsilon - \langle\varepsilon\rangle)^2\rangle = \frac{\partial^2 q}{\partial\beta^2} = \frac{15q}{4\beta^2}, \tag{6.616}$$

or, using Eqs. (6.607) and (6.615),

$$\frac{\langle(\varepsilon - \langle\varepsilon\rangle)^2\rangle}{\langle\varepsilon\rangle^2} = \frac{5}{3\langle\nu\rangle}. \tag{6.617}$$

That the fluctuations in the energy are now larger than those given by Eq. (5.0804) is due to the fact that there is an additional source of fluctuations, namely, the number of particles.

6.7 THE PERFECT BOSE-EINSTEIN GAS

Apart from the calculations of the fluctuations the preceding section brought no new results. This could have been expected, since exclusion effects had been neglected and finally even the sum in Eq. (6.606) replaced by an integral, thus carefully removing any chance we may have had to obtain new results. In the present and the following section we shall see how the exclusion principle will change the considerations of the preceding section. The results will be practically the same as those obtained in Chapter 4, but our results will be less restricted, just as Eq. (6.606) was more general than Eq. (4.607). We shall assume for the sake of simplicity that all energy levels ε_s are non-degenerate, that is, that all g_s are equal to one.

The case of a perfect Bose-Einstein gas is rather simple. We can use Eq. (6.605) for the grand potential, but γ is now not given by Eq. (6.604) but is equal to one for all values of ε, or

$$\gamma = 1. \tag{6.701}$$

We have then for q the equation

$$e^q = \sum_{\nu=0}^{\infty} e^{\alpha\nu} \sum_{\sum n_s = \nu} e^{-\beta \sum n_s \varepsilon_s}, \tag{6.702}$$

or

$$e^q = \sum \mathop{\cdots}_{n_s=0}^{\infty} \sum e^{\sum n_s(\alpha - \beta\varepsilon_s)}$$

$$= \prod_s (1 - e^{\alpha - \beta\varepsilon_s})^{-1},$$

or

$$q = -\sum_s \ln\,[1 - e^{\alpha - \beta\varepsilon_s}]. \tag{6.703}$$

Again, this equation is practically the same as Eq. (4.707), but it was not necessary in deriving Eq. (6.703) to assume the energy levels to lie densely, and the ε_s are the energy levels and not a representative average of a group of energy levels as the E_j in Eq. (4.707) were.

We find for $\langle v \rangle$ and the fluctuations in v

$$\langle v \rangle = \frac{\partial q}{\partial \alpha} = \sum_s (e^{-\alpha+\beta\varepsilon_s} - 1)^{-1}, \tag{6.704}$$

and

$$\langle (v - \langle v \rangle)^2 \rangle = \frac{\partial^2 q}{\partial \alpha^2} = \langle v \rangle + \sum_s (e^{-\alpha+\beta\varepsilon_s} - 1)^{-2}. \tag{6.705}$$

We see here that the fluctuations are larger than in the case of a Boltzmann gas.

Of special interest again are the $\langle n_s \rangle$ and the fluctuations in the n_s, and we have

$$\langle n_s \rangle = -\frac{1}{\beta}\frac{\partial q}{\partial \varepsilon_s} = \frac{1}{e^{-\alpha+\beta\varepsilon_s} - 1}, \tag{6.706}$$

and

$$\langle (n_s - \langle n_s \rangle)^2 \rangle = \frac{1}{\beta^2}\frac{\partial^2 q}{\partial {\varepsilon_s}^2} = \langle n_s \rangle + \langle n_s \rangle^2 \tag{6.707}$$

We note here that the fluctuations in n_s may become quite large. Let us consider for a moment the case of a perfect boson gas below its condensation temperature T_0 (see Section 4.7). If we are well below T_0, practically all bosons will be in the ground state so that $\langle n_0 \rangle \approx N$, and we get then from Eq. (6.707)

$$\langle (n_0 - \langle n_0 \rangle)^2 \rangle = \langle n_0 \rangle + \langle n_0 \rangle^2 \approx N^2. \tag{6.708}$$

This means that we can expect large fluctuations in density at very low temperatures. The fact that they have never been observed in liquid helium is a strong argument against identifying liquid helium with a perfect boson gas.

Let us now consider the situation well above T_0. In that case, α will be large and negative so that $\langle n_s \rangle \ll 1$. There is thus little point in discussing the fluctuations in n_s. However, we can group together a number of levels in such a way that the quantity

$$N_i = \sum{}' n_s, \tag{6.709}$$

where the summation is over the ith group of levels, is large compared to unity. We can now consider the fluctuations in N_i. We have, first of all,

$$N_i = \sum{}' \frac{\partial \varepsilon}{\partial \varepsilon_s}, \tag{6.710}$$

and thus

$$\langle N_i \rangle = \frac{1}{\beta} \sum' \frac{\partial q}{\partial \varepsilon_s} = \sum' \frac{1}{e^{-\alpha+\beta\varepsilon_s} - 1}. \tag{6.711}$$

As the fluctuations in the n_s are independent, we have for the fluctuations in the N_i

$$\langle (N_i - \langle N_i \rangle)^2 \rangle = \frac{1}{\beta^2} \sum' \frac{\partial^2 q}{\partial {\varepsilon_s}^2}, \tag{6.712}$$

or

$$\begin{aligned} \langle (N_i - \langle N_i \rangle)^2 \rangle &= \langle N_i \rangle + \sum' (e^{-\alpha+\beta\varepsilon_s} - 1)^2 \\ &= \langle N_i \rangle + \sum' \langle n_s \rangle^2. \end{aligned} \tag{6.713}$$

As the $\langle n_s \rangle$ are small, the second term on the right-hand side of Eq. (6.713) will be small compared to the first one, and we see that, at sufficiently high temperatures, the relative fluctuations in the N_i are small.

Let us now restrict ourselves to the case of a perfect monatomic gas. We can then use the results of Chapter 4 for q, since the sum in Eq. (6.703) can be replaced by an integral, and we have

$$q = \left(\frac{2\pi m}{\beta h^2} \right)^{3/2} V g(\alpha), \tag{6.714}$$

where

$$g(\alpha) = \sum_{n=1}^{\infty} \frac{e^{n\alpha}}{n^{5/2}}. \tag{6.715}$$

Using this formula for q we get easily from Eqs. (6.419)

$$\frac{\langle (\nu - \langle \nu \rangle)^2 \rangle}{\langle \nu \rangle^2} = \frac{g''(\alpha)}{g'(\alpha)} \cdot \frac{1}{\langle \nu \rangle}, \qquad \left(g'(\alpha) = \frac{dg}{d\alpha}, \qquad g''(\alpha) = \frac{d^2 g}{d\alpha^2} \right) \tag{6.716}$$

and

$$\frac{\langle (\varepsilon - \langle \varepsilon \rangle)^2 \rangle}{\langle \varepsilon \rangle^2} = \frac{5}{3\langle \nu \rangle} \frac{g'(\alpha)}{g(\alpha)}, \tag{6.717}$$

where we used the fact that

$$\langle \nu \rangle = \frac{\partial q}{\partial \alpha} = q \frac{g'(\alpha)}{g(\alpha)}. \tag{6.718}$$

Since $g''(\alpha)$ diverges as $\alpha \to 0$, the fluctuations in ν can become arbitrarily large as soon as $\alpha \to 0$.

6.8 THE PERFECT FERMI-DIRAC GAS

The calculations are similar to those of the preceding section. We now have for γ

$$\gamma = 1, \qquad \sum n_s^2 = \nu; \qquad \gamma = 0, \qquad \sum n_s^2 > \nu, \tag{6.801}$$

and we thus have

$$e^q = \sum_{\nu=0}^{\infty} e^{\alpha\nu} \sum_{\substack{\sum n_s = \nu \\ \sum n_s^2 = \nu}} e^{-\beta \sum n_s \varepsilon_s}, \tag{6.802}$$

or

$$e^q = \sum_{n_s=0}^{1} \cdots \sum e^{\sum n_s(\alpha - \beta\varepsilon_s)}$$
$$= \prod_s (1 + e^{\alpha - \beta\varepsilon_s})$$

or

$$q = \sum_s \ln\,[1 + e^{\alpha - \beta\varepsilon_s}]. \tag{6.803}$$

This equation corresponds to Eq. (4.802), and again Eq. (6.803) is less restricted than its counterpart of Chapter 4.

Since the calculations are completely analogous to those of Section 6.7, we shall give only the more important results. We have for the fluctuations in ν

$$\langle(\nu - \langle\nu\rangle)^2\rangle = \langle\nu\rangle - \sum_s (e^{-\alpha+\beta\varepsilon_s} + 1)^{-2} < \langle\nu\rangle, \tag{6.804}$$

which shows that the fluctuations are now smaller than in the Boltzmann case.

We have for $\langle n_s\rangle$ and the fluctuations in the n_s

$$\langle n_s\rangle = \frac{1}{e^{-\alpha+\beta\varepsilon_s} + 1}, \tag{6.805}$$

$$\langle(n_s - \langle n_s\rangle)^2\rangle = \langle n_s\rangle - \langle n_s\rangle^2. \tag{6.806}$$

Since the n_s can only be zero or one, n_s is certainly not a large number, and although the fluctuations in the n_s are smaller than in the Boltzmann case, the relative fluctuations $\langle(n_s - \langle n_s\rangle)^2\rangle/\langle n_s\rangle^2$ can never be very small—unless $\langle n_s\rangle$ is practically equal to one, that is, unless we are dealing with the lowest energy levels which are practically completely occupied as long as the temperature is not extremely high. However, it is possible to consider again a group of levels so that the average number of particles in this group of levels is large compared to unity. For the number of fermions in the ith group of levels we again have Eqs. (6.709) and (6.710). Instead of Eqs. (6.711) and (6.713) we have now

$$\langle N_i \rangle = \frac{1}{\beta} \sum \frac{\partial q}{\partial \varepsilon_s} = \sum' \frac{1}{e^{-\alpha+\beta\varepsilon_s} + 1}, \tag{6.807}$$

$$\begin{aligned}\langle (N_i - \langle N_i \rangle)^2 \rangle &= \frac{1}{\beta^2} \sum' \frac{\partial^2 q}{\partial {\varepsilon_s}^2} \\ &= \langle N_i \rangle - \sum' [e^{-\alpha+\beta\varepsilon_s} + 1]^2 = \langle N_i \rangle - \sum' \langle n_s \rangle^2,\end{aligned} \tag{6.808}$$

and we see that the relative fluctuations in the N_i are, indeed, small compared to unity, as soon as $\langle N_i \rangle$ is large.

Considering once more the case of a perfect monatomic gas, we can replace the sum in Eq. (6.803) by an integral and we get for q the equation

$$q = \left(\frac{2\pi m}{\beta h^2}\right)^{3/2} V h(\alpha), \tag{6.809}$$

where

$$h(\alpha) = -\sum_{n=1}^{\infty} \frac{(-e^{\alpha})^n}{n^{5/2}}. \tag{6.810}$$

After an easy calculation we get for the fluctuations in the energy

$$\frac{\langle (\varepsilon - \langle \varepsilon \rangle)^2 \rangle}{\langle \varepsilon \rangle^2} = \frac{5}{3\langle \nu \rangle} \frac{h'(\alpha)}{h(\alpha)}, \tag{6.811}$$

where we have used for $\langle \nu \rangle$ the equation

$$\langle \nu \rangle = q \frac{h'(\alpha)}{h(\alpha)}. \tag{6.812}$$

6.9 THE SAHA EQUILIBRIUM

In the preceding three sections we considered a system consisting of one kind of particle only. In the present section, we shall consider systems consisting of three kinds of particles, A, B, and C, between which the following reaction can take place:

$$\mathrm{A} + \mathrm{B} \leftrightarrows \mathrm{C}. \tag{6.901}$$

In reaction (6.901), A and B may be atoms and C the diatomic molecule AB, or C may be an atom, A an electron, and B the ion C^+, and so on.

We shall use Eq. (6.403) for the grand potential; this equation can now be written in the form

$$e^q = \sum_{\nu_A=0}^{\infty} \sum_{\nu_B=0}^{\infty} \sum_{\nu_C=0}^{\infty} e^{\alpha_A \nu_A + \alpha_B \nu_B + \alpha_C \nu_C} \sum_k \gamma_k e^{-\beta E_k(\nu_A, \nu_B, \nu_C)}, \tag{6.902}$$

where the last summation extends over all energy levels $E_k(\nu_A,\nu_B,\nu_C)$ of a system consisting of ν_A particles A, ν_B particles B, and ν_C particles C, and where γ_k is the multiplicity of the level $E_k(\nu_A,\nu_B,\nu_C)$. It must be borne in mind that α_A, α_B, and α_C are not independent, but satisfy the equilibrium condition (compare Eq. (5.1117))

$$\alpha_C = \alpha_A + \alpha_B. \tag{6.903}$$

We now make the first of a number of assumptions. This first assumption is that we may treat all particles in the system to be independent. In that case we can write

$$E_k(\nu_A,\nu_B,\nu_C) = E_{kA}(\nu_A) + E_{kB}(\nu_B) + E_{kC}(\nu_C), \tag{6.904}$$

and

$$\gamma_k = \gamma_{kA} \cdot \gamma_{kB} \cdot \gamma_{kC}, \tag{6.905}$$

so that Eq. (6.902) can be written in the form

$$e^q = \{A\} \cdot \{B\} \cdot \{C\}, \tag{6.906}$$

with

$$\{A\} = \sum_{\nu_A=0}^{\infty} e^{\alpha_A \nu_A} \sum_{kA} \gamma_{kA} e^{-\beta E_{kA}(\nu_A)}, \cdots \tag{6.907}$$

Since all particles are assumed to be noninteracting, we have further (compare Eq. (6.601)):

$$E_{kA}(\nu_A) = \sum_{sA} n_{sA}\varepsilon_{sA}, \qquad \sum_{sA} n_{sA} = \nu_A. \tag{6.908}$$

We now make our second assumption: that we may neglect all symmetry effects. In that case we get for γ_{kA} the expression (compare Eq. (6.604))

$$\gamma_{kA} = \prod_{sA} \frac{g_{sA}{}^{n_{sA}}}{n_{sA}!}. \tag{6.909}$$

Substituting Eqs. (6.908) and (6.909) into Eq. (6.906), we find

$$q = e^{\alpha_A} \cdot Z_A + e^{\alpha_B} \cdot Z_B + e^{\alpha_C} \cdot Z_C, \tag{6.910}$$

where $Z_A, \cdots$ are the partition functions,

$$Z_A = \sum_{sA} g_{sA} e^{-\beta\varepsilon_{sA}}, \cdots. \tag{6.911}$$

We note that it follows from Eqs. (6.910) and (6.414) that the total pressure is the sum of the partial pressures due to A, B, and C, as we should have expected for noninteracting particles.

From Eqs. (6.910), (6.415), and (6.903) we get

$$\frac{\langle\nu_A\rangle\langle\nu_B\rangle}{\langle\nu_C\rangle} = \frac{Z_A Z_B}{Z_C}. \tag{6.912}$$

Let us now remind ourselves that the energies ε_{sA}, ε_{sB}, and ε_{sC} can be split into at least two independent parts: the kinetic energy of the center of mass and the internal energy. As far as the kinetic energy is concerned, we can perform the summation involved, and the result is (see Eq. (4.609)) a factor

$$\left(\frac{2\pi m}{\beta h^2}\right)^{3/2} V, \tag{6.913}$$

where V is the volume occupied by the system and m the mass of the particle concerned. If we now reckon the internal energies of the particles from their ground states as zero, we have (ε_{kin} denotes the kinetic energies)

$$\varepsilon_{sA}' = \varepsilon_{sA} - \varepsilon_{\text{kinA}}, \qquad \varepsilon_{sB}' = \varepsilon_{sB} - \varepsilon_{\text{kinB}}, \qquad \varepsilon_{sC}' = \varepsilon_{sC} + \chi - \varepsilon_{\text{kinC}}, \tag{6.914}$$

where we have primed the internal energies reckoned from the ground state as zero and where χ is the dissociation energy of the reaction (6.901). If $c_A = \langle \nu_A \rangle / V, \cdots$ denote the concentrations, we find, combining the various results (compare Problem 3 at the end of Chapter 4)

$$\frac{c_A c_B}{c_{AB}} = \left(\frac{2\pi\mu}{\beta h^2}\right)^{3/2} e^{-\beta\chi} \frac{Z_A' Z_B'}{Z_C'}, \tag{6.915}$$

where $Z_A', \cdots$ are the internal partition functions,

$$Z_A' = \sum \gamma_{sA}' e^{-\beta\varepsilon_{sA}'}, \ldots, \tag{6.916}$$

with $\gamma_{sA}', \cdots$ referring to the internal energies only, and μ is the reduced mass,

$$\mu = \frac{m_A m_B}{m_C}. \tag{6.917}$$

Equation (6.915) is essentially Saha's formula. Special cases of this formula are considered in Problems 31 and 32 at the end of this chapter.

PROBLEMS

1. Prove that Eq. (6.115) is invariant under a unitary transformation.

   In the Schrödinger representation the wave function ψ_S satisfies the wave equation

   $$i\hbar\dot{\psi}_S = \mathscr{H}\psi_S,$$

   and, if we exclude explicit time dependence, the operators $\hat{G}_S$ are time-independent operators.

In the Heisenberg representation the system is described by a wave function ψ_H, which is related to ψ_S by the relation

$$\psi_H = e^{i\hat{\mathscr{H}}t/\hbar}\psi_S,$$

from which it follows that ψ_H is time-independent, while the operators $\hat{G}_H$ in the Heisenberg representation are related to those in the Schrödinger representation $\hat{G}_S$ through the equation

$$\hat{G}_H = e^{i\hat{\mathscr{H}}t/\hbar}\hat{G}_S e^{-i\hat{\mathscr{H}}t/\hbar}.$$

In the so-called interaction representation one is dealing with a Hamiltonian $\hat{\mathscr{H}}$ which can be split into two parts:

$$\hat{\mathscr{H}} = \hat{\mathscr{H}}_0 + \hat{\mathscr{H}}'.$$

Often $\hat{\mathscr{H}}_0$ corresponds to a system of noninteracting particles and $\hat{\mathscr{H}}'$ to the interaction between the particles—hence the name interaction representation—but other situations may also occur (compare Problems 8 and 9). The wave function ψ_{int} is now related to ψ_S by the equation

$$\psi_{int} = e^{i\hat{\mathscr{H}}_0 t/\hbar}\psi_S,$$

so that ψ_{int} satisfies the equation of motion

$$i\hbar\dot{\psi}_{int} = \hat{\mathscr{H}}_{int}'\psi_{int},$$

where $\hat{\mathscr{H}}_{int}'$ is $\hat{\mathscr{H}}'$ in the interaction representation in which the operators $\hat{G}_{int}$ are related to $\hat{G}_S$ by the equation

$$\hat{G}_{int} = e^{i\hat{\mathscr{H}}_0 t/\hbar}\hat{G}_S e^{-i\hat{\mathscr{H}}_0 t/\hbar}.$$

2. Prove that Eq. (6.134) holds in the Heisenberg representation.
3. Prove that Eq. (6.134) holds in the interaction representation.
4. Consider a system of many degrees of freedom. Denote those degrees of freedom in which we are interested collectively by x and those that we wish to ignore collectively by q. The total normalized wave function of our system will then be a function of both q and x : $\psi(q,x)$. Let $\hat{G}$ be an operator that is a function of x only. Introduce the density matrix $\hat{\rho}$ in the Dirac notation* by the equation

$$\langle x|\hat{\rho}|x'\rangle = \int \psi^*(q,x')\psi(q,x)\,dq;$$

then prove that $\hat{\rho}$ is normalized and find an expression for the average value of $\hat{G}$. Find also the equation of motion for $\hat{\rho}$ assuming that the Hamiltonian can be split into two parts, one pertaining to the q and one pertaining to the x.

* See, for instance, A. S. Davydov, *Quantum Mechanics*, Oxford, 1965, Chap. IV.

5. Show that, if $\psi(q,x)$ can be written as the product $\Phi(q)\chi(x)$, we are dealing with a pure case.
6. Introduce the density matrix in the coordinate representation by using the eigenfunctions $\varphi_n(x)$ (which in the Dirac notation can be written $\langle n|x\rangle$) as transformation matrix.*
7. Write down the general expression in terms of the φ_n for the density matrix in the coordinate representation for the case where the ensemble is a macrocanonical ensemble.
8. Consider a Hamiltonian given by

$$\mathscr{H} = \mathscr{H}_0 + \mathscr{H}', \qquad \mathscr{H}' = \hat{V}e^{-i\omega t}.$$

If at $t = -\infty$, the density matrix $\hat{\rho}(-\infty)$ is given by expression (6.301) with $\mathscr{H} = \mathscr{H}_0$, find and solve the equation of motion for $\Delta\hat{\rho}(=\hat{\rho} - \hat{\rho}(-\infty))$ in the interaction representation, neglecting the second order term involving the product of $\Delta\hat{\rho}$ and $\hat{V}$. Hence prove the following expression for the average value of the operator $\hat{G}$ at time t (Kubo formula†):

$$\langle G(t)\rangle = \langle G\rangle_0 - \frac{i}{\hbar}\int_{-\infty}^{t} \langle[\hat{G}_{\text{int}}(t), \hat{V}_{\text{int}}(\tau)]_-\rangle_0 e^{-i\omega\tau}\, d\tau,$$

where $\langle\cdots\rangle_0$ indicates an average over $\hat{\rho}(-\infty)$.
9. Use the result of the preceding problem to find an expression for the electrical conductivity tensor, putting

$$\hat{G} = \hat{\jmath}_k, \quad k = x, y, z, \qquad \text{and} \qquad \hat{V} = -e\sum_l (\mathbf{E}\cdot\mathbf{r}_l),$$

where $\mathbf{E}$ is an external electric field, the $\hat{\jmath}_k$ are the components of the electrical current operator, and the summation over l is over all charge carriers in the system.
10. Find the density matrix of the macrocanonical ensemble in the coordinate representation for the case of a single free particle in a volume V.
11. Find the density matrix of the macrocanonical ensemble in the coordinate representation for the case of a system of N identical free particles, distinguishing carefully between bosons, fermions, and Boltzmann particles.‡
12. If $\hat{\sigma}_x$, $\hat{\sigma}_y$, and $\hat{\sigma}_z$ are the Pauli spin matrices,

$$\hat{\sigma}_x = \begin{pmatrix} 0 & 1 \\ 1 & 0 \end{pmatrix}, \qquad \hat{\sigma}_y = \begin{pmatrix} 0 & -i \\ i & 0 \end{pmatrix}, \qquad \hat{\sigma}_z = \begin{pmatrix} 1 & 0 \\ 0 & -1 \end{pmatrix},$$

* Compare the discussion by Davydov (*loc. cit.*) of change of representation.
† R. Kubo, *J. Phys. Soc. Japan*, **12**, 570, 1957.
‡ See, for example, D. ter Haar, *Elements of Statistical Mechanics*, 1st ed., New York, 1954, Section VIII.4.

one can for the case of a system of spin-$\frac{1}{2}$ particles define the polarization vector **P** by the equation

$$\mathbf{P} = \langle \boldsymbol{\sigma} \rangle.$$

Use this definition of **P** to find an expression for the spin density matrix, which is in this case a two-by-two matrix in terms of the three components of $\boldsymbol{\sigma}$ and the unit matrix, and the value of **P**.

Using the equation of motion (6.134) for the average value of an operator and the expression

$$\mathscr{H}_s = -\tfrac{1}{2}\gamma\hbar(\hat{\boldsymbol{\sigma}} \cdot \mathbf{H})$$

for the spin Hamiltonian $\mathscr{H}_s$, where γ is the magnetogyric ratio and **H** an external magnetic field, find the equation of motion of **P**.

13. Evaluate the fluctuations in the energy for a system of independent localized harmonic oscillators. Consider the asymptotic expressions for $\beta\hbar\omega \ll 1$ and $\beta\hbar\omega \gg 1$.
14. Evaluate, up to terms of order β, the high-temperature expression for the fluctuations in the energy of a system of independent localized three-dimensional rigid rotators. (Use the series expansion given in Problem 16 at the end of Chapter 3.)
15. By comparing the average value of the number of particles in the sth single-particle state in a canonical ensemble for a system of N independent particles, with the same quantity in a canonical ensembles for a system of $N + 1$ independent particles, find an expression for that average number for the case of Boltzmann particles, bosons, or fermions. Use the fact that the two averages should be the same in the limit of large N.*
16. A microcanonical quantum-mechanical ensemble describes a system with a fixed number of particles, N, and with a fixed total energy, E. Let Ω be the total number of quantum states compatible with these conditions. According to the principle of equal a priori probabilities for all quantum states, the probability P_l of finding a system of the ensemble in the lth state will be equal to $1/\Omega$. This corresponds to a density matrix $\hat{\rho}$ of the following form:

$$\rho_{kl} = \frac{\delta_{kl}}{\Omega}, \text{ if the } k\text{th state is one of the } \Omega \text{ states compatible with } N \text{ and } E;$$

$$= 0, \text{ otherwise,}$$

where we use the energy representation.

Prove that the entropy S is given by the equation

$$S = k \ln \Omega.$$

* F. Ansbacher and W. Ehrenberg, *Phil. Mag.*, **40,** 626, 1949; H. Schmidt, *Z. Physik*, **134,** 430, 1953.

17. Consider a system* of N distinguishable one-dimensional harmonic oscillators of frequency ω. Neglect the zero-point energy and write $E = N_0\hbar\omega$; then find an expression for Ω, and hence for the entropy S, using Stirling's formula for the factorial. Use the thermodynamic relation (compare Eq. (4.411)):

$$\beta = \left(\frac{\partial(S/k)}{\partial E}\right)_{V,N},$$

to find a relation between β and E, and thus derive the Planck expression for the energy of a system of N oscillators.

18. Discuss why α vanishes in the case of a gas of photons or of phonons.

19. Use the method of Section 5.5 to derive the following relation between the energy level density $\rho(\langle\varepsilon\rangle,\langle\nu_i\rangle)$ and the entropy

$$\rho(\langle\varepsilon\rangle,\langle\nu_i\rangle) \approx e^{S/k}\left[-\frac{\pi}{2}\frac{\partial\langle\varepsilon\rangle}{\partial\beta}\prod_i\left(\frac{\pi}{2}\frac{\partial\langle\nu_i\rangle}{\partial\alpha_i}\right)\right]^{-1/2}$$

where $\rho(\varepsilon,\nu_i)\,d\varepsilon\,\Pi\,d\nu_i$ is the number of levels with energies between ε and $\varepsilon + d\varepsilon$ for a system with the number of particles of the ith kind between ν_i and $\nu_i + d\nu_i$, and where we have used Eq. (6.409).

20. Consider a nucleus, which for the sake of simplicity is considered to be a system of $Z(=\frac{1}{2}A)$ protons and $N(=\frac{1}{2}A)$ neutrons which move without interactions in a volume V equal to $Ar_0^{\,3}$, where r_0 is a length of nuclear dimensions of the order of a fermi ($=10^{-13}$ cm). First, verify that as long as the nuclear temperature which can be defined by the equation $(kT)^{-1} = \beta = \partial S/\partial E$, expressed in energy units, does not exceed 100 meV, the proton and neutron gases are degenerate.
Secondly, bear in mind the fact that microcanonical ensembles lead to the same results as canonical grand ensembles; use the results of Section 4.8 to evaluate the energy level density, from the result of the preceding problem.

21. Repeat the calculations of the preceding problem for the case where the nucleons, while being independent, have a single-particle energy spectrum which is not that of free point-particles in a volume V, but is such that the number of levels dZ with energies between ε and $\varepsilon + d\varepsilon$ is given by the equation

$$dZ = f(\varepsilon)\,d\varepsilon.$$

22. Prove that if all systems consisting of particles of one kind in an ensemble have the same number of particles, N, and an energy lying within a given interval E, $E + \delta E$, $\langle\eta\rangle$ is minimum for the density matrix $\hat{\rho}$ given by the

* A. V. Tobolsky, *Am. J. Phys.*, **32**, 799, 1964.

equation (we use the Dirac notation):

$$\langle k, v|\hat{\rho}|l,v'\rangle = c\delta_{kl}\delta_{vN}\delta_{v'N} \qquad \text{for} \qquad E \leqslant E_k(v) \leqslant E + \delta E;$$

$$\langle k, v|\hat{\rho}|l,v'\rangle = 0, \qquad \text{otherwise.}$$

23. Prove that if all systems consist of one kind of particle in an ensemble have the same number of particles, N, while only the average energy is given, so that Eq. (6.503) holds, $\langle \eta \rangle$ is a minimum for the density matrix $\hat{\rho}$ given by the equation

$$\langle k, v|\hat{\rho}|l,v'\rangle = \delta_{kl}\delta_{vN}\delta_{v'N}e^{\beta[\psi - E_k(v)]}.$$

24. Prove for the two cases considered in the preceding problems that $d\langle \text{H} \rangle/dt < 0$.
25. Prove that the canonical ensembles are the proper representative ensembles, following the line of argument of Section 5.13.
26. By considering a unitary transformation that brings $\hat{\mathbf{A}}$ onto diagonal form, prove Eq. (6.510).
27. By considering the unitary transformation that transforms the probability amplitudes at t' to those at t'', prove Klein's lemma.
28. If w_{kl} is the probability for a transition per unit time from a state k to a state l with the same energy, and if $w_{kl} = w_{lk}$, prove that the P_i given in Problem 16 are attained exponentially (quantum-mechanical ergodic theorem in its simplest form).
29. Use the Darwin-Fowler method to prove Eqs. (6.612), (6.707), and (6.806).
30. If we consider a system of N bosons, which can be treated as independent point particles, below the transition temperature T_0, we can treat the part of the system consisting of particles in excited states by using a canonical grand ensemble, the particles in the ground state acting as a "particle bank." Prove that the fluctuations in the total number of particles in excited states, N_1, satisfy the equation

$$\langle (N_1 - \langle N_1 \rangle)^2 \rangle = C\langle N_1 \rangle^{4/3},$$

where C is a constant.
31. Apply formula (6.912) to the case of an ionization equilibrium, considering especially the case where the temperature is relatively low.
32. Apply Eq. (6.912) to the case of the dissociation equilibrium of a diatomic molecule AB into its constituent atoms A and B. Assume that the electronic, vibrational, and rotational energies of the molecule are completely independent.

33. At a given temperature and volume a gas of AB molecules is 50 percent dissociated into A and B atoms. If at constant temperature the volume is doubled, evaluate the relative change in pressure.

34. Discuss the changes to be made in Eq. (6.912) if symmetry effects are taken into account.

35. Consider the reaction

$$A + 2B \rightarrow AB_2,$$

where A, B, and AB_2 are all bosons. We introduce N_A atoms A and N_B atoms B into a volume V at temperature T. Find the fraction of A atoms in molecular form in the high-temperature limit.

BIBLIOGRAPHICAL NOTES

The density matrix was first introduced independently by Landau, who used the quantum-mechanical approach, and von Neumann:

1. L. D. Landau, *Z. Physik*, **45,** 430, 1927; *Collected Papers*, Oxford, 1965, p. 8.
2. J. von Neumann, *Göttinger Nachr.*, **1927,** 1, 24, 273.

Dirac discussed both the quantum-mechanical approach:

3. P. A. M. Dirac, *Proc. Camb. Phil. Soc.*, **26,** 376, 1930; **27**, 240, 1931,

and the statistical approach:

4. P. A. M. Dirac, *Proc. Camb. Phil. Soc.*, **25,** 62, 1929; **26,** 361, 1930,

without, however, mentioning the connection between the two kinds of density matrices and normalizing the density matrix in different ways.

For general surveys of the density matrix and its applications, see:

5. K. Husimi, *Proc. Phys.-Math. Soc. Japan*, **22,** 264, 1940.
6. D. ter Haar, *Repts. Progr. Phys.*, **24,** 304, 1961.

Section 6.2. See Reference 2 and also

7. J. von Neumann, *Mathematical Foundations of Quantum Mechanics*, Princeton, N. J., 1955.

Section 6.4. Pauli was the first to introduce canonical grand ensembles in quantum statistics:

8. W. Pauli, *Z. Physik*, **41,** 81, 1927.

Our treatment is largely based upon Kramers' work:

9. H. A. Kramers, *Proc. Kon. Ned. Akad. Wet.* (*Amsterdam*), **41,** 10, 1938; *Collected Scientific Papers*, Amsterdam, 1956, p. 738.

Section 6.5. See

10. R. C. Tolman, *The Principles of Statistical Mechanics*, Oxford, 1938.
11. D. ter Haar, *Elements of Statistical Mechanics*, New York, 1954, 1st ed., Appendix I.
12. D. ter Haar, *Rev. Mod. Phys.*, **27,** 289, 1955.
13. R. Jancel, *Les Fondements de la Mécanique Statistique Classique et Quantique*, Paris, 1963 (English translation, Oxford, in course of publication).
14. I. E. Farquhar, *Ergodic Theory in Statistical Mechanics*, New York, 1964.

These references all contain extensive bibliographies.

Section 6.9. Saha's formula was first published in 1920:

15. M. N. Saha, *Phil. Mag.*, **40,** 472, 1920.

See also

16. M. N. Saha, *Z. Physik*, **6,** 40, 1921.

For a bibliography and a discussion of the various assumptions involved, see:

17. D. ter Haar, *Am. J. Phys.*, **23**, 326, 1955.

7

Relativistic Statistics

7.1 THE ELECTRON GAS; GENERAL FORMULAS

Although so far we have used only nonrelativistic formulas, it can easily be shown that most formulas can immediately be adapted to the case of relativistic statistics. We wish to discuss in this chapter how such an adaptation can be made and we have chosen as the case to be discussed in some detail the case of a gas of free electrons, since this is the case that has been of most interest in the application of relativistic statistics. These applications have been in the field of astrophysics and especially in the discussion of the theory of white dwarfs by Fowler, Chandrasekhar, and other authors.* We are thus

* We refer for an account of the application of relativistic statistics to the theory of white dwarfs to S. Chandrasekhar, *An Introduction to the Study of Stellar Structure*, Chicago, 1939, Chap. X and XI. The reader will also find extensive bibliographies there.

confining ourselves to the case of a system of independent Fermi-Dirac particles. On inspection it will be seen that we can use most of the results of Chapter 6 and we have, for instance, for the grand potential q from Eq. (6.803)

$$q = \sum_s \ln (1 + e^{\alpha - \beta \varepsilon_s}), \tag{7.101}$$

where the summation is over all energy levels of one particle.

In the derivation of Eq. (7.101) no use was made of the fact that in Chapter 6 we were interested only in the nonrelativistic case, but only of the fact that we are dealing with a system of independent particles, since the ε_s are the energy levels of a single particle.

If the energy levels are lying sufficiently densely, the sum in Eq. (7.101) may be replaced by an integral and we have

$$q = \int \ln (1 + e^{\alpha - \beta \varepsilon})\, dZ, \tag{7.102}$$

where dZ is once again the number of energy levels between ε and $\varepsilon + d\varepsilon$.

It is at this juncture that the relativistic case differs from the nonrelativistic one. We can no longer use Eq. (4.604) for dZ but we must derive its relativistic counterpart. We shall again assume the gas to be enclosed in a cube of edge length L and volume $V = L^3$. The wave functions of a free particle which satisfy the Dirac equations with periodic boundary conditions can be written as*

$$\psi_\lambda = a_\lambda e^{i[(\mathbf{p} \cdot \mathbf{r}) - \varepsilon t]}, \tag{7.103}$$

where λ runs from 1 to 4, where the ψ_λ are the four components of the Dirac wave function, and where the momentum $\mathbf{p}$ and the energy ε are related to each other as follows:

$$\frac{\varepsilon^2}{c^2} = p^2 + m^2c^2. \tag{7.104}$$

Moreover, for a given $\mathbf{p}$ there are two linearly independent solutions of the Dirac equations, which means that each energy level is twofold degenerate. This degeneracy is the usual spin degeneracy ($g_s = 2s + 1$, $s = \frac{1}{2}$).

The fact that the particle is enclosed in a cube of edge length L entails that not all values of p_x, p_y, and p_z are permissible, but only those which satisfy the equations

$$p_x = \frac{k_1 h}{L}, \qquad p_y = \frac{k_2 h}{L}, \qquad p_z = \frac{k_3 h}{L}, \tag{7.105}$$

where the k_i can take on the values $\pm 1, \pm 2, \cdots$.†

* See, for example, A. S. Davydov, *Quantum Mechanics*, Oxford, 1965, Section 61.

† The situation is slightly different here from the case discussed in Section 4.6. In the nonrelativistic case the wave functions with negative values of the k_i are not linearly independent of those with only positive values of the k_i, but in the relativistic case they are.

Substituting expressions (7.105) into Eq. (7.104), we have

$$\frac{\varepsilon^2}{c^2} = \frac{h^2k^2}{L^2} + m^2c^2, \tag{7.106}$$

where

$$k^2 = k_1{}^2 + k_2{}^2 + k_3{}^2. \tag{7.107}$$

In terms of k we have for dZ (compare Eq. (4.603) and the footnote on p. 249)

$$dZ = 8\pi k^2\, dk. \tag{7.108}$$

From Eqs. (7.106) and (7.108) we now get

$$dZ = \frac{8\pi}{h^3} V \left(\frac{\varepsilon^2}{c^2} - m^2c^2\right)^{1/2} \frac{\varepsilon d\varepsilon}{c^2}. \tag{7.109}$$

If the energy ε is only just larger than the rest energy mc^2, we can write $\varepsilon = mc^2 + \varepsilon'$, where ε' is now the usual nonrelativistic energy. Substituting this expression into Eq. (7.109) and neglecting all terms but the most important one, we get

$$dZ = 4\pi\left(\frac{2m}{h^2}\right)^{3/2} V\sqrt{\varepsilon'}\, d\varepsilon', \tag{7.110}$$

which is Eq. (4.604) apart from a factor 2 due to the spin.

If we introduce as variable instead of ε the absolute magnitude p of the momentum, we can write Eq. (7.109) by using Eq. (7.104) in the form*

$$dZ = 2 \cdot \frac{V \cdot 4\pi p^2\, dp}{h^3}. \tag{7.111}$$

Since $4\pi p^2 dp \cdot V$ is the volume in phase space corresponding to momenta of absolute magnitude between p and $p + dp$, Eq. (7.111) expresses the fact that to each volume h^3 in μ-space there correspond two stationary states, the factor 2 arising from spin degeneracy.

From Eqs. (7.102) and (7.109) we now have for the grand potential

$$q = \frac{8\pi}{h^3c^3} V \int \varepsilon(\varepsilon^2 - m^2c^4)^{1/2} \ln\left(1 + e^{\alpha - \beta\varepsilon}\right) d\varepsilon, \tag{7.112}$$

where the integration extends from mc^2 to ∞.

We remarked that the energy ε contains the rest energy. As a consequence, the α in the equations of this chapter differs from the α encountered, for instance, in Chapter 6 by a term βmc^2. If we denote the α of Chapter 6 by α',

* It can easily be verified by using the nonrelativistic relation $\varepsilon' = p^2/2m$ that Eq. (7.111) is valid both in the nonrelativistic case and in the relativistic case. That such should be the case follows also from the discussion of Eq. (7.111).

we have $\alpha - \beta\varepsilon = \alpha' - \beta\varepsilon'$, or

$$\alpha = \alpha' + \beta mc^2. \tag{7.113}$$

Our equations will simplify if we use p instead of ε, as integration variable. Equation (7.112) then becomes

$$q = \frac{8\pi}{h^3} V \int_0^\infty \ln(1 + e^{\alpha - \beta\varepsilon}) p^2\, dp, \tag{7.114}$$

where ε is a function of p given by Eq. (7.104).

Using Eqs. (6.415), (6.416) and (6.414) we now have for the total number of particles, N, the total energy (including the rest energy Nmc^2), E, and the pressure P the following equations

$$N = \frac{8\pi}{h^3} V \int_0^\infty \frac{p^2\, dp}{e^{-\alpha+\beta\varepsilon} + 1}, \tag{7.115}$$

$$E = \frac{8\pi}{h^3} V \int_0^\infty \frac{\varepsilon p^2\, dp}{e^{-\alpha+\beta\varepsilon} + 1}, \tag{7.116}$$

$$P = \frac{kT}{V} q = \frac{8\pi}{h^3} kT \int_0^\infty \ln(1 + e^{\alpha+\beta\varepsilon}) p^2\, dp. \tag{7.117}$$

It is more convenient for the further discussion to introduce with Jüttner a variable θ defined by the equation

$$\theta = \operatorname{arc\,sinh} \frac{p}{mc}. \tag{7.118}$$

From Eqs. (7.104) and (7.118) we get

$$\varepsilon = mc^2 \cosh\theta, \tag{7.119}$$

and Eqs. (7.115) to (7.117) can be written in the form

$$N = 8\pi \frac{m^3c^3}{h^3} V \int_0^\infty \frac{\sinh^2\theta \cosh\theta\, d\theta}{e^{-\alpha+\beta mc^2\cosh\theta} + 1}, \tag{7.120}$$

$$E = 8\pi \frac{m^4c^5}{h^3} V \int_0^\infty \frac{\sinh^2\theta \cosh^2\theta\, d\theta}{e^{-\alpha+\beta mc^2\cosh\theta} + 1}, \tag{7.121}$$

$$P = 8\pi \frac{m^3c^3kT}{h^3} \int_0^\infty \sinh^2\theta \cosh\theta \ln(1 + e^{\alpha-\beta mc^2\cosh\theta})\, d\theta. \tag{7.122}$$

Integrating by parts, we can write Eq. (7.122) in the form

$$P = \frac{8\pi}{3} \frac{m^4c^5}{h^3} \int_0^\infty \frac{\sinh^4\theta\, d\theta}{e^{-\alpha+\beta mc^2\cosh\theta} + 1}. \tag{7.123}$$

7.2 THE DEGENERATE ELECTRON GAS

As was remarked in Chapter 4 a Fermi-Dirac gas is degenerate if $e^\alpha \gg 1$. Let us now consider the case where $e^\alpha \to \infty$. We can split the integrals in Eqs. (7.120), (7.121), and (7.123) in two parts. We have in that way, for instance, for $N(\mu = \alpha/\beta)$:

$$N = 8\pi \frac{m^3c^3}{h^3} V \left\{ \int_0^{\theta_0} \frac{\sinh^2\theta \cosh\theta \, d\theta}{e^{-\alpha[1-\mu mc^2 \cosh\theta]} + 1} + \int_{\theta_0}^{\infty} \frac{\sinh^2\theta \cosh\theta \, d\theta}{e^{\alpha[\mu mc^2 \cosh\theta - 1]} + 1} \right\}, \tag{7.201}$$

where θ_0 is given by the equation

$$\alpha = \beta mc^2 \cosh\theta_0. \tag{7.202}$$

In the limit $\alpha \to \infty$ the second integral will be equal to zero, and in the first integral the denominator reduces to 1 (compare the discussion of Eq. (4.820)). In a first approximation we get thus for large α the following equations for N, E, and P.

$$N = 8\pi \frac{m^3c^3}{h^3} V \int_0^{\theta_0} \sinh^2\theta \cosh\theta \, d\theta, \tag{7.203}$$

$$E = 8\pi \frac{m^4c^5}{h^3} V \int_0^{\theta_0} \sinh^2\theta \cosh^2\theta \, d\theta, \tag{7.204}$$

$$P = \frac{8\pi}{3} \frac{m^4c^5}{h^3} \int_0^{\theta_0} \sinh^4\theta \, d\theta. \tag{7.205}$$

These equations are exact for a completely degenerate relativistic Fermi-Dirac gas. We shall discuss presently under what circumstances they represent a reasonable approximation for a not completely degenerate gas. Before doing this, however, we shall evaluate the integrals in Eqs. (7.203) to (7.205). Introducing a quantity x by the equation

$$x = \sinh\theta_0, \tag{7.206}$$

we have easily

$$N = \frac{8\pi}{3} \frac{m^3c^3}{h^3} Vx^3, \tag{7.207}$$

$$E = \frac{\pi}{3} \frac{m^4c^5}{h^3} Vg(x), \tag{7.208}$$

$$P = \frac{\pi}{3} \frac{m^4c^5}{h^3} f(x), \tag{7.209}$$

where $f(x)$ and $g(x)$ are given by the equations*

$$f(x) = (2x^3 - 3x)(x^2 + 1)^{1/2} + 3 \operatorname{arc\,sinh} x, \tag{7.210}$$

$$g(x) = 8x^3(x^2 + 1)^{1/2} - f(x). \tag{7.211}$$

The physical meaning of the variable x can most easily be seen as follows. From Eqs. (7.121) and (7.206) it follows that we can also write

$$x = \frac{p_0}{mc}, \tag{7.212}$$

and that by using Eq. (7.112) we can write Eq. (7.207) in the form

$$N = \frac{8\pi}{3} \frac{V}{h^3} p_0{}^3 = \int_0^{p_0} dZ. \tag{7.213}$$

This equation is exact in the case of complete degeneracy. It expresses the fact that all the lowest energy levels up to a limiting energy are fully occupied, as should be expected at the absolute zero, which is the temperature at which complete degeneracy exists (compare the considerations in Section 4.8 and also formula (7.217)). The limiting energy corresponds to a threshold momentum of absolute magnitude p_0.

Equations (7.207), (7.209), and (7.210) give us together the equation of state of a completely degenerate (relativistic) electron gas in parameter form.

It can be shown (see Problem 5 at the end of this chapter) that Eqs. (7.203) to (7.205) are a fair approximation as long as

$$4\pi^2 \frac{x(x^2 + 1)^{1/2}}{f(x)} \ll (\beta mc^2)^2. \tag{7.214}$$

The left-hand side of this inequality depends only on the mean density $n(=N/V)$ of the electron gas, since we have from Eq. (7.207)

$$x = \frac{h}{mc}\left(\frac{3n}{8\pi}\right)^{1/3}. \tag{7.215}$$

Since h/mc, the Compton wavelength of the electron, is only $2\cdot 10^{-10}$ cm, we see that unless n becomes of the order of 10^{30} cm^{-3} (corresponding to densities of at least 10^6 g cm^{-3}), x will always be small compared to one. The function $f(x)$ behaves for small x as $\frac{8}{5}x^5$, and inequality (7.214) can thus nearly always be written in the form

$$c\frac{h^2}{mk}\frac{n^{2/3}}{T} \gg 1, \tag{7.216}$$

* It must be noted that our $g(x)$ differs from the function $g(x)$ introduced by Chandrasekhar (*loc. cit.*), $g_{\text{here}} = g_{\text{Chand}} + 8x^3$. This difference is due to the fact that Chandrasekhar's U does not include the rest mass energy, while our E does.

where c is a numerical const ($c = 10.2$). Expressing T in °K and n in cm^{-3}, we have from inequality (7.216)

$$3.6 \cdot 10^{-9} \frac{n^{2/3}}{T} \gg 1 \tag{7.217}$$

as the condition for the validity of Eqs. (7.207) to (7.209).*

We remarked a moment ago that x is practically always small compared to one; we can in that case simplify Eqs. (7.208) and (7.209) by using the asymptotic expressions for $f(x)$ and $g(x)$,

$$f(x) \approx \tfrac{8}{5} x^5, \qquad g(x) \approx 8x^3 + \tfrac{12}{5} x^5. \tag{7.218}$$

In that way we get for E and P the expressions

$$E = Nmc^2 + \frac{4\pi}{5} \frac{m^4 c^5}{h^3} V x^5, \qquad (x \ll 1), \tag{7.219}$$

$$P = \frac{8\pi}{15} \frac{m^4 c^5}{h^3} x^5, \qquad (x \ll 1). \tag{7.220}$$

Denoting the kinetic energy $E - Nmc^2$ by $\mathscr{T}$ and using Eq. (7.215) for x, we get the equations

$$\mathscr{T} = \frac{3}{40} \left(\frac{3}{\pi}\right)^{2/3} \frac{h^2}{m} V n^{5/3}, \qquad (x \ll 1), \tag{7.221}$$

$$PV = \tfrac{2}{3} \mathscr{T}, \qquad (x \ll 1). \tag{7.222}$$

Equation (7.222) is the same as Eq. (4.715). The derivation of Eq. (4.715) is now no longer always applicable since the energy ε of one electron now includes the rest energy, so that it is not always true that ε' is proportional to $V^{-2/3}$. Indeed, we shall presently see a case where Eq. (7.222) is not valid.

In some astrophysical applications we may encounter cases where $x \gg 1$, for instance, in stellar interiors where n may exceed $10^{30}\ \text{cm}^{-3}$. In that case we have for $f(x)$ and $g(x)$ the asymptotic expressions

$$f(x) \approx 2x^4, \qquad g(x) \approx 6x^4. \tag{7.223}$$

Inequality (7.214) reduces also in this case to an inequality of the form (7.216) or (7.217), the only difference being a factor $\frac{1}{2}\sqrt{5}$ in the numerical constant.

We get easily for E, which is now in a first approximation equal to $\mathscr{T}$,

* It can easily be verified that this condition is equivalent to the condition $e^\alpha \gg 1$ (compare Eq. (4.703)) or to the condition $\eta \ll 1$ which we met in Section 4.8 as the condition for the applicability of the formulas for complete degeneracy in the nonrelativistic case.

and for P the expressions

$$\mathscr{T} \cong E = 2\pi \frac{m^4c^5}{h^3} Vx^4, \qquad (x \gg 1) \qquad (7.224)$$

$$P = \frac{2\pi}{3} \frac{m^4c^5}{h^3} x^4, \qquad (x \gg 1) \qquad (7.225)$$

or

$$\mathscr{T} = \frac{3}{8}\left(\frac{3}{\pi}\right)^{1/3} hcVn^{4/3}, \qquad (x \gg 1) \qquad (7.226)$$

$$PV = \tfrac{1}{3}\mathscr{T}, \qquad (x \gg 1) \qquad (7.227)$$

and we see that now, indeed, Eq. (4.715) is no longer valid. The reason is that while in the case $x \ll 1$ the kinetic energy is small compared to the rest energy so that the nonrelativistic case is a good approximation, in the case $x \gg 1$ the rest energy is small compared to the kinetic energy and the non-relativistic approximation breaks down.

It must, however, be mentioned that one can derive Eqs. (7.222) and (7.227) in an elementary way completely similar to the derivation of Eq. (4.715). The energy levels ε_k now satisfy the Eq. (7.107). If the kinetic energy is small compared to the rest mass energy, we have approximately $\varepsilon_k = mc^2 + \varepsilon_k'$ where ε_k' is proportional to $V^{-2/3}$ and hence P will be equal to $2\mathscr{T}/3V$. If the rest mass, on the other hand, is small compared to the kinetic energy, ε_k is proportional to $V^{-1/3}$ and P will be equal to $E/3V$, or $\mathscr{T}/3V$. The question whether or not the gas is degenerate does not enter into these considerations and we can thus expect to find a similar situation in the next section (see Eqs. (7.313) and (7.320)).

7.3 THE NONDEGENERATE ELECTRON GAS

In this last section we shall consider the case where the temperature is so high or the density is so low that $e^\alpha \ll 1$. In that case the term 1 in the denominator of the integrands in Eqs. (7.120) and (7.121) can be neglected and the logarithm in the integrand in Eq. (7.122) can be expanded. We then get

$$N = 8\pi \frac{m^3c^3}{h^3} Ve^\alpha \int_0^\infty e^{-\beta mc^2 \cosh\theta} \sinh^2\theta \cosh\theta \, d\theta, \qquad (7.301)$$

$$E = 8\pi \frac{m^4c^5}{h^3} Ve^\alpha \int_0^\infty e^{-\beta mc^2 \cosh\theta} \sinh^2\theta \cosh^2\theta \, d\theta, \qquad (7.302)$$

$$P = 8\pi \frac{m^3c^3}{h^3} kTe^\alpha \int_0^\infty e^{-\beta mc^2 \cosh\theta} \sinh^2\theta \cosh\theta \, d\theta. \qquad (7.303)$$

First of all, we note that the perfect gas law is identically true, since from Eqs. (7.301) and (7.303) it follows that

$$PV = NkT. \tag{7.304}$$

We see that, irrespective of the inclusion or exclusion of relativistic effects, the Boyle–Gay-Lussac formula applies in the nondegenerate case.

The integrals in Eqs. (7.301) and (7.302) can be expressed in terms of Bessel functions (see Chandrasekhar, *loc. cit.*), but it would be beyond the scope of this chapter to enter into a detailed discussion. However, we shall discuss briefly the two limiting cases $\beta mc^2 \ll 1$ and $\beta mc^2 \gg 1$.

Putting

$$\beta mc^2 = a \tag{7.305}$$

and introducing as a new integration variable the quantity u defined by the equation

$$u = a \cosh \theta, \tag{7.306}$$

we have from Eqs. (7.301) to (7.303):

$$N = \frac{PV}{kT} = A \int_a^\infty e^{-u}\sqrt{u^2 - a^2}\, u\, du, \tag{7.307}$$

$$E = AkT \int_a^\infty e^{-u}\sqrt{u^2 - a^2}\, u^2\, du, \tag{7.308}$$

where

$$A = 8\pi \frac{m^3c^3}{h^3a^3} Ve^{\alpha}. \tag{7.309}$$

We now consider the two limiting cases $a \to 0$ and $a \to \infty$.

1. In the limit as $a \to 0$ we have from Eqs. (7.307) and (7.308)

$$N = \frac{PV}{kT} \cong A \int_0^\infty e^{-u}\, u^2\, du = 2A, \tag{7.310}$$

$$E \cong AkT \int_0^\infty e^{-u}\, u^3\, du = 6AkT. \tag{7.311}$$

Since it follows from Eqs. (7.304) and (7.305) that

$$\frac{Nmc^2}{PV} = a, \tag{7.312}$$

we see that for $a \ll 1$ $Nmc^2 \ll PV = 2AkT = \frac{1}{3}E$. This means that the rest energy is small compared to E. We get thus for the relation between the kinetic energy $\mathcal{T}$ and PV

$$PV = \tfrac{1}{3}E \cong \tfrac{1}{3}\mathcal{T} \tag{7.313}$$

We could have expected a deviation from the nonrelativistic relation, since we are now in the region of the very high temperatures and energies where relativistic effects make Eq. (7.222) inapplicable.

2. In the limit $a \to \infty$ it is more advantageous to introduce a new variable v by the equation

$$v^2 = u - a, \tag{7.314}$$

and we get from Eqs. (7.307) and (7.308)

$$N = \frac{PV}{kT} = B \int_0^\infty e^{-v^2} \sqrt{(v^4 + 2av^2)}\,(v^2 + a)\, v\, dv, \tag{7.315}$$

$$E = BkT \int_0^\infty e^{-v^2} \sqrt{(v^4 + 2av^2)}\,(v^2 + a)^2 v\, dv, \tag{7.316}$$

where

$$B = 2Ae^{-a} \tag{7.317}$$

For large values of a we may expand the integrands in powers of v^2/a, in that way we get

$$N = \frac{PV}{kT} = Ba\sqrt{2a} \int_0^\infty e^{-v^2} v^2\, dv = \tfrac{1}{2}\, Ba\sqrt{2\pi a}. \tag{7.318}$$

$$\begin{aligned} E &= BkT\left[a^2\sqrt{2a}\int_0^\infty e^{-v^2}v^2\,dv + a\sqrt{2a}\int_0^\infty e^{-v^2}v^4\,dv\right] \\ &= Nmc^2 + \tfrac{3}{4}\,BkTa\sqrt{2\pi a}, \end{aligned} \tag{7.319}$$

and we get for the relation between $\mathcal{T}$ and PV

$$E - Nmc^2 = \mathcal{T} = \tfrac{3}{2}PV, \tag{7.320}$$

as we should expect, since now $\mathcal{T} \ll Nmc^2$.

PROBLEMS

1. Discuss the changes to be made in the formalism of this chapter, if the particles obeyed Boltzmann or Bose-Einstein rather than Fermi-Dirac statistics.
2. Prove Eqs. (7.208) and (7.209).
3. Prove Eq. (7.223).

4. Use the methods of Section 4.8 to find the first corrections to Eqs. (7.207) to (7.209).
5. Use the results of the preceding problem to prove inequality (7.214).
6. Find expressions for the Fermi level, up to the first correction term, and for the specific heat of a nearly completely degenerate relativistic fermion gas.
7. Find an expression for the paramagnetic susceptibility of a nearly completely degenerate strongly relativistic electron gas.
8. Prove that the perfect gas law $PV = NkT$ holds for a perfect Boltzmann gas, independent of whether it is relativistic or not.
9. For a nonrelativistic Boltzmann gas the average kinetic energy equals $\frac{3}{2}kT$. Find the corresponding expression that equals $\frac{3}{2}kT$ for a relativistic Boltzmann gas.
10. Prove that for a Boltzmann gas of noninteracting particles the ultrarelativistic specific heat is twice that of the nonrelativistic one.
11. Prove that a two-dimensional ultrarelativistic boson gas shows the phenomenon of the Einstein condensation, while a one-dimensional ultrarelativistic boson gas does not condense.*
12. Prove that the specific heats of two-dimensional ultrarelativistic perfect boson and fermion gases are identical.†
13. Show that if the temperature is sufficiently high that pair production can take place, in the limit as $T \to \infty$, the particle density will be proportional to T^3, independent of statistics.
14. Consider a system with an electron number N_0 at $T = 0$. Find the ratio N^+/N^- of the number of positrons to the number of electrons, as a function of temperature.

BIBLIOGRAPHICAL NOTES

Planck and his pupil Jüttner developed relativistic statistics for the classical case:

1. M. Planck, *Ann. Physik*, **26,** 1, 1908.
2. F. Jüttner, *Ann. Physik*, **34,** 856, 1911; **35,** 145, 1911.

The treatment in this chapter is mainly based on the following reference to which we also refer for a more extensive bibliography:

3. S. Chandrasekhar, *An Introduction to the Study of Stellar Structure*, Chicago 1939, Chap. X.

* See, R. M. May, *Phys. Rev.*, **135,** A1515, 1964.
† R. M. May, *ibid.*

See also

4. R. H. Fowler, *Statistical Mechanics*, Cambridge 1936, Sections 16.3–16.5.

For a discussion of recent work in this field, see

5. J. Meixner (ed.), *Statistical Mechanics of Equilibrium and Non-equilibrium*, Amsterdam, 1965; especially the articles by P. Havas, I. Prigogine, and P. T. Landsberg and J. Dunning-Davies.
6. P. T. Landsberg and J. Dunning-Davies, *Phys. Rev.*, **138,** A1049, 1965.

Section 7.1. Formula (7.110) appears for the first time in

7. P. A. M. Dirac, *Proc. Roy. Soc.* (*London*), **A112,** 671, 1926.

See also

8. P. Jordan, *Statistische Mechanik auf Quantentheoretischer Grundlage*, Braunschweig 1933, p. 20.
9. C. Møller and S. Chandrasekhar, *Monthly Notices Roy. Astron. Soc.*, **95,** 673, 1935.
10. R. Peierls, *Monthly Notices Roy. Astron. Soc.*, **96,** 780, 1936.
11. E. K. Broch, *Phys. Rev.*, **51,** 586, 1937, and Reference 3, Section 2.

The general formulas for the case of an electron gas can be found in Section 5 of Reference 3. Our representation differs slightly from Chandrasekhar's since we have used ε and not ε' as variable. This entails that α includes a term βmc^2 and that our E differs from Chandrasekhar's U by a term Nmc^2.

Formulas (7.123) to (7.125) are due to Jüttner:

12. F. Jüttner, *Z. Physik*, **47,** 542, 1928.

Section 7.2. See Reference 3, Section 6.

Section 7.3. Reference 3, Section 7; Reference 2; and also

13. R. C. Tolman, *Phil. Mag.*, **28,** 583, 1914.

This last reference also contains other applications of relativistic statistics. General remarks on relativistic statistics can also be found in Pauli's article on the theory of relativity in

14. W. Pauli, *Enzyklopädie der mathematischen Wissenschaften*, Vol. V. Part 19, Leipzig-Berlin, 1921, Sections 48 and 49.

8

Fluctuations and Irreversible Processes

8.1 FLUCTUATIONS

In Chapters 5 and 6 we briefly showed how one can evaluate fluctuations and correlations, once the Helmholtz free energy (in the case of a macrocanonical ensemble) or the q-potential (in the case of a grand canonical ensemble) are known. In the present chapter we shall give a slightly more detailed discussion of various aspects of fluctuation theory.

We shall consider, to begin with, the approach to fluctuations that was first developed by Einstein in his considerations of critical opalescence. Although Einstein based his discussion upon the relation between entropy and probability (compare Eq. (4.413) and see Problem 1 at the end of this chapter), we shall use ensemble theory. We first of all note that in a grand canonical ensemble, the quantity

$$P(\nu;p,q;a_k)\,d\Omega = e^{-q+\alpha\nu-\beta\varepsilon}\,d\Omega \tag{8.101}$$

gives us the probability that there are in the ensemble—for which we have taken to simplify the discussion a classical, one-component ensemble—systems with ν particles and with p- and q-values lying within the extension in phase $d\Omega$ (compare Eq. (5.0906)). Let us now look at the fluctuations in some variable, λ say. To find the probability that a certain value of λ is realized, we must take the sum of all $P(\nu;p,q;a_k)$ compatible with that value of λ. We thus find

$$P(\lambda)\,d\lambda = \int_{\lambda}^{\lambda+d\lambda} d\lambda \sum_{\nu} \int d\Omega\, e^{-q+\alpha\nu-\beta\varepsilon}, \tag{8.102}$$

where the summation over ν and the integration over phase space is performed only over those regions that are compatible with λ lying between λ and $\lambda + d\lambda$. If we introduce a quantity $q(\lambda)$ by the equation

$$e^{q(\lambda)} = \left[\sum_{\nu} \int d\Omega\, e^{\alpha\nu-\beta\varepsilon}\right]_{\lambda<\lambda<\lambda+d\lambda}, \tag{8.103}$$

we can write Eq. (8.102) in the form

$$P(\lambda) = e^{-\Delta q}, \quad \Delta q = q - q(\lambda), \tag{8.104}$$

where q is the equilibrium value of q.* If $\langle\lambda\rangle$ is the equilibrium value of λ, and $\Delta\lambda = \lambda - \langle\lambda\rangle$, it follows from the thermodynamical meaning of q that $\Delta q/\beta$ is equal to the minimum work ΔA required to produce the fluctuation $\Delta\lambda$. We can therefore rewrite Eq. (8.104) in the form

$$P(\lambda) = e^{-\beta\Delta A}, \tag{8.105}$$

and one can easily convince oneself that this formula is valid independent of which kind of ensemble we are considering. If we are considering a macro-canonical rather than a grand canonical ensemble, ΔA is equal to the change in the Helmholtz free energy.†

In most cases of interest—the main exception being the case of critical fluctuations (see Section 8.2)—$\beta\Delta A \gg 1$ as soon as λ becomes appreciably different from $\langle\lambda\rangle$; this is, of course, stating in a different way that almost all fluctuations are negligibly small, a fact appealed to repeatedly in earlier

* We should note that in our derivation we have neglected a possible factor in $P(\lambda)$ depending on λ, which follows from the fact that we are integrating over a range of λ-values. For a discussion of this factor see, for instance, R. C. Tolman, *The Principles of Statistical Mechanics*, Oxford, 1938, Section 141.

† Although there is a superficial resemblance between expression (8.105) and the Maxwell-Boltzmann distribution, it must be emphasized that ΔA is here a free energy and not an internal energy (compare also Problem 5 of Chapter 5).

chapters. As it moreover follows from the fact that q corresponds to the equilibrium situation that $[\partial q/\partial\lambda]_{\lambda=\langle\lambda\rangle}=0$, we can write to a fair approximation

$$\Delta q=\frac{1}{2}\frac{\partial^2 q}{\partial\lambda^2}(\lambda-\langle\lambda\rangle)^2, \tag{8.106}$$

and we see that to this approximation the probability distribution (8.104) is a Gaussian one. It also follows easily that the mean square deviation in λ satisfies the equation

$$\langle(\lambda-\langle\lambda\rangle)^2\rangle=\left[\frac{\partial^2 q}{\partial\lambda^2}\right]^{-1}. \tag{8.107}$$

Although this formula looks a simple one, it is often difficult to apply, since it is not at all obvious, in general, which variables must be kept constant when the derivative is taken with respect to λ.

For many applications it is advantageous to use the following thermodynamic expression for the minimum work ΔA:*

$$\Delta A=\tfrac{1}{2}(\Delta S\,\Delta T-\Delta P\,\Delta V), \tag{8.108}$$

and Eq. (8.105) becomes

$$P(\lambda)=e^{-\beta(\Delta S\Delta T-\Delta P\Delta V)/2}. \tag{8.109}$$

To conclude this section we wish to discuss briefly the various kinds of probability that can be applied in statistical mechanics. When discussing Eq. (8.101), we called $P(\nu;p,q;a_k)$ the probability to find a system in the grand canonical ensemble with ν particles and with p,q within well-defined limits. On the other hand, we have often discussed (see, for instance, Section 5.13) that the probability to find p,q within an extension of phase $d\Omega$ should be proportional to $d\Omega$. The answer to this apparent paradox is that in defining probabilities we must state the conditions under which these probabilities are defined. Equal probabilities for equal volumes in phase space is a valid assumption, provided either no external conditions are imposed or the total energy, E, total number of particles, N, and all the a_k are given. This means that we have, for a microcanonical ensemble,

$$P=1, \tag{8.110}$$

where we have tacitly assumed that we are only considering those systems for which E and N have their selected values.

In a macrocanonical ensemble, we have instead of Eq. (8.110)

$$P=e^{-\beta\varepsilon}, \tag{8.111}$$

as follows, for instance, from the formalism of Section 5.3. This means that

* See, for instance, L. D. Landau and E. M. Lifshitz, *Statistical Physics*, Oxford, 1958, p. 351.

Eq. (8.111) holds when the a_k and the total number of particles is given, but instead of E the value of β is given.

If instead of E and N we give β and α, we get for the probability Eq. (8.101):

$$P = e^{\alpha\nu - \beta\varepsilon}, \tag{8.112}$$

where now P, as in Eqs. (8.110) and (8.111), is no longer normalized.

The thermodynamic properties of a system represented by the various ensembles are determined by the appropriate partition functions

$$Z = \Sigma P, \tag{8.113}$$

as follows either directly from the fact that the P are the probabilities, or from Eqs. (5.0302), (5.0535), and (5.0907).

We note also that the thermodynamic potentials corresponding to ln Z are for the microcanonical, the macrocanonical, and the grand canonical ensembles, respectively, S/k, $-\beta F$, and q, and that we can get from each one of them to each of the others through a Legendre transformation, since

$$\frac{S}{k} = q - \alpha N + \beta E = \beta E - \beta F. \tag{8.114}$$

So far we have treated the external parameters as given. We know from thermodynamics that apart from S, and F, the enthalpy H and the Gibbs free energy G can also be used as thermodynamic potentials. This suggests that there may be other ensembles that could be considered. To simplify our discussion, we shall only consider the case where a_k is the volume, and we can then generalize by considering the possibility that not V is given, but the pressure P. This corresponds to considering systems that are *mechanically open*.* We shall consider systems which are also thermally open, but which may be open with respect to particle exchange with their surroundings or closed.

Since we have the thermodynamic relations

$$\frac{S}{k} = \beta E + PV - \beta G \tag{8.115}$$

and

$$q = \alpha N + PV - \beta G, \tag{8.116}$$

we expect that for a system in thermal and mechanical contact with its surroundings, but with fixed N, we should have

$$P = e^{-\beta\varepsilon - PV}, \tag{8.117}$$

where we have to bear in mind that now different systems in the ensemble—which is called an isothermal-isobaric ensemble by Hill†‡—may correspond to different values of V.

* See, for instance, D. ter Haar and H. Wergeland, *Elements of Thermodynamics*, Reading, Mass., 1966, Chap. III.

† T. L. Hill, *Statistical Mechanics*, New York, 1956. See also: A. Münster, *Statistische Thermodynamik*, Berlin, 1956.

‡ The isothermal-isobaric ensemble is called the *T-P*-distribution by Kubo (*Statistical Mechanics*, Amsterdam, 1965) and is also known as the pressure ensemble.

For a system in thermal, mechanical, and material contact with its surroundings, we have what Hill* calls a generalized ensemble, and P now is given by the equation

$$P = e^{\alpha\nu - \beta\varepsilon - PV}. \tag{8.118}$$

Corresponding to the partition function following from Eq. (8.117) we have

$$-\beta G = \ln Z, \tag{8.119}$$

but it follows from Eqs. (8.116) and (5.0926) that there is no thermodynamic potential corresponding to the generalized ensemble.

8.2 SPECIAL CASES

As illustrations of the theory of Chapters 5 and 6 and of the preceding chapter, we shall in the present chapter consider (1) fluctuations in energy, (2) fluctuations in density, (3) fluctuations in pressure, and (4) critical fluctuations. A few more examples are discussed in Problems 12 and 14 of Chapter 8.

We first of all consider energy fluctuations. From Eq. (5.103) we have

$$\langle(\varepsilon - \langle\varepsilon\rangle)^2\rangle = -\frac{\partial\langle\varepsilon\rangle}{\partial\beta} = kT^2C_v, \tag{8.201}$$

where C_v is the specific heat at constant volume. Usually, C_v will be of the order of magnitude of the number of particles in the system, and as $\langle\varepsilon\rangle$ is of the same order of magnitude, the relative fluctuations in energy will be of the order of $\langle\nu\rangle^{-1}$, and thus negligibly small (compare Eq. (5.1210)).

There are, however, cases where the relative fluctuations may become large. One example is provided by the case of Problem 19 at the end of Chapter 8. Another case was suggested by Gibbs. Consider a box containing a liquid. Over the liquid is saturated vapor and on top of that rests a piston (see Fig. 8.1). If this system is placed in a thermostat, it will have an infinite specific heat, as any amount of heat taken up or given off by the system will result in evaporation or condensation of vapor, but not in a change of temperature. Energy fluctuations may thus become arbitrarily large, provided not all of the substance is either evaporated or condensed.

Let us now turn to density fluctuations. From Eq. (5.101) we have

$$\frac{\langle(\nu - \langle\nu\rangle)^2\rangle}{\langle\nu\rangle^2} = \frac{1}{\langle\nu\rangle^2}\frac{\partial\langle\nu\rangle}{\partial\alpha}. \tag{8.202}$$

From Eq. (5.0922) and the fact that for a homogeneous system $\beta G = \alpha\,\langle\nu\rangle$, we find that

$$\langle\nu\rangle\delta\alpha = \beta V\delta P \quad (\beta\text{: const}) \tag{8.203}$$

* T. L. Hill, *op. cit.*

and thus

$$\langle v \rangle \left(\frac{\partial \alpha}{\partial \langle v \rangle}\right)_{\beta,V} = -\frac{\beta V^2}{\langle v \rangle}\left(\frac{\partial P}{\partial V}\right)_{\beta,\langle v \rangle}, \tag{8.204}$$

where we have used the relation

$$\left(\frac{\partial y}{\partial x}\right)_z = -\left(\frac{\partial y}{\partial z}\right)_x \left(\frac{\partial z}{\partial x}\right)_y,$$

which follows from writing

$$\delta y = \left(\frac{\partial y}{\partial x}\right)_z \delta x + \left(\frac{\partial y}{\partial z}\right)_x \delta z,$$

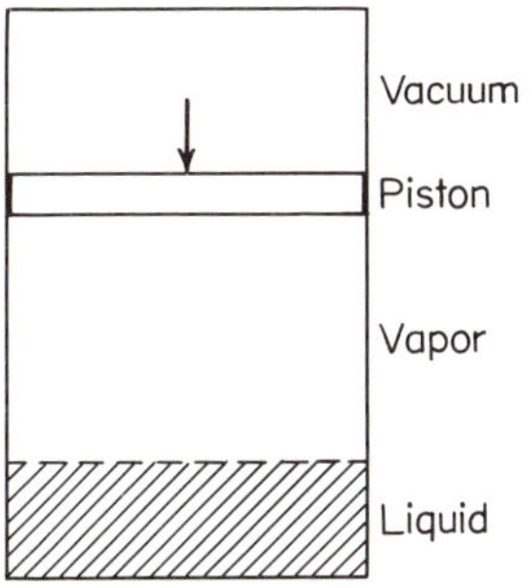

FIG. 8.1 A system in which very large energy fluctuations will occur.

and substituting in it a similar relation for δz in terms of δy and δx, and the fact that for a homogeneous system

$$\left(\frac{\partial \langle v \rangle}{\partial V}\right)_P = \frac{\langle v \rangle}{V}.$$

We can thus rewrite Eq. (8.202) in the form

$$\frac{\langle (v - \langle v \rangle)^2 \rangle}{\langle v \rangle^2} = -\frac{1}{\beta V^2}\frac{\partial V}{\partial P} = \frac{\kappa}{\beta V}, \tag{8.205}$$

where κ is the compressibility,

$$\kappa = -\frac{1}{V}\frac{\partial V}{\partial P}. \tag{8.206}$$

An interesting problem is posed by the fluctuations in the density of bosons.

We found for this Eq. (6.707):

$$\langle (n_s - \langle n_s \rangle)^2 \rangle = \langle n_s \rangle + \langle n_s \rangle^2, \tag{8.207}$$

or Eq. (6.713):

$$\langle (N_i - \langle N_i \rangle)^2 \rangle = \langle N_i \rangle + \sum{}' \langle n_s \rangle^2, \tag{8.208}$$

where n_s is the occupation number for the state ε_s and $N_i = \sum' n_s$, where the summation is over the ith group of levels. The first term on the right-hand side of Eqs. (8.207) and (8.208) is the term that also occurs in the case of Boltzmann statistics. Einstein in his paper on the perfect boson gas pointed out that this term arises from the particle nature of the bosons. The second term, on the other hand, arises from the wave nature of the bosons. To illustrate this, let us consider the case of black-body radiation so that we can identify n_s and N_i, respectively, with the number of oscillators of frequency ω (compare Section 3.3) and the number of oscillators with frequencies between ω and $\omega + d\omega$. Multiplying Eq. (8.208) by $(\hbar\omega)^2$, we get from it the relation

$$\begin{aligned} \langle (\rho(\omega)\, d\omega - \langle \rho(\omega) \rangle\, d\omega)^2 \rangle &= \hbar\omega \langle \rho(\omega) \rangle\, d\omega + \frac{V \langle \rho(\omega) \rangle^2\, d\omega}{D(\omega)} \\ &= \hbar\omega \langle \rho(\omega) \rangle\, d\omega + \frac{\pi^2 c^3 \langle \rho(\omega) \rangle^2\, d\omega}{\omega^2} \end{aligned} \tag{8.209}$$

where $\rho(\omega)$ is the radiation energy density, and $D(\omega)$ is given by Eq. (3.307). The first term is again the "particle term." On the other hand, in the limit as $\hbar \to 0$, when we should regain the classical wave picture of electromagnetic radiation, only the second term remains.*

These results are of importance for photon fluctuation experiments which can now be performed much more easily as a result of the high intensity laser beams that are available for such experiments. Hanbury Brown and Twiss† have reported such large fluctuations. To see simply why there should be large fluctuations, Purcell considered the interference of two coherent wave trains. If each of the wave trains contains one photon, when they overlap, a packet may emerge with anything between 0 and 4 photons (the amplitudes can be added, and the number of photons is proportional to the intensity).

We now come to the subject of pressure fluctuations. We discussed these briefly in Section 5.8 for the case of a perfect gas where we could invoke Eq. (5.0811). In the discussion we then emphasized, as we had done in Section

* Heisenberg (*Physical Principles of Quantum Theory*, Chicago, 1930, Section 5.7) has given a quantum-mechanical derivation of Eq. (8.208). This equation can also be derived directly from Eq. (8.201); see Problem 8 at the end of this chapter.

† R. Hanbury Brown and R. Q. Twiss, *Nature*, **177**, 27, 1956; R. Q. Twiss, A. D. Little, and R. Hanbury Brown, *Nature*, **180**, 324, 1957. For a discussion, see E. M. Purcell, *Nature*, **178**, 1449, 1956, and also G. J. Troup, *Bull. Inst. Phys. and Phys. Soc.*, **16**, 257, 1965.

5.4 that the physical meaning of a pressure naturally involves the idea that one is dealing with a time-averaged quantity. The question of pressure fluctuations in a perfect gas has been the subject of an extensive discussion between Klein, Münster, and Wergeland. As this topic provides us both with an instructive example of the application of fluctuation theory and with an occasion to clear up a few difficult conceptual points, we shall discuss it here in some detail.

To begin with, we remind ourselves that according to Eq. (5.102) the fluctuations in the pressure are given by the formula

$$\langle (P - \langle P \rangle)^2 \rangle = \frac{1}{\beta}\frac{\partial \langle P \rangle}{\partial V} + \frac{1}{\beta}\left\langle \frac{\partial^2 \varepsilon}{\partial V^2} \right\rangle, \tag{8.210}$$

where, of course, P is here not a truly macroscopic quantity—as was pointed out by Münster—but the microscopic quantity $-\partial\varepsilon/\partial V$. We have earlier in Section 5.4 pointed out that the second term on the right-hand side of Eq. (8.210) becomes infinite for an infinitely steep wall potential. This is not surprising, since it corresponds to the instantaneous infinite forces exerted by the wall on the gas particles when they collide with the wall. The fact that the fluctuations depend crucially upon the nature of the potential field near the wall by itself shows that we must tread extremely warily. One can evaluate the right-hand side of Eq. (8.210) for particular choices of the wall potential. Fowler* used a potential that varies as an inverse power of the distance of the molecule from the wall and found that

$$\frac{\langle (P - \langle P \rangle)^2 \rangle}{\langle P \rangle^2} = A \cdot N^{-2/3}, \tag{8.211}$$

that is, small fluctuations, but fluctuations that are larger than normal ones which are proportional to N^{-1}. Fowler also found that the constant A tended to infinity when the potential became infinitely steep. A different potential was investigated by Wergeland (see Problem 19 at the end of Chapter 8); he found normal fluctuations, but again fluctuations that tended to infinity as the steepness of the wall potential increased.

To see why it is reasonable to expect Eq. (8.210) to give nonphysical results, we must make clear to ourselves what kind of fluctuations we are discussing. Strictly speaking, Eq. (8.210) tells us the spread in the values of P we can expect when we are looking successively at the different systems which together make up our ensemble. As $\langle P \rangle$ itself is finite, we know that the chance of picking out of the ensemble a system for which we get an infinite pressure is infinitesimal. In an actual physical situation we are looking at the fluctuations in the pressure observed *in one system*. If we treat an ensemble as

* R. H. Fowler, *Statistical Mechanics*, 2d edition, Cambridge, 1936, p. 756.

representing the time evolution of a single system, that is, as mathematically equivalent to a microcanonical ensemble and hence, through the ergodic theorem, as determining the time averages of phase functions, we are back to the approach used in Section 5.8 and incidentally to the approach favored by Münster. However, there is a different way of looking at Eq. (8.210). When we are considering fluctuations in a system represented by an ensemble, and when we are then using the formalism of ensemble theory to derive Eq. (8.210), we are comparing two situations with slightly different values of V (compare the discussion in Section 5.9). That means that we are investigating how small reversible changes in the a_k affect the system represented by the ensemble. In order that we can throughout use the ensemble formalism, it is necessary—as was emphasized by Wergeland—that the change $a_k \to a_k + \delta a_k$ must be quasistatic as well as infinitesimal, which means that it must be an *adiabatic* change. This means, in turn, that we should have specified in writing $\partial^2\varepsilon/\partial V^2$ *not* that this derivative was taken for const $p,q,$—this is shown by Wergeland to correspond to infinitely fast changes in V—but for constant entropy. This means that, strictly speaking, Eq. (5.102) should be written in the form

$$\langle(A_k - \langle A_k\rangle)^2\rangle = \frac{1}{\beta}\frac{\partial\langle A_k\rangle}{\partial a_k}\bigg|_{\alpha,\beta} + \frac{1}{\beta}\left\langle\frac{\partial A_k}{\partial a_k}\bigg|_S\right\rangle. \tag{8.212}$$

If we apply the corresponding equation for the macrocanonical ensemble, we can use thermodynamic relations for the perfect gas to derive again Eq. (5.0812). On the other hand, for a perfect gas in a cubical vessel one finds easily for the action integral J_i, which is given by the formula*

$$J_i = \oint p_i \, dq_i, \tag{8.213}$$

which for a particle moving with a kinetic energy $\mathscr{T}_i$ corresponding to q_i leads easily to

$$J_i = 2V^{1/3}\sqrt{2m\mathscr{T}_i}, \tag{8.214}$$

so that the energy of the system, expressed in terms of the J_i is given by the equation

$$\varepsilon = \sum_{i=1}^{3N} \mathscr{T}_i = \sum_i \frac{J_i^2}{8mV^{2/3}}, \tag{8.215}$$

which leads to $(\partial^2\varepsilon/\partial V^2)_s = 10\varepsilon/9V^2$, and this together with expression (5.0803) for the average value of the energy in a macrocanonical ensemble and the relation $P = 2\bar{\varepsilon}/3V$ again leads to Eq. (5.0812).

* See, for instance, D. ter Haar, *Elements of Hamiltonian Mechanics*, Amsterdam, 1961, Section 6.2.

To conclude this section we shall consider critical fluctuations. We saw earlier that the fluctuations in density are given by Eq. (8.205). In a critical point $\partial P/\partial V$ vanishes and we should therefore expect large density fluctuations. Smoluchowski suggested that this was the reasons for the Tyndall effect of critical opalescence and the theory was worked out by Einstein. Although we now know that a proper theory of the phenomenon of critical opalescence need take into account the correlation between different regions of the system, we shall first show what happens to the density fluctuations if we treat them simple-mindedly and apply ensemble theory to the whole system, using the formalism of the grand canonical ensemble.

From the considerations of Section 5.9 we find for $\langle(\nu - \langle\nu\rangle)^2\rangle$

$$\langle(\nu - \langle\nu\rangle)^2\rangle = \sum_{\nu} \frac{e^{-q+\alpha\nu}}{\nu!} (\nu - \langle\nu\rangle)^2 \int e^{-\beta\varepsilon} d\Omega. \tag{8.216}$$

Changing the sum over ν to an integral, we can write the normalization condition,

$$1 = \sum_{\nu} \frac{e^{-q+\alpha\nu}}{\nu!} \int e^{-\beta\varepsilon}\, d\Omega, \tag{8.217}$$

in the form

$$1 = \int d\nu \int d\varepsilon\, e^{-\Phi(\nu,\varepsilon)-\beta\varepsilon}, \tag{8.218}$$

where $\Phi(\nu,\varepsilon)$ is a generalization of the $\phi(\varepsilon)$ of Section 5.5. From Eq. (8.218) we get now, instead of Eq. (5.0930) the relation

$$\beta P = -\left\langle \left(\frac{\partial \Phi}{\partial V}\right)_{\beta} \right\rangle. \tag{8.219}$$

From the general behavior of statistical quantities we may expect that the integrand in Eq. (8.218), and in other equations, will be steeply peaked around the value $\nu = \langle\nu\rangle$. Expanding Φ as function of ν in the neighborhood of this value, we have

$$\Phi(\nu) = \Phi(\langle\nu\rangle) + (\nu - \langle\nu\rangle) \left.\frac{\partial \Phi}{\partial \nu}\right|_{\nu=\langle\nu\rangle} + \frac{1}{2}(\nu - \langle\nu\rangle)^2 \left.\frac{\partial^2 \Phi}{\partial \nu^2}\right|_{\nu=\langle\nu\rangle} + \frac{1}{3!}(\nu - \langle\nu\rangle)^3 \left.\frac{\partial^3 \Phi}{\partial \nu^3}\right|_{\nu=\langle\nu\rangle} + \frac{1}{4!}(\nu - \langle\nu\rangle)^4 \left.\frac{\partial^4 \Phi}{\partial \nu^4}\right|_{\nu=\langle\nu\rangle} + \cdots. \tag{8.220}$$

The partial derivatives of Φ occurring in Eq. (8.220) can be expressed in terms of partial derivatives with respect to V, and thus in terms of the partial derivatives of P with respect to V, as follows:

$$\left.\frac{\partial^2\Phi}{\partial v^2}\right|_{v=\langle v\rangle} = \frac{V^2}{\langle v\rangle^2}\left.\frac{\partial^2\Phi}{\partial V^2}\right|_{v=\langle v\rangle} = -\frac{V^2}{\beta\langle v\rangle^2}\frac{\partial P}{\partial V}, \tag{8.221}$$

$$\left.\frac{\partial^3\Phi}{\partial v^3}\right|_{v=\langle v\rangle} = \frac{3V^2}{\beta\langle v\rangle^3}\frac{\partial P}{\partial V} + \frac{V^3}{\beta\langle v\rangle^3}\frac{\partial^2 P}{\partial V^2}, \tag{8.222}$$

$$\left.\frac{\partial^4\Phi}{\partial v^4}\right|_{v=\langle v\rangle} = -\frac{12V^2}{\beta\langle v\rangle^4}\frac{\partial P}{\partial V} - \frac{8V^3}{\beta\langle v\rangle^4}\frac{\partial^2 P}{\partial V^2} - \frac{V^4}{\beta\langle v\rangle^4}\frac{\partial^3 P}{\partial V^3}. \tag{8.223}$$

In the critical point both $\partial P/\partial V$ and $\partial^2 P/\partial V^2$ vanish so that we need the last term written out on the right-hand side of Eq. (8.220). Substituting Eq. (8.220) into Eq. (8.216), after we have changed the sum over v to an integration, and bearing in mind that $\partial\Phi/\partial v$ vanishes for $v = \langle v\rangle$, we find after a simple integration that

$$\frac{\langle(v-\langle v\rangle)^2\rangle}{\langle v\rangle^2} = \frac{B}{\sqrt{\langle v\rangle}}\left(\beta n^2\frac{\partial^3 P}{\partial n^3}\right)^{-1/2}, \tag{8.224}$$

where we have used the fact that $\langle v\rangle$ is proportional to V, and where the constant B is of order of magnitude unity while n is the particle density,

$$n = \frac{\langle v\rangle}{V}. \tag{8.225}$$

We note that, indeed, the fluctuations are large.

We have already mentioned that the above calculation does not really explain the phenomenon of critical opalescence. In order to explain the Tyndall effect it is necessary to show—as was first done by Ornstein and Zernike—that at the critical point there is a long-range correlation between the density at different points in the system. We do not want to enter into a detailed discussion, but we wish to derive a relation between this long-range correlation and the fluctuations in the number of particles ΔN contained in a macroscopic volume ΔV in the system. If $g(\mathbf{r})$ is the so-called radial distribution function, which means that $ng(\mathbf{r})\,d^3\mathbf{r}$ is the number of particles in the volume element $d^3\mathbf{r}$ around $\mathbf{r}$, if there is a particle at the origin, we have the relation

$$\langle(\Delta N - \langle\Delta N\rangle)^2\rangle = \langle\Delta N\rangle\left[1 + n\int\{g(\mathbf{r})-1\}\,d^3\mathbf{r}\right], \tag{8.226}$$

where n is again the average particle density, given by Eq. (8.225). Equation

(8.226) can be proved easily, if we write ΔN in the form

$$\Delta N = \sum_{i=1}^{\nu} \Delta(\mathbf{r}_i), \tag{8.227}$$

where $\Delta(\mathbf{r})$ is a function which is unity, if $\mathbf{r}$ lies within the volume ΔV, and which vanishes otherwise. One proves first of all that

$$\langle \Delta N \rangle = n\Delta V. \tag{8.228}$$

We then obtain Eq. (8.226), using the fact that

$$[\Delta(\mathbf{r})]^2 = \Delta(\mathbf{r}), \tag{8.229}$$

neglecting surface effects, and bearing in mind that one obtains the probability for the ith particle to be at $\mathbf{r}_i$ and the jth particle to be at $\mathbf{r}_j$ by integrating the normalized ensemble density over all variables except $\mathbf{r}_i$ and $\mathbf{r}_j$.

We note that from our definition of $g(\mathbf{r})$ it follows that for large values of $\mathbf{r}$, when normally we would not expect any correlations, $g(\mathbf{r}) \approx 1$.

8.3 THE RETURN TO EQUILIBRIUM

We have several times considered in the preceding chapters the approach to equilibrium of nonequilibrium situations, and a few times we have also discussed transport phenomena, which is certainly one aspect of nonequilibrium thermodynamics. In 1931, Onsager derived the so-called *Onsager relations*; since the war an extensive literature has appeared on the *thermodynamics of irreversible processes*, based upon the Onsager relations. The theory can be based either upon a discussion of fluctuations and their regression, or directly upon ensemble theory. In the present section we shall briefly consider the return to equilibrium. In the next section we shall derive the Onsager relations, and in the concluding section of this chapter we shall discuss a few applications.

Let us consider a system containing a number of different constituents. We saw in Chapters 5 and 6 how such a system can be described by a canonical grand ensemble and how the equilibrium situation is completely determined once we know the grand potential q as a function of β, the α_i, and the external parameters a_k. Once β, α_i, and a_k are fixed, the relative occurrence of a situation where the number of particles of the ith kind is ν_i, while the total energy of the system is E_k, is given by the following expression (compare the discussion in Sections 6.4 and 8.1):

$$\rho(\nu_i;E_k) = e^{-q+\sum \alpha_i \nu_i - \beta E_k}, \tag{8.301}$$

The partial derivatives of q with respect to β, the α_i, and the a_k give us, respectively, minus the average energy, the average number of particles of the

ith kind, and β times the average generalized forces. For the sake of simplicity we shall denote by b_j any of the variables β, α_i, or a_k and the corresponding "forces," ε, v_i, or βA_k, we shall denote by B_j so that Eqs. (6.415) to (6.417) can be written as

$$\langle B_j \rangle = \frac{\partial q}{\partial k_j}. \tag{8.302}$$

Let us now suppose that the values of the b_j are changed so fast that they have taken on their new values, $\tilde{b}_j = b_j + \Delta b_j$, before the density function $\rho(v_i, E_k)$ can change. If we wait long enough, the density function $\tilde{\rho}$ of the new ensemble which represents the system under the changed conditions will again be of the canonical form, but will be different from the old density function:

$$\tilde{\rho}(v_i, E_k) = \rho(v_i, E_k) + \Delta\rho, \tag{8.303}$$

where

$$\Delta\rho = \sum_j \Delta b_j \frac{\partial \rho}{\partial b_j} = \sum_j \Delta b_j \frac{\partial \tilde{\rho}}{\partial \tilde{b}_j}, \tag{8.304}$$

provided the Δb_j are sufficiently small. Bearing in mind that the B_j can be written as

$$B_j = \frac{\partial \Psi}{\partial k_j}, \tag{8.305}$$

where

$$\Psi = \sum_i \alpha_i v_i - \beta E_k, \tag{8.306}$$

and using Eq. (8.302) we can rewrite Eq. (8.304) in the following form

$$\Delta\rho = \tilde{\rho} \sum_j (B_j - \langle B_j \rangle) \Delta b_j, \tag{8.307}$$

where $\langle B_j \rangle$ is the equilibrium value of the "force" corresponding to the density function $\tilde{\rho}$, while B_j stands for its value defined by Eq. (8.305), that is, the value in a situation where the v_i and E_k are given, but are not necessarily equal to $\langle v_i \rangle$ and $\langle \varepsilon \rangle$.

Immediately after the disturbance the density function differs from the equilibrium density function by $\Delta\rho$. Let us now consider the rate of change, $d\rho/dt$, of the density function. We have for this rate of change*

$$\frac{d\rho_m}{dt} = \sum_n \left[\rho_n \, e^{-\Sigma \tilde{\alpha}_i v_i' + \tilde{\beta} E_l} - \rho_m \, e^{-\Sigma \tilde{\alpha}_i v_i + \tilde{\beta} E_k}\right] d_{mn}, \tag{8.308}$$

where ρ_m stands for $\rho(v_i, E_k)$ and ρ_n for $\rho(v_i', E_l)$, where the summation extends

* We exclude from our considerations the case when a magnetic field is present, when various changes must be made in the equations.

over all possible values of all the ν_i' and of E_l, and where $d_{mn}(=d_{nm})$ is related to the probability per unit time $a_{m\to n}$ that a transition takes place from m to n,

$$d_{mn} = a_{m\to n}e^{\Sigma\tilde{\alpha}_i\nu_i - \tilde{\beta}E_k}. \tag{8.309}$$

Equation (8.308) can be derived as follows.* Our system is in contact with a heat bath and a particle bath which fix the values of β (temperature) and of the α_i (thermal potentials). The density function $\bar{\rho}$ of the system which forms the heat and particle bath will be given by the equation

$$\bar{\rho}(\nu_i,\bar{E}_k) = e^{-q+\Sigma\tilde{\alpha}_i\nu_i - \tilde{\beta}E_k}, \tag{8.310}$$

The bath system will be so large that it will always be in thermodynamic equilibrium. Moreover, the fact that it acts as a heat and thermal bath entails that the α_i and β in Eq. (8.310) are the same as those occurring in $\tilde{\rho}$.

Let us consider a transition from a state where the situation of the original system is characterized by ν_i and E_k and the situation of the bath system by $\bar{\nu}_i$ and $\bar{E}_k$ to a state where we have ν_i', E_l and $\bar{\nu}_i'$, $\bar{E}_l$. We shall denote the corresponding density functions by $\rho_m, \bar{\rho}_m$, ρ_n and $\bar{\rho}_n$, and indicate the transition by $m\bar{m} \to n\bar{n}$. The probability per unit time for the transition $m\bar{m} \to n\bar{n}$ may be denoted by $a_{m\bar{m}\to n\bar{n}}$. We know that $a_{m\bar{m}\to nn}$ will be equal to zero, unless the following conditions are fulfilled.:

$$E_k + \bar{E}_k = E_l + \bar{E}_l, \tag{8.311}$$

$$\nu_i + \nu_i = \nu_i' + \bar{\nu}_i'. \tag{8.312}$$

Furthermore, from the principle of microscopic reversibility it follows that

$$a_{m\bar{m}\to n\bar{n}} = a_{n\bar{n}\to m\bar{m}}. \tag{8.313}$$

We are interested in a transition where the original system goes from a state specified by ρ_m to a state specified by ρ_n, irrespective of the states of the bath system. Let $a_{m\to n}$ be the probability per unit time that such a transition takes place. It can be shown that

$$a_{m\to n} = \sum_{m}\sum_{n}\bar{\rho}_m a_{m\bar{m}\to n\bar{n}}, \tag{8.314}$$

or, using Eq. (8.310),

$$a_{m\to n} = \sum_{m}\sum_{\bar{n}} a_{m\bar{m}\to n\bar{n}}e^{-\bar{q}+\Sigma\tilde{\alpha}_i\bar{\nu}_i - \tilde{\beta}E_k}. \tag{8.315}$$

Similarly, we have for $a_{n\to m}$ the equation

$$a_{n\to m} = \sum_{\bar{m}}\sum_{\bar{n}} a_{m\bar{m}\to n\bar{n}}e^{-\bar{q}+\Sigma\tilde{\alpha}_i\bar{\nu}_i' - \tilde{\beta}\bar{E}_l}, \tag{8.316}$$

where Eq. (8.313) has been used.

From Eqs. (8.315), (8.316), the the fact that $a_{m\bar{m}\to n\bar{n}}$ is zero unless conditions (8.311) and (8.312) are fulfilled it follows that the quantity d_{mn} given by Eq. (8.309) is symmetrical in its two indices,

$$d_{mn} = d_{nm}. \tag{8.317}$$

* Compare R. T. Cox, *Revs. Modern Phys.*, **22**, 238, 1950, Section 2.

We have for the rate of change of ρ_m the equation

$$\frac{d\rho_m}{dt} = \sum_n (\rho_n a_{n\to m} - \rho_m a_{m\to n}), \tag{8.318}$$

and using Eqs. (8.309) and (8.317), Eq. (8.308) follows.

It may be remarked here that it follows from Eq. (8.308) that a nonequilibrium situation will change until ρ satisfies Eq. (8.302).

Writing

$$\exp\left[-\sum \tilde{\alpha}_i v_i + \tilde{\beta} E_k\right]\left(\exp\left[-\sum \tilde{\alpha}_i v_i' + \tilde{\beta} E_l\right]\right)$$

as

$$e^q \tilde{\rho}_m (e^q \tilde{\rho}_n)$$

and using Eq. (8.305) for the B_j, we get from Eqs. (8.308), (8.303), and (8.307)

$$\frac{d\rho_m}{dt} = e^{-q} \sum_n \sum_j (B_j^{(n)} - B_j^{(m)}) \Delta b_j \, d_{mn}, \tag{8.319}$$

where the superscripts (n) and (m) are self-explanatory.

From this equation we can calculate the rate of change of the B_j, and we have

$$\frac{dB_j}{dt} = \frac{d}{dt} \sum_m \rho_m B_j^{(m)} = \sum_m B_j^{(m)} \frac{d\rho_m}{dt} \tag{8.320}$$

$$= \frac{\partial T}{\partial \Delta b_j}, \tag{8.321}$$

where

$$T = \tfrac{1}{4} e^{-q} \sum_j \sum_m \sum_n (B_j^{(m)} - B_j^{(n)})^2 \Delta b_j^{\,2} \, d_{mn}. \tag{8.322}$$

One obtains Eq. (8.321) by substituting Eq. (8.319) for $d\rho_m/dt$ into Eq. (8.320) and taking the arithmetic mean of that equation and the equation obtained by exchanging m and n and using Eq. (8.317).

One can use Eqs. (8.307), (8.319), (8.309), and (8.302) to obtain an order of magnitude estimate of relaxation times. The relaxation time τ will be of the order of magnitude of the ratio of $\Delta\rho_m$ to $-d\rho_m/dt$ and we get from the aforementioned equations

$$\tau = \frac{\langle B_j \rangle - B_j^{(m)}}{\sum_n a_{m\to n}(B_j^{(n)} - B_j^{(m)})} \tag{8.323}$$

which means that τ will be of the order of magnitude of $[a_{m\to n}]^{-1}$ (with $m \approx n$), which might have been expected. The actual value of $a_{m\to n}$ will vary from example to example, but in the case of a perfect gas we can make a rough estimate by assuming $a_{m\to n}$ to be of the order of $n \cdot a_{12\to 1'2'}$ where $a_{12\to 1'2'}$ is the quantity encountered in Section 1.4 and where n is the number of particles per cm^3. We then get

$$\tau \sim \frac{\lambda}{\bar{c}}, \tag{8.324}$$

where λ is the mean free path and $\bar{c}$ the average velocity of a particle in the gas. Equation (8.324) might also have been derived from simple dimensional considerations.

Taking as an example helium gas under normal conditions, we get

$$\tau \sim 10^{-7} \text{ sec.} \tag{8.325}$$

8.4 THE ONSAGER RELATIONS

In many instances when we are interested in irreversible processes these will be described by phenomenological laws, for instance, Ohm's law giving a proportionality between electric current and potential gradient, or Fourier's law giving a proportionality between heat flow and temperature gradient. In such cases the phenomenological law is of the kind

$$J_j = L_{jj} X_j, \tag{8.401}$$

where J_j is the *flux* (electric current and heat flow in the two examples given above), X_j the affinity* (potential gradient and temperature gradient in our examples), and L_{jj} the coefficients of proportionality (coefficients of electric and thermal conductivity).

Sometimes two or more affinities are present and cross phenomena may occur, such as the well-known thermoelectric effects, the Peltier effect which is the production of heat as the result of the existence of an electric current and the thermoelectric force, that is, a potential drop as the result of the existence of a temperature difference (see Section 8.5). In such cases Eq. (8.401) must be generalized to read

$$J_j = \sum_l L_{jl} X_l. \tag{8.402}$$

It can be shown that if a proper choice is made for the affinities and the fluxes, the L_{jl} satisfy the equation

$$L_{jl} = L_{lj} \tag{8.403}$$

* We do not wish to use the term, "force," since we have already used it in another sense.

which was first derived by Onsager as a general principle. We shall see presently how one can choose the J_j and the X_j in a proper way so that Eq. (8.403) holds.

Let $\lambda_j (j = 1,2,\cdots)$ be a number of macroscopic variables which together describe the macroscopic state of the system, and let $\lambda_j^{(m)}$ be the values of these variables in a situation where the number of particles of the ith kind is ν_i and the energy of the system E_k. If the state of the system can be represented by an ensemble with a density function $\rho(\nu_i,E_k)$, we have the equation (compare the discussion of Eq. (8.101)):

$$\lambda_j = \sum_m \lambda_j^{(m)} \rho^{(m)}, \tag{8.404}$$

where $\rho^{(m)}$ is the ρ_m of the preceding section, that is, it is one of the diagonal elements of the density matrix in the number-energy representation.

Consider now a situation where the values of the λ_j are not those corresponding to equilibrium, but are slightly different. The system will then try to regain equilibrium and we shall assume that a state of steady flow has set in. In that case we may expect that at any time the situation present, though not in equilibrium, will correspond to a state where a suitably defined entropy has the largest value consistent with the momentary values of the λ_j. By analogy with the situation at equilibrium, we shall define the entropy in a nonequilibrium state by the equation

$$S = -k\langle \ln \rho \rangle = -k \sum_m \rho^{(m)} \ln \rho^{(m)}. \tag{8.405}$$

The condition that S should be maximum for given values of the λ_j leads to the following expression for $\rho^{(m)}$ (by a reasoning completely similar to the one which in Section 1.5 led to Eq. (1.522)),

$$\rho^{(m)} = \frac{\exp\left[-\frac{1}{k}\sum X_j \lambda_j^{(m)}\right]}{\sum_m \exp\left[-\frac{1}{k}\sum X_j \lambda_j^{(m)}\right]}, \tag{8.406}$$

where the X_j are Lagrangian multipliers. Substituting Eq. (8.406) into Eq. (8.405) and expressing δS in terms of the $\delta\lambda_j$, we have

$$X_j = \frac{\partial S}{\partial \lambda_j}. \tag{8.407}$$

From Eq. (8.406) it follows that

$$\frac{\partial \rho^{(m)}}{\partial X_j} = \frac{1}{k}(\lambda_j - \lambda_j^{(m)})\rho^{(m)}, \tag{8.408}$$

a result which we shall need presently.

Let us now consider the rate of change of the λ_j. We have from Eqs. (8.404) and (8.308),

$$\frac{d\lambda_j}{dt} = \sum_m \lambda_j^{(m)} \frac{d\rho^{(m)}}{dt} = \sum_m \sum_n \lambda_j^{(m)} \left[\rho^{(n)} e^{-\Sigma \alpha_i v_i' + \beta E_l} - \rho^{(m)} e^{-\Sigma \alpha_i v_i + \beta E_k} \right] d_{mn}$$

$$= \sum_m \sum_n (\lambda_j^{(n)} - \lambda_j^{(m)}) \rho^{(m)} e^{-\Sigma \alpha_i v_i + \beta E_k} d_{mn}. \tag{8.409}$$

In order to find out how this expression depends on the X_l we take the derivative with respect to X_l and using Eq. (8.408) we have

$$\frac{\partial}{\partial X_l} \frac{d\lambda_j}{dt} = \frac{1}{k} \sum_m \sum_n (\lambda_j^{(n)} - \lambda_j^{(m)})(\lambda_l - \lambda_l^{(m)}) \rho^{(m)} e^{-\Sigma \alpha_i v_i + \beta E_k} d_{mn}. \tag{8.410}$$

In the first approximation we may use for $\rho^{(m)}$ the equilibrium expression given by Eq. (8.301); after interchanging m and n in the summation, taking the arithmetical mean of the two expressions, and using Eq. (8.317), we get

$$\frac{\partial}{\partial X_l} \frac{d\lambda_j}{dt} = \frac{e^{-q}}{2k} \sum_m \sum_n (\lambda_j^{(n)} - \lambda_j^{(m)})(\lambda_l^{(n)} - \lambda_l^{(m)}) d_{mn}, \tag{8.411}$$

or

$$\frac{d\lambda_j}{dt} = \sum_l X_l L_{jl}, \tag{8.412}$$

where

$$L_{jl} = \frac{e^{-q}}{2k} \sum_m \sum_n (\lambda_j^{(n)} - \lambda_j^{(m)})(\lambda_l^{(n)} - \lambda_l^{(m)}) d_{mn}. \tag{8.413}$$

In deriving Eq. (8.412) we have used the fact that at equilibrium we have $\partial S/\partial \lambda_j = X_j = 0$ and also $d\lambda_j/dt = 0$.

Identifying the X_j with the affinities and the $d\lambda_j/dt$ with the fluxes J_j, we see from Eq. (8.413) that Eq. (8.403) holds.

In order to choose in a given situation the proper fluxes and affinities, one usually considers the *entropy source strength*, which is the time derivative of the entropy and given by the equation

$$\frac{dS}{dt} = \sum_j X_j \frac{d\lambda_j}{dt} = \sum_j X_j J_j. \tag{8.414}$$

We may remark here that although our derivation of the Onsager relations was based on the use of grand canonical ensembles, which implies that we are considering systems that can exchange energy and particles with their surroundings, this restriction can easily be removed. If we are interested in a system which, although still able to exchange energy with its surroundings, cannot exchange particles, we must use macrocanonical ensembles and this results only in some changes in Eqs. (8.308) and (8.409), but not in Eq. (8.413). Finally, when we are dealing with an adiabatically insulated system, again slight changes must be made in some equations, but not in the final results.

The change to be made in the last case, for instance, is that instead of letting the situation n range over all possible values of all the ν_i and of the energy, we must restrict these situations to those with fixed numbers of particles and with fixed energy.

In the preceding derivation of Onsager's relation we have used ensemble theory and an essential part was played by Eq. (8.317), which followed from Eq. (8.313). Equation (8.313) does not hold when magnetic fields are present. If we exclude this case—as we have done so far—Eq. (8.313) holds as a consequence of the fact that all equations of motion are invariant under time reversal. To see, that, indeed, time reversal leads to the Onsager relations, we shall present a different derivation of Eq. (8.403).

Before doing this we want to say something about the decay of fluctuations. Let λ be one of the fluctuating variables and let its average value be 0. If λ has at a certain moment a value different from zero, it will tend to approach its equilibrium value according to some equation of motion:

$$\frac{d\lambda}{dt} = f(\lambda), \tag{8.415}$$

and if λ is "small," that is, if there is only a slight deviation from equilibrium, we can substitute for $f(\lambda)$ the first term, $-\gamma\lambda$, of a Taylor expansion, and we can then solve Eq. (8.415) and obtain

$$\lambda(t) = \lambda(0)\, e^{-\gamma t}. \tag{8.416}$$

Consider now the following time-correlation function:

$$\langle \lambda(t)\lambda(t + \tau)\rangle. \tag{8.417}$$

If Eq. (8.416) were valid for all values of λ, we would find for the correlation function (8.417):

$$\langle \lambda(t)\lambda(t + \tau)\rangle = \langle \lambda^2 \rangle\, e^{-\gamma\tau}. \tag{8.418}$$

The question arises for what values of τ the above equations hold. From the fact that, in deriving Eq. (8.418) we have in an essential way used the assumption that the time derivative of λ is proportional to λ, it follows that τ must be smaller than the time over which appreciable changes in λ can be expected, while on the other hand if τ is too small we can not expect a smooth decay of λ. This imposing of a lower limit for τ below which no steady flow condition can have been established is a condition that is normally encountered in the discussion of transport phenomena and is in fact the basis of many applications of the phenomenological, macroscopic equations of the kind (8.402).

Let us now consider instead of the correlation function (8.417) the more general functions

$$\langle \lambda_j(t)\lambda_l(t+\tau)\rangle. \tag{8.419}$$

Because of invariance under time reversal it should not make any odds whether we consider λ_j a time interval τ later or earlier than λ_l, and we have thus

$$\langle \lambda_j(t)\lambda_l(t+\tau)\rangle = \langle \lambda_j(t+\tau)\lambda_l(t)\rangle, \tag{8.420}$$

or, subtracting from both sides of this equation $\langle \lambda_j(t)\lambda_l(t)\rangle$,

$$\langle \lambda_j(t)[\lambda_l(t+\tau) - \lambda_l(t)]\rangle = \langle [\lambda_j(t+\tau) - \lambda_j(t)]\lambda_l(t)\rangle. \tag{8.421}$$

Assuming that τ is both sufficiently large and sufficiently small so that we may put

$$\lambda_j(t+\tau) - \lambda_j(t) = \tau\frac{d\lambda_j}{dt} \tag{8.422}$$

and using Eq. (8.412) for $d\lambda_j/dt$, we get from Eq. (8.421):

$$\sum_k L_{lk}\langle \lambda_j(t)X_k(t)\rangle = \sum_k L_{jk}\langle \lambda_l(t)X_k(t)\rangle. \tag{8.423}$$

We now use the following relation for the averages occurring in Eq. (8.423):

$$\langle \lambda_j X_l\rangle = -k\delta_{jl}, \tag{8.424}$$

where k is Boltzmann's constant. Using Eq. (8.424), we again find Eq. (8.403).

The proof of Eq. (8.424) follows from Eq. (8.407) and the fact that the probability for finding a set of values of the λ_j is again given by the expression $e^{-S/k}$ (compare Eq. (8.406) and the discussion in Section 8.1).

8.5 APPLICATIONS

In this concluding section we shall consider the following two applications of the Onsager relations: (1) a perfect gas enclosed in a reservoir consisting of two connected vessels, and (2) thermoelectric processes.

1. If we maintain a temperature difference between the two vessels that together form the reservoir in which our perfect gas finds itself, there will result a flow of matter leading to a pressure difference. The ratio of this pressure difference to the temperature difference is the so-called *thermo-molecular pressure difference*. Similarly, a pressure difference at constant

temperature will result in a heat flow, the so-called thermomechanical effect.*

We shall assume that the volumes of the two vessels are the same and that the gas contains only one constituent. Let N_1 and N_2 be the number of particles in the two vessels and T_1 and T_2 the respective temperatures. At equilibrium we have

$$N_1 = N_2(=N), \quad T_1 = T_2(=T). \tag{8.501}$$

We shall assume that the system is adiabatically insulated so that the total energy, which in the case of a perfect gas is the total kinetic energy, and that the total number of particles are constant, or

$$N_1 T_1 + N_2 T_2 = 2NT, \tag{8.502}$$

$$N_1 + N_2 = 2N. \tag{8.503}$$

We choose for our λ_j the two quantities

$$\lambda_1 = N_1 T_1 - NT(=NT - N_2 T_2), \tag{8.504}$$

$$\lambda_2 = N_1 - N(=N - N_2). \tag{8.505}$$

We have for the entropy the following equation (compare Eq. (4.618)),

$$\begin{aligned} S = S_1 + S_2 = k\Big[&\frac{3}{2}(N_1 \ln T_1 + N_2 \ln T_2) \\ &- \frac{5}{2}(N_1 \ln N_1 + N_2 \ln N_2) - 2N \ln \frac{k}{V} + 2N\mathfrak{C} + 3N \ln M + 5N\Big] \\ = S_{\text{eq}} &- \frac{3k}{2N}\frac{\lambda_1{}^2}{T^2} + \frac{3k\lambda_1\lambda_2}{NT} - \frac{5k\lambda_2{}^2}{2N}, \end{aligned} \tag{8.506}$$

where only terms up to the second degree in the λ_j are included.

From Eq. (8.506) we have

$$\dot{S} = \left(-\frac{3k}{NT^2}\lambda_1 + \frac{3k}{NT}\lambda_2\right)\dot{\lambda}_1 + \left(\frac{3k}{NT}\lambda_1 - \frac{5k}{N}\lambda_2\right)\dot{\lambda}_2, \tag{8.507}$$

and thus

$$X_1 = -\frac{3k}{NT^2}\lambda_1 + \frac{3k}{NT}\lambda_2, \tag{8.508}$$

$$X_2 = \frac{3k}{NT}\lambda_1 - \frac{5k}{N}\lambda_2. \tag{8.509}$$

* Although we restrict our discussion to the case of a perfect gas, it is of interest to note that the same effects also occur in helium II. The thermomolecular pressure difference is then called the *fountain effect* (J. F. Allen and H. Jones, *Nature*, **141**, 243, 1938).

These equations could also have been obtained from Eq. (8.506) by using Eq. (8.407).

We can express the affinities in terms of the pressure difference ΔP and the temperature difference ΔT between the two vessels, where

$$\Delta P = \frac{k}{V}(N_1 T_1 - N_2 T_2) = \frac{2k\lambda_1}{V}, \tag{8.510}$$

$$\Delta T = T_1 - T_2 \cong \frac{2\lambda_1}{N} - \frac{2\lambda_2 T}{N}, \tag{8.511}$$

and where V is the volume of one vessel.

From Eqs. (8.508) to (8.511) we get

$$X_1 = -\frac{3}{2}k\frac{\Delta T}{T^2}, \tag{8.512}$$

$$X_2 = -\frac{V\Delta P}{NT} + \frac{5}{2}k\frac{\Delta T}{T}. \tag{8.513}$$

Taking $J_j = d\lambda_j/dt$, we see that J_1 is proportional to the energy flow and that J_2 is the flow of particles. From Eqs. (8.412), (8.512), and (8.513) we have

$$J_1 = \left[-\frac{3k}{2T}L_{11} + \frac{5k}{2}L_{12}\right]\frac{\Delta T}{T} - L_{12}\frac{V\Delta P}{NT}, \tag{8.514}$$

$$J_2 = \left[-\frac{3k}{2T}L_{21} + \frac{5k}{2}L_{22}\right]\frac{\Delta T}{T} - L_{22}\frac{V\Delta P}{NT}. \tag{8.515}$$

Let us first consider the case where $\Delta T = 0$. We then have from Eqs. (8.514) and (8.515)

$$J_1 = \frac{L_{12}}{L_{22}}J_2 = \frac{2\mathscr{E}}{3k}J_2, \tag{8.516}$$

and on inspection it is seen that $\mathscr{E}$ is the average energy transferred per particle. Equation (8.516) describes the thermomechanical effect, transfer of energy, although $\Delta T = 0$.

We can consider two limiting cases, the Knudsen case and the ordinary case. The difference between these two cases lies in the geometry of the communicating opening between the two vessels. Whereas in the Knudsen case we are dealing with an opening the width of which is small compared to the mean free path of the particles in the gas, in the ordinary case the width of the opening is large compared to the mean free path.

In the second case, $\mathscr{E}$ is equal to the average kinetic energy per particle plus the average work done against the pressure, or

$$\mathscr{E} = \frac{3}{2}kT + \frac{PV}{N} = \frac{5}{2}kT. \tag{8.517}$$

In the Knudsen case we must calculate the average energy per particle, averaged over all the particles arriving at the hole, since these particles will pass freely through it. The average energy is then calculated in the usual manner from the velocity distribution function $f(u,v,w)$,

$$\mathscr{E} = \frac{\int_{-\infty}^{+\infty} du \int_{-\infty}^{+\infty} dv \int_0^{\infty} dw f(u,v,w)\, \tfrac{1}{2}mc^2}{\int_{-\infty}^{+\infty} du \int_{-\infty}^{+\infty} dv \int_0^{\infty} dw f(u,v,w)}, \tag{8.518}$$

where $c^2 = u^2 + v^2 + w^2$ and where we have supposed the z-coordinate to be perpendicular to the hole. Using for $f(u,v,w)$ the Maxwell distribution (see Eq. (1.104)), we get

$$\mathscr{E} = 2kT. \tag{8.519}$$

We can now study the so-called stationary state, which occurs when we wait until there is no net flow of particles from one vessel to the other, so that $J_2 = 0$. From Eq. (8.515) we then have

$$\frac{\Delta P}{\Delta T} = \frac{N}{V}\left[\frac{5k}{2} - \frac{\mathscr{E}}{T}\right], \tag{8.520}$$

which gives us the thermomolecular pressure difference. In deriving Eq. (8.520) we have used the fact that, according to the Onsager relations, $L_{12} = L_{21}$.

In the Knudsen case, Eqs. (8.519) and (8.520) lead to

$$\frac{\Delta P}{\Delta T} = \frac{1}{2}\frac{kN}{V} = \frac{1}{2}\frac{P}{T}, \tag{8.521}$$

or

$$\frac{P_1}{\sqrt{T_1}} = \frac{P_2}{\sqrt{T_2}}. \tag{8.522}$$

In the ordinary case we have

$$\frac{\Delta P}{\Delta T} = 0, \qquad \text{or} \qquad P_1 = P_2. \tag{8.523}$$

2. We now consider the case of a thermocouple made of two metals A and B. The two junctions are at temperatures T and $T + \Delta T$, and, in the wire made

of metal A there is a condenser over which there exists a potential drop $\Delta\varphi$ (see Fig. 8.2). In order to treat this case we must once more find an expression for the entropy production. We can use the equation

$$\delta S = \frac{1}{T}\delta U + \frac{1}{T}\sum A_k \delta a_k. \tag{8.524}$$

In our case, energy will be transported from P to Q and the only external parameter is the charge e on the condenser, so that A_k will be $-\Delta\varphi$. The

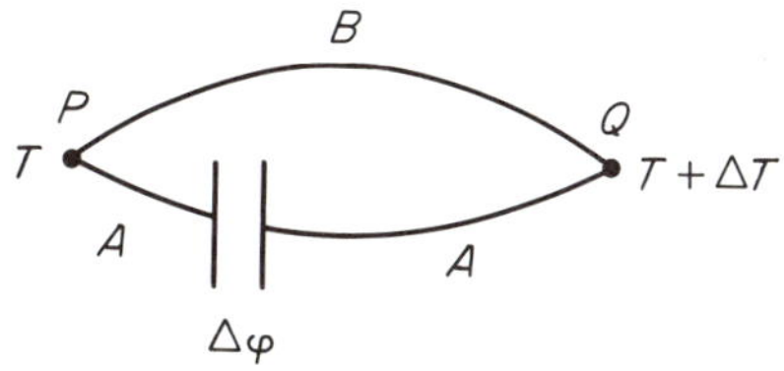

FIG. 8.2 A thermocouple.

change in the entropy of the total system, when δU is transported from P to Q and a charge δe is brought on the condenser, will be given by the equation

$$\delta S = -\frac{\delta U}{T} + \frac{\delta U}{T + \Delta T} - \frac{\Delta\varphi}{T}\delta e, \tag{8.525}$$

or

$$\frac{dS}{dt} = -\frac{dU}{dt}\frac{\Delta T}{T^2} - \frac{de}{dt}\frac{\Delta\varphi}{T}. \tag{8.526}$$

From Eqs. (8.526) and (8.414) it follows that we can choose

$$J_1 = \frac{dU}{dt}, \qquad J_2 = \frac{de}{dt}, \tag{8.527}$$

$$X_1 = -\frac{\Delta T}{T^2}, \qquad X_2 = -\frac{\Delta\varphi}{T}. \tag{8.528}$$

Equation (8.412) now reads

$$J_1 = -L_{11}\frac{\Delta T}{T^2} - L_{12}\frac{\Delta\varphi}{T} \tag{8.529}$$

$$J_2 = -L_{21}\frac{\Delta T}{T^2} - L_{22}\frac{\Delta\varphi}{T}. \tag{8.530}$$

Let us first consider the stationary case, where $J_2 = 0$. We then get from Eq. (8.530)

$$\frac{\Delta\varphi}{\Delta T} = -\frac{L_{21}}{L_{22}}\frac{1}{T}, \tag{8.531}$$

giving the thermoelectric power of the thermocouple (Seebeck effect), that is, the potential difference per unit temperature difference when there is no electric current.

A second case is that where $\Delta\varphi$ is fixed and $\Delta T = 0$. We then get the so-called *Peltier effect.* The Peltier heat Π is defined as the flow of energy per unit electric current, and we get from Eqs. (8.529) and (8.530), putting $\Delta T = 0$

$$\frac{J_1}{J_2} = \prod = \frac{L_{12}}{L_{22}}. \tag{8.532}$$

From Eqs. (8.531) and (8.532) and the Onsager relations (8.403), we then get the following relation between the thermoelectric power and the Peltier heat:

$$\frac{\Delta\varphi}{\Delta T} = -\frac{\prod}{T}. \tag{8.533}$$

This relation goes back to Thomson, who derived it in 1854 by a different method.*

PROBLEMS

1. Using the relation $S = k \ln W$ between the entropy S and the probability W for a situation, and the thermodynamic relation between a change in entropy and a change in free energy for a closed system at temperature equilibrium, derive Eq. (8.105).†
2. Use Eq. (8.105) and the fact that, for small deviations from equilibrium, ΔA can be expanded in a Taylor series (compare Eq. (8.106)) to prove that the average value of ΔA equals $\frac{1}{2}kT$.
3. Use thermodynamic relations and Eq. (8.109) to find expressions for the average values of ΔV^2, $\Delta V \Delta T$, and ΔT^2.‡
4. Do the same for ΔP^2, ΔS^2, and $\Delta S \Delta P$.

* W. Thomson, *Proc. Roy. Soc.* (Edinburgh), **3,** 255, 1854.

† A. Einstein, *Ann. Physik*, **33,** 1275, 1910.

‡ For problems 3 to 6, compare L. D. Landau and E. M. Lifshitz, *Statistical Physics*, Oxford, 1958, Section 111, and R. Kubo, *Statistical Mechanics*, Amsterdam, 1965, pp. 381ff.

5. Do the same for ΔE^2, ΔV^2, and $\Delta E \Delta V$.
6. Do the same for $\Delta S \Delta V$, $\Delta V \Delta P$, and $\Delta S \Delta T$.
7. Generalize Eq. (8.109) to take into account possible fluctuations in occupation numbers.
8. Use the formula obtained in the preceding problem to derive a relation between the fluctuations in the number of solute molecules in a dilute solution and the osmotic pressure.
9. Evaluate the fluctuations in the energy of a system of N harmonic oscillators.
10. Evaluate the fluctuations in the energy of a Debye solid.
11. Express $\langle(\varepsilon - \langle\varepsilon\rangle)^3\rangle$ in terms of C_v and its temperature derivative. Evaluate this expression for the case of a perfect gas.
12. Express the fluctuations in the magnetization of a system in terms of the magnetic susceptibility.
13. Consider a system of N spin-$\frac{1}{2}$ particles which have a magnetic moment μ in a magnetic field $\mathbf{B}$. Calculate the fluctuations in the total magnetic moment of the system both directly and by applying the results of the preceding problem.
14. Use a pressure ensemble to evaluate the fluctuations in the enthalpy.
15. As in the case of photons the chemical potential does not occur in the ensemble density, one cannot apply Eq. (8.202) directly. Prove that Eq. (8.205) is also valid for the case of a system of photons.
16. Use a pressure ensemble to prove Eq. (8.205).
17. Derive Eq. (8.209) from Eq. (8.201).
18. By a direct calculation of number of microstates compatible with a given macrostate, evaluate the fluctuations in the number of particles per cell, if N particles are distributed amongst Z cells, for the case where neither Z nor N are large. Consider both Boltzmann and Bose-Einstein statistics.*
19. Use Eq. (8.210), taking the derivatives to be meant as being evaluated at constant p,q, and a wall potential of the form

$$U(r) = 0, \quad r < R;$$
$$= C(r - R), \quad r \geqslant R,$$

where C is a constant, and where we consider a perfect gas in a spherical container, to evaluate the pressure fluctuations.†

* B. Decomps and A. Kastler, *Compt. Rend. Acad. Sci.*, **256**, 1087, 1963.
† H. Wergeland, *Proc. Norwegian Acad. Sci.*, **28**, 106 1955.

20. Prove Eqs. (8.221) to (8.223) by using the fact that q will be a homogeneous function of degree 1 in V and v.
21. Find an expression for the density fluctuations in the neighborhood of the critical point. Assume that Eq. (8.205) may still be used.
22. Complete the proof of Eq. (8.226).
23. Prove Eq. (8.424).

BIBLIOGRAPHICAL NOTES

Sections 8.1 and 8.2. We may refer to the following general references:

1. L. D. Landau and E. M. Lifshitz, *Statistical Physics*, Oxford, 1958.
2. R. Kubo, *Statistical Mechanics*, Amsterdam, 1965.
3. A. Münster, *Statistische Thermodynamik*, Berlin, 1956.
4. A. Münster, *Handb. Physik*. $\mathbf{3}_2$, 176, 1959.
5. R. C. T. da Costa, *Nuovo Cimento*, **32,** 654, 1964.
6. R. C. Tolman, *The Principles of Statistical Mechanics*, Oxford, 1938.

Section 8.1. Fluctuation theory really started with the classical papers of Einstein and Smoluchowski:

7. M. von Smoluchowski, *Ann. Physik*, **25,** 205, 1908.
8. A. Einstein, *Ann. Physik*, **33,** 1275, 1910.

For a discussion of different kinds of ensembles and distributions see References 3 and 4, and also

9. T. L. Hill, *Statistical Mechanics*, New York, 1956.
10. D. ter Haar and H. Wergeland, *Elements of Thermodynamics*, Reading, Mass., Chaps. III and IV.

Section 8.2. See References 3, 4, and 6. For a discussion of fluctuations in the radiation field see also

11. A. Einstein, *Physik. Z.* **10,** 185, 1909.
12. L. Rosenfeld, *Osiris*, **2,** 149, 1936.
13. M. J. Klein, *Proc. Roy. Acad. Sc., Amsterdam*, **B62,** 41, 1959.
14. D. ter Haar, *The Old Quantum Theory*, Oxford, 1966, Chap. 2.

For a discussion of pressure fluctuations, see also

15. H. Wergeland, *Proc. Norwegian Acad. Sc.*, **28,** 106, 1955.
16. M. J. Klein, *Physica*, **26,** 1073, 1960.
17. A. Münster, *Physica*, **26,** 1117, 1960.
18. H. Wergeland, in *Fundamental Problems in Statistical Mechanics* (E. G. D. Cohen, ed.) Amsterdam, 1962.

For a discussion of critical fluctuations see References 7 and 8, and also

19. M. J. Klein and L. Tisza, *Phys. Rev.*, **76,** 1861, 1949.

20. A. Münster, in *Statistical Mechanics of Equilibrium and Non-Equilibrium* (J. Meixner, ed.), Amsterdam, 1965.

21. A. Münster, in *Fluctuation Phenomena in Solids* (R. E. Burgess, ed.) New York, 1965.

For a discussion of the importance of the radial distribution function see Reference 2, but especially

22. F. Zernike, *Thesis*, Amsterdam, 1915.

23. L. Ornstein and F. Zernike, *Proc. Kon. Akad. Wet. Amsterdam*, **17,** 793, 1914; **18,** 1520, 1916; **19,** 1321, 1917; *Physik. Z.*, **19,** 134, 1918; **27,** 761, 1926.

Section 8.3. For general references see References 1 and 2, and also

24. S. R. de Groot, *Thermodynamics of Irreversible Processes*, Amsterdam, 1951.

25. S. R. de Groot and P. Mazur, *Non-equilibrium Thermodynamics*, Amsterdam, 1962.

26. C. Kittel, *Elementary Statistical Physics*, New York, 1958, Part 2.

27. R. T. Cox, *Revs. Modern, Phys.* **22,** 238, 1950, especially Sections 2 and 5. Cox's treatment is based on petit ensembles, while we have used the theory of the grand ensembles.

28. T. L. Hill and I. W. Plesner, *J. Chem. Phys.*, **41,** 1359, 1964; **43,** 267, 1965.

Section 8.4. Equation (8.403) was derived by Onsager in 1931:

29. L. Onsager, *Phys. Rev.*, **37,** 405, 1931.

30. L. Onsager, *Phys. Rev.*, **38,** 2265, 1931.

See also Reference 24, Chap. II; Reference 27, Section 6; and

31. H. B. G. Casimir, *Revs. Modern Phys.*, **17,** 343, 1945.

32. H. B. Callen, *Phys. Rev.*, **73,** 1349, 1948.

Section 8.5. See Reference 24, Sections 9, 14, and 57; Reference 31, Section 3; and Reference 32. We refer especially to Reference 24 for a discussion of more applications.

9

The Third Law of Thermodynamics

9.1 THE UNATTAINABILITY OF THE ABSOLUTE ZERO

"The greatest advance in thermodynamics since van der Waals' time is the enunciation of a new general law of thermodynamics in 1905 by Nernst, the Third Law." This is the opening sentence of a paper by Simon on Nernst's heat theorem. A textbook on statistical mechanics is not the place to enter into a detailed discussion of Nernst's heat theorem and we must refer the reader to the literature given in the bibliographical notes at the end of this chapter for such a discussion. We shall be concerned only with those aspects of this theorem that are most closely connected with statistical considerations.

First of all, let us state the third law in the now generally accepted form given by Simon: *The contribution to the entropy of a system due to each component which is in internal equilibrium disappears at absolute zero.* A slightly different though equivalent way of stating the third law is the following one, given by Fowler and Guggenheim: *For any isothermal process involving only phases in internal equilibrium or, alternatively, if any phase is in frozen equilibrium, provided the process does not disturb this frozen equilibrium,*

$$\lim_{T\to 0} \Delta S = 0. \tag{9.101}$$

As Simon has emphasized in his discussion of the third law, it should not be surprising that there are some restricting clauses in the formulation of the third law, but these clauses do not mean that the third law is not a general thermodynamical law. Indeed, one sees immediately that the restrictions are equivalent to the statement that the third law is valid as long as we are dealing with systems to which thermodynamics applies at all.

It can be shown that the two formulations of the third law that we have just given are equivalent to the following one: *It is impossible by any procedure, no matter how idealized, to reduce any system to the absolute zero in a finite number of operations.* This means that the third law of thermodynamics is equivalent to the principle of the unattainability of the absolute zero.*

We may briefly sketch the proof of the equivalence of the two principles.† Consider a process where a transition from a situation 1 to a situation 2 takes place. This process may, for instance, be the result of a change in one of the external parameters. Denoting by $S_i(T)$ the entropy per unit mass in situation i at a temperature T, we have from the thermodynamical formula

$$dS = \frac{C\,dT}{T} \tag{9.102}$$

the equations

$$S_1(T) - S_1(0) + \int_0^T \frac{C_1}{T}\,dT, \tag{9.103}$$

$$S_2(T) = S_2(0) + \int_0^T \frac{C_2}{T}\,dT, \tag{9.104}$$

where C_i is the specific heat per unit mass in situation i. It follows from general quantum-mechanical principles that the integrals in Eq. (9.103) and (9.104) converge since the specific heats go sufficiently rapidly to zero for $T \to 0$ (compare the discussion in Section 9.2).

* It should be mentioned that in classical physics the unattainability of the absolute zero is a triviality and becomes meaningful only if we remind ourselves that according to quantum mechanics all specific heats must vanish at the absolute zero.

† F. E. Simon, *Science Museum Handbook*, **3**, 61, 1937.

1. Let us assume that Eq. (9.101) holds. If the transition $1 \to 2$ takes place adiabatically (no change of entropy during the transition) and reversibly,* and if the initial temperature is T_1 and the final temperature T_2, we have from Eqs. (9.101), (9.103), and (9.104)

$$\int_0^{T_1} \frac{C_1}{T} \, dT = \int_0^{T_2} \frac{C_2}{T} \, dT. \tag{9.105}$$

If the process could be used to reach the absolute zero ($T_2 = 0$), we should have

$$\int_0^{T_1} \frac{C_1}{T} \, dT = 0, \tag{9.106}$$

but this is clearly impossible, since C_1 is always larger than zero, as follows from thermodynamic stability conditions.† It is also clear that the inverse transition ($2 \to 1$) cannot be used to reach the absolute zero.

2. Let us now start from the principle of the unattainability of the absolute zero and prove that $S_1(0)$ and $S_2(0)$ are equal. Consider again an adiabatic, reversible process. We then have

$$S_2(0) - S_1(0) = \int_0^{T_1} \frac{C_1}{T} \, dT - \int_0^{T_2} \frac{C_2}{T} \, dT. \tag{9.107}$$

Let us assume first that $S_2(0) > S_1(0)$. We can then choose T_1 in such a way that

$$\int_0^{T_1} \frac{C_1}{T} \, dT = S_2(0) - S_1(0). \tag{9.108}$$

From Eqs. (9.107) and (9.108) we see that it is now possible to reach the absolute zero by an adiabatic, reversible process. Since this is in contradiction to the principle of the unattainability of the absolute zero, we must have the condition

$$S_2(0) \leqslant S_1(0). \tag{9.109}$$

Similarly, we can prove that

$$S_2(0) \geqslant S_1(0), \tag{9.110}$$

from which Eq. (9.101) follows.

Fowler and Guggenheim's formulation of the third law deals only with entropy differences, but Simon's formulation opens the door to the introduction of an *absolute entropy*. We can introduce an absolute entropy only

* If the process is not reversible, our conclusions hold a fortiori (see Simon, *loc. cit.*).

† See, for instance, D. ter Haar and H. Wergeland, *Elements of Thermodynamics*, Reading, Mass., 1966, Chap. III.

when we have a means to determine the entropy at the absolute zero (or at any other temperature) absolutely, corresponding to the fixing of the value of the const $S_1(0)$ in Eq. (9.103). This can be done by using the statistical definition of the entropy (see Eq. (4.413) and Eq. (6.426) with the subsequent discussion),

$$S = k \ln W, \tag{9.111}$$

where W is the number of ways (microsituations) in which a given thermodynamical situation (macrosituation) can be realized. In the next section we shall discuss how this equation leads to

$$\lim_{T\to 0} S = 0, \tag{9.112}$$

which fixes the constant $S_1(0)$ as zero. Equation (9.101) then follows as a consequence.

In our discussion up to this moment we have tacitly assumed that we are always dealing with situations that are in stable equilibria. Often, however, one is dealing with a state of metastable equilibrium, and Eq. (9.112) will no longer hold, but the third law as formulated at the beginning of the present section still holds. Indeed, the existence of glasses and other metastable equilibria was the reason why many scientists doubted the general validity of Nernst's heat theorem, until Simon discussed in great detail the domain of applicability of the third law and gave the formulation that we quoted. The problem of metastability really lies outside the normal thermodynamical or statistical discussion; therefore we must refer the reader to Simon's papers on this subject. The only case where we met with an example of metastable equilibrium was in the discussion of the specific heat of hydrogen (Section 3.5); otherwise we have disregarded the existence of frozen equilibria.

9.2 MICROSCOPIC AND MACROSCOPIC ASPECTS OF THE THIRD LAW

In this section we shall discuss the implications of Eq. (9.111). From a purely microscopic, statistical point of view one can say that at the absolute zero the system will be in its ground state, and since many people believe that the ground state of any system will be nondegenerate (this has never actually been proven), it follows that at the absolute zero we would have $W = 1$, and hence $S = 0$.* In this approach we base the decrease of entropy to zero upon the

* It must be remembered that even if the degree of degeneracy of the ground state is larger than one, then if that degree of degeneracy is independent of the total number of particles in the system, N, the entropy at the absolute zero would only be of the order of k, which would be negligibly small compared to the usual values of entropies at finite temperatures which are of the order kN (compare also a discussion by O. Stern (*Ann. Physik* **49**, 823, 1916)).

discreteness of the quantum-mechanical energy levels. This approach fails, however, to take properly into account that the third law—as all laws of thermodynamics—are dealing with experimentally accessible phenomena. In Section 3.5 we noted that the spacing of the translational energy levels corresponds to a temperature of the order of 10^{-15} °K, while the spacing of vibrational levels in a Debye solid (see Problem 6 at the end of Chapter 3) is of the order of 10^{-6} °K. We see thus that at experimentally accessible temperatures many levels will be occupied and that the experimentally observed approach to zero of the entropy must be due to some other factor. In fact, as we shall see in a moment, the density of energy levels determines the asymptotic behavior of the entropy.*

We remind ourselves of the connection between the entropy S and the density of states $\rho(\varepsilon)$. We shall consider a system containing particles of one kind only, whose total number N, is fixed, at a temperature $T\,(=(k\beta)^{-1})$, so that we can use a macrocanonical ensemble to describe the system. From Eq. (6.302) we have in the energy representation

$$e^{-\beta\psi} = \sum_k g_k e^{-\beta E_k}. \tag{9.201}$$

For the sake of simplicity we shall assume that the energy spectrum can be split rigorously into two parts: one part with an average distance between the levels of the order of or larger than β_r^{-1} and a second part with a spacing well below β_r^{-1}, where $(k\beta_r)^{-1}$ is room temperature. We have just seen that the second part is the one that concerns us. As far as the first part is concerned, its contribution to the entropy will be vanishingly small at low temperatures and we shall not consider it in our present discussion; its contribution to the free energy will be a constant multiple of the temperature. We are thus concerned with those energy levels E_k which are spaced so closely that we can safely replace the sum in Eq. (9.201) by an integral. We can now use an analysis similar to the one in Section 5.5 (see also Section 6.4 and Problem 19 at the end of Chapter 6) and find finally that

$$S = k \ln[\rho(\bar{\varepsilon})\Delta\varepsilon]. \tag{9.202}$$

We note, by the way, that the right-hand side of Eq. (9.202) contains $\Delta\varepsilon$, the spread around the average energy, under the sign of the logarithm, showing that special care must be exercised when fluctuations are large. We remind ourselves that $\Delta\varepsilon$ satisfies the equation (compare Eq. (5.0407))

$$(\Delta\varepsilon)^2 = -\frac{\partial\bar{\varepsilon}}{\partial\beta}. \tag{9.203}$$

* This was apparently first pointed out by Kramers. The author is indebted to H. B. G. Casimir for correspondence about this point.

In Section 9.4 we shall consider a few special cases, but here we are concerned with the general theory. We shall now consider three possible expressions for $\rho(\varepsilon)$ and we shall assume that in Eq. (9.202) we may neglect $k \ln \Delta\varepsilon$ in comparison with $k \ln \rho$. The possibilities to be considered are:

(1) $$\rho(\varepsilon) = \exp[K\varepsilon^p], \tag{9.204}$$

(2) $$\rho(\varepsilon) = C\varepsilon^K, \tag{9.205}$$

and

(3) $$\rho(\varepsilon) = C\left(\frac{\varepsilon}{\Delta}\right)e^{-\varepsilon/\Delta} \tag{9.206}$$

where the C's are constants and the K's constants times N.

Before discussing these various possibilities for ρ we remind ourselves that $\bar{\varepsilon}$ is the value of ε for which the integral that replaces the sum in Eq. (9.201) has a maximum, which means that $\bar{\varepsilon}$ is the solution of the equation

$$\frac{\partial}{\partial\varepsilon} \ln \rho(\varepsilon) - \beta = 0, \tag{9.207}$$

which is equivalent to the thermodynamic relation $\beta = \partial[S/k]/\partial\varepsilon$.

From Eqs. (9.207) and (9.202) we now get for case (1):

$$\bar{\varepsilon} = C_1\beta^{1/(p-1)}, \tag{9.208}$$

$$S = C_2\beta^{p/(p-1)}, \tag{9.209}$$

and we see that p must satisfy the inequality

$$0 < p < 1, \tag{9.210}$$

in order that $\bar{\varepsilon}$ and S behave properly at $T = 0$.

For case (2) we get:

$$\bar{\varepsilon} = \frac{C_3}{\beta}, \tag{9.211}$$

$$S = -C_4 \ln \beta, \tag{9.212}$$

and we see that for this case Eq. (9.112) is not satisfied.

For case (3) we get:

$$\bar{\varepsilon} = C_5\Delta e^{-\beta/\Delta}, \tag{9.213}$$

$$S = C_6\beta\Delta e^{-\beta\Delta}, \tag{9.214}$$

and we see here an approach to zero in the same way as for the case where the level spacing is large compared to β^{-1}. In Eqs. (9.208) to (9.214) the C_i are temperature-independent constants.

We note in conclusion that, in the cases where the third law holds, the behavior of the entropy as $T \to 0$ is governed by the behavior of $\rho(\varepsilon)$ as $\varepsilon \to 0$.

In Section 9.4 we shall consider a few special cases and verify that they fall into the categories discussed here.

9.3 THE CHEMICAL CONSTANTS

In Section 4.6 we introduced a constant $\mathfrak{C}$ by the equation

$$\mathfrak{C} = \ln\left[\left(\frac{2\pi m_{\rm H} k}{h^2}\right)^{3/2} k\right], \tag{9.301}$$

where $m_{\rm H}$ is the molecular weight unit. We mentioned then that $\mathfrak{C}$ is connected with the so-called chemical constants. This relationship will now be discussed.

Let us consider the equilibrium between a vapor and its condensed state, especially at low temperatures. At low temperatures the vapor pressure will be so low that we may treat it as a perfect Boltzmann gas. We have then from Eq. (4.618) for the entropy per particle of the vapor, $S_{\rm v}$, the equation

$$S_{\rm v} = k[\tfrac{5}{2}\ln T - \ln P + \mathfrak{C} + \tfrac{3}{2}\ln M + \tfrac{5}{2}], \tag{9.302}$$

where P is the vapor pressure and M the molecular weight of the vapor. Let $S_{\rm c}$ be the entropy per particle of the condensed state. Since the condensation takes place at constant temperature, we have from the second law of thermodynamics

$$S_{\rm v} - S_{\rm c} = \frac{\chi}{T}, \tag{9.303}$$

where χ is the sublimation heat per particle at temperature T.

We can use Eqs. (9.302) and (9.303) to write down an equation for the vapor pressure, and the result is

$$\ln P = -\frac{\chi}{kT} + \tfrac{5}{2}\ln T + \mathfrak{C} + \tfrac{3}{2}\ln M + \tfrac{5}{2} - \frac{S_{\rm c}}{k}. \tag{9.304}$$

We now want to compare Eq. (9.304) with the vapor pressure equation obtained from classical thermodynamics. That equation is obtained from the Clausius-Clapeyron equation in the form

$$\frac{d\ln P}{dT} = \frac{\chi}{kT^2}, \tag{9.305}$$

and Kirchhoff's equation for the latent heat of sublimation

$$\frac{d\chi}{dT} = c_{\rm v} - c_{\rm c}. \tag{9.306}$$

In Eq. (9.305) we have assumed the vapor to be a perfect gas and we neglected the specific volume of the condensed phase as compared with that of the vapor. Assuming c for the specific heat per particle of the vapor, c_v, its classical expression, $\frac{5}{2}k$ (we are dealing with c_P rather than c_v), we get by integration from Eqs. (9.305) and (9.306)

$$\ln P = -\frac{\chi_0}{kT} + \frac{5}{2}\ln T - \frac{1}{k}\int_0^T \frac{dT'}{T'^2}\int_0^{T'} c_c\, dT'' + i, \tag{9.307}$$

where c_c is the specific heat per particle of the condensed phase, where χ_0 is the value of χ at the absolute zero, and where i is an integration constant which is called the *chemical constant* of the substance under consideration.

Comparing Eqs. (9.302), (9.301), and (9.307), we see that in order to get agreement, i should theoretically satisfy the equation (m = mass of one particle)*

$$i = \ln\left(\frac{2\pi m}{h^2}\right)^{3/2} k^{5/2}. \tag{9.308}$$

One can now compare the theoretical value of i given by Eq. (9.308) with the value following from Eq. (9.307) by inserting experimental values. If this is done, it is found that there is excellent agreement. This can be interpreted as meaning either that Nernst's heat theorem is satisfied in this case, or that the fundamental assumptions, which were introduced in deriving Eq. (9.301) (which is the same as Eq. (4.616)) for $\mathfrak{C}$, are sound.

Tetrode used Eq. (9.307) to derive from the experimental data the size of the cells in Γ-space. It must be noted that he had introduced the extra factor $1/N!$ (compare Eqs. (4.208) and (4.209)). His reason for the introduction of this factor was that otherwise the entropy would not be an extensive quantity, and he stated explicitly that one should use generic, and not specific, phases in calculating the entropy (compare the discussion in Section 5.12). He then put the elementary cell in Γ-space equal to σ and, on dimensional grounds, wrote

$$\sigma = (zh)^n, \tag{9.309}$$

where n is the number of degrees of freedom, z a numerical factor, and h Planck's constant. From the experimental data about mercury, Tetrode determined z, and the result was

$$z = 1.0_5, \tag{9.310}$$

from which he arrived at the conclusion: "Es erscheint plausibel, dasz z genau 1 zu setzen ist, was schon von O. Sackur angenommen wurde."† In

* We do not wish to discuss here the influence of the statistical weights of the atoms on the chemical constants, but refer the reader to the literature given in the bibliographical notes at the end of this chapter.

† "It seems plausible that z must be put exactly equal to 1, as has been already assumed by O. Sackur."

this way Tetrode found an experimental justification for the choice of h^n as the size of the elementary cell in Γ-space.

We wish to conclude this section with a brief discussion of the importance of the chemical constants and of Eq. (9.307). In the derivation of Eq. (9.307) use was made of Nernst's heat theorem by putting S_c equal to zero at the absolute zero. From Eq. (9.307) it follows that we can determine i and thus the entropy constant S_{v0} of the gas, which is defined by the equation

$$S_v = k \ln \frac{T^{5/2}}{P} + S_{v0}, \tag{9.311}$$

from a measurement of the vapor pressure. Once S_{v0} is determined, all thermodynamic properties of the gas can be calculated. Since these properties play an important role in the computation of chemical equilibria, Nernst gave i the name of chemical constant. We must refer to the literature for a discussion of the calculation of chemical constants and their importance in the discussion of chemical reactions and equilibria.

9.4 SPECIAL CASES

In this last section we wish to consider a few special cases and see whether or not the third law is satisfied. Before doing that, we wish to point out that, although we saw in Section 9.2 that the low energy behavior of the density of states determines the low temperature behavior of the entropy, there are very few cases where one can actually directly evaluate the density of states; usually one obtains the density of states by writing Eq. (9.201)—or the equivalent expression for the case of a canonical grand ensemble—in the form

$$e^{-\beta\psi} = \int e^{-\beta\varepsilon}\rho(\varepsilon)\,d\varepsilon, \tag{9.401}$$

noting that it follows from this equation that $\rho(\varepsilon)$ is the Laplace transform of $e^{-\beta\psi}$, and then writing the inverse equation:*

$$\rho(\varepsilon) = \frac{1}{2\pi i}\int_{\beta'-i\infty}^{\beta'+i\infty} e^{-\beta\psi(\beta)}e^{\beta\varepsilon}\,d\beta, \tag{9.402}$$

where the path of integration lies to the right of the imaginary β axis. This means that only in a few cases will we be able to find S from ρ and verify the validity or otherwise of the third law. Usually we can find S and from S (or rather from ψ) infer the behavior of ρ.

* See, for instance, I. N. Bronshtein and K. A. Semendyayev, *A Guidebook to Mathematics*, Oxford, 1963, p. 541.

The first case we shall consider is a perfect gas. In the case of a Boltzmann gas we can actually evalute ρ. The total number of states available to a system of N Boltzmann particles for energies not exceeding ε, $\Omega(\varepsilon)$, is given by the expression

$$\sum 1 = \Omega(\varepsilon), \tag{9.403}$$

where the summation is over all points in the $3N$-dimensional space with coordinates equal to a positive integer within a sphere of radius $[2V^{1/3}\sqrt{(2m)}/h]\sqrt{\varepsilon}$ (compare Eq. (4.601)). From the formula for the volume of a $3N$-dimensional sphere we find thus

$$\Omega(\varepsilon) = C\varepsilon^{3N/2}, \tag{9.404}$$

where C is a constant that is independent of ε but depends on the volume. The density $\rho(\varepsilon)$ now follows from the relation

$$\rho(\varepsilon) = \frac{d\Omega}{d\varepsilon}. \tag{9.405}$$

We see that we are here in case (2) of Section 9.2, and the entropy will contain a term proportional to ln T. This is, of course, in agreement with Eq. (4.618), and if we had written down the explicit expression for C in Eq. (9.404) and used Eqs. (9.202) and (9.207), we would have found exactly Eq. (4.618) times N, as we are now dealing with the entropy of the whole system.

Let us now consider the case of a boson gas. In Sections 4.7 and 6.7 we found that the q-potential—or the grand potential—is related to the entropy, on the one hand, and to the pressure, on the other hand, by the equations

$$q = \frac{PV}{kT} = \frac{S}{k} + \alpha N - \frac{E}{kT}, \tag{9.406}$$

where V is the volume occupied by the gas, E its total energy, and α the thermal potential divided by kT. In the case of a perfect Bose-Einstein gas this potential is given by Eq. (4.709), or

$$\frac{PV}{kT} = \left(\frac{2\pi mk}{h^2}\right)^{3/2} T^{3/2} V \sum_{n=1}^{\infty} \frac{e^{n\alpha}}{n^{5/2}}. \tag{9.407}$$

Since we are interested in the value of S at $T = 0$, we are in the condensation region (see Section 4.7) where $\alpha = 0$, and we have thus from Eqs. (9.406) and (9.407)

$$S = \frac{PV}{T} + \frac{E}{T} = \frac{5}{2}\frac{PV}{T} = \left(\frac{2\pi mk}{h^2}\right)^{3/2} kVT^{3/2}\zeta\left(\frac{5}{2}\right), \tag{9.408}$$

where we have used the equation

$$PV = \tfrac{2}{3}E, \tag{9.409}$$

and where $\zeta(\frac{5}{2})$ is Riemann's ζ-function ($\zeta(n) = \sum_{k=1}^{\infty} k^{-n}$). From Eq. (9.408) it is seen immediately that Eq. (9.112) is satisfied. This case corresponds to case (1) of Section 9.2 with $p = \frac{3}{5}$.

Let us finally consider the case of a perfect gas of fermions. We cannot invoke Eq. (4.843) here directly, because it was derived assuming that $S = 0$ for $T = 0$. However, using Eqs. (9.406) and (9.409), we find

$$S = \frac{5}{3}\frac{E}{T} - k\alpha N = \frac{\frac{5}{3}E - \mu N}{T}, \tag{9.410}$$

and using Eqs. (4.839) and (4.840), we find easily

$$S = \frac{\pi^2}{2}\frac{\eta^2 N\mu_0}{T}, \tag{9.411}$$

in accordance with Eq. (4.843), if we bear in mind the definition (4.829) of η. As η is proportional to T, we see that, indeed, S vanishes as $T \to 0$ proportional to T itself. This case corresponds to case (1) of Section 9.2 with $p = \frac{1}{2}$.

The next case to be considered is that of a system of N particles with a permanent magnetic moment. In Section 2.4 we considered the classical case, and found for the magnetization in the direction of the field (see Eq. (2.412))

$$M = N\mu\left[\coth x - \frac{1}{x}\right], \tag{9.412}$$

where

$$x = \beta\mu B, \tag{9.413}$$

with B the magnitude of the applied magnetic field, and μ the magnitude of the magnetic moment. We can now use one of the thermodynamic Maxwell relations*

$$\frac{\partial S}{\partial B} = \frac{\partial M}{\partial T}, \tag{9.414}$$

and we see from Eqs. (9.412) and (9.414) that $\partial S/\partial B$ does not vanish as $T \to 0$, as it should if the third law were valid for the system (compare the third law in the form (9.101)).

* See, for instance, D. ter Haar and H. Wergeland, *Elements of Thermodynamics*, Reading, Mass., 1966, Section V.3.

Let us now consider the case of N spin-$\frac{1}{2}$ particles with a permanent magnetic moment. In that case the magnetization is not given by Eq. (9.412), but by the quantum-mechanical equivalent (compare Problem 13 of Chapter 3)

$$M = N\mu \tanh x, \tag{9.415}$$

and $\partial S/\partial B \to 0$ as $T \to 0$.

The last case to be considered is that of liquid helium. We first consider the melting curve of helium. According to the Clausius-Clapeyron equation we have

$$\frac{dP}{dT} = \frac{\Delta S}{\Delta V}, \tag{9.416}$$

where ΔS and ΔV are, respectively, the change of entropy and the change of volume during melting, while the left-hand side of this equation gives us the slope of the melting curve in a P-T-diagram. Measurements by Simon and Swenson show that indeed the slope of the melting curve of He^4 becomes zero for $T \to 0$, as should follow from Eqs. (9.101) and (9.416). Below 1.4 °K dP/dT vanishes as T^7, a most striking example of the validity of the third law.

Another interesting feature is the behavior of the surface tension, σ, of He^3. Applying the relevant Maxwell relation, we find that the third law requires that

$$\lim_{T \to 0} \frac{\partial \sigma}{\partial T} = 0, \tag{9.417}$$

a relation verified experimentally by Zinov'eva.

We may also mention the fact that while, above 1 °K, He^3 and He^4 mix in all proportions, below about 0.8 °K phase separation takes place; as $T \to 0$, the two phases tend to become pure He^3 and pure He^4, thus leading to a vanishing of the mixing entropy in that limit.

Finally, a last point of interest is: If ρ be the heat of melting, one has, on the one hand, the equation

$$\rho = T\Delta S, \tag{9.418}$$

and, on the other hand, the equation

$$\rho = \Delta E + P\Delta V, \tag{9.419}$$

where ΔE is the difference in energy content between the liquid and the solid state. From Eqs. (9.418) and (9.101) it follows that $\rho \to 0$ for $T \to 0$. Since in the case of helium ΔV does not vanish at the absolute zero and remains positive, it follows that at sufficiently low temperatures the energy of the liquid is lower than that of the solid, a most unusual situation.

PROBLEMS

1. Falk* has formulated the third law as follows: When the energy has its minimum value, the entropy also has its minimum value. Show that this statement is equivalent to the statement: The absolute minimum of the energy cannot be reached with a finite number of processes.†
2. Prove that the third law in the strictest sense of Eq. (9.112) is violated, if the density of state contains a factor that is a function of N and V only.
3. Prove that if a system consists of N independent subsystems with a discrete energy spectrum, we are dealing with case (3) of Section 9.2.
4. Consider the case of N distinguishable one-dimensional harmonic oscillators (see Problem 17 at the end of Chapter 6). Show that this system belongs to case (3) of Section 9.2.
5. Prove, by explicitly evaluating the density of states, that a system of N spin-$\frac{1}{2}$ particles with a magnetic moment in an external magnetic field belongs to case (3) of Section 9.2. Reckon all energies from the ground state energy as zero.
6. Consider the case of N independent classical one-dimensional harmonic oscillators. Prove that this system belongs to case (2) of Section 9.2.
7. By using the thermodynamic relations

$$N\chi = E_v - E_c + P(V_v - V_c),$$

and

$$E(T) = E(0) + \int_0^T C_v \, dT,$$

and making suitable assumptions, derive Eq. (9.306).

8. Prove that $\alpha_v \to -\infty$ as $T \to 0$ so that the assumption that the vapor can be treated as a perfect gas is the more justified, the lower the temperature.
9. Prove that the only phase present at the absolute zero is the condensed phase, so that the fact that the latent heat of sublimation remains finite as $T \to 0$ does not imply a violation of the third law.‡
10. Derive the vapor pressure equation for the case where we can use the Einstein model of a crystal for the condensed state, that is, where we may assume that the crystal consists of N independent one-dimensional harmonic oscillators, and that it costs an energy χ to remove one oscillator from the crystal into the vapor.

* G. Falk, *Phys. Rev.*, **115,** 249, 1959.

† J. Kvasnica, *Czech. J. Phys.*, **B10,** 883, 1960.

‡ See S. J. Glass and M. J. Klein, "*Investigations on the Third Law of Thermodynamics*," *AEC Technical Report, AT (II-I)-623*, August 1958.

BIBLIOGRAPHICAL NOTES

It would be outside the scope of this book to enter into a detailed discussion of Nernst's heat theorem, its applications, and its foundations, which was published in 1906:

1. W. Nernst, *Göttinger Nachr.* **1906,** 1.

For a general discussion and extensive bibliographies we refer to

2. F. Simon, *Ergeb. exakt. Naturwiss.*, **9,** 222, 1930.
3. J. Wilks, *The Third Law of Thermodynamics*, Oxford, 1961.

We may also refer to the following papers:

4. F. Simon, *Physica*, **4,** 1089, 1937.
5. G. O. Jones and F. E. Simon, *Endeavour*, **8,** 175, 1949.
6. F. E. Simon, *Z. Naturf.*, **6a,** 397, 1951.

For a discussion of some special aspects see also

7. S. J. Glass and M. J. Klein, "Investigations on the Third Law of Thermodynamics," *AEC Technical Report No.* 2, *Contract AT* (II-I)-623, August, 1958 (unpublished).
8. D. ter Haar and H. Wergeland, *Elements of Thermodynamics*, Reading, Mass., 1966, Chap. VIII.

For a more axiomatic approach see

9. P. T. Landsberg, *Rev. Mod. Phys.*, **28,** 363, 1956; *Thermodynamics*, New York, 1961, Chap. III.

Section 9.1. See Reference 2 and

10. P. Jordan, *Statistische Mechanik auf Quantentheoretischer Grundlage*, Braunschweig, 1933, Section 3.7.
11. R. H. Fowler and E. A. Guggenheim, *Statistical Thermodynamics*, Cambridge, 1939, p. 219.
12. E. Schrödinger, *Statistical Thermodynamics*, Cambridge, 1948, Chap. III.

Section 9.2. See References 7 and 8, and especially

13. H. B. G. Casimir, *Z. Physik*, **171,** 246, 1963.

Section 9.3. See References 2, 3, 10, and

14. O. Sackur, *Ann. Physik*, **36,** 958, 1911.
15. H. Tetrode, *Ann. Physik*, **38,** 434, 1912 (corrections to this paper in *Ann. Physik*, **39,** 255, 1912).
16. O. Sackur, *Ann. Physik*, **40,** 67, 1912.
17. H. Tetrode, *Proc. Kon. Ned. Akad. Wet.* (*Amsterdam*), **17,** 1167, 1915.
18. P. Ehrenfest and V. Trkal, *Proc. Kon. Ned. Akad.* (*Amsterdam*), **23,** 162, 1920.
19. P. Ehrenfest and V. Trkal, *Ann. Physik*, **65,** 609, 1921.

Section 9.4. See References 2, 3, 7, and 9. For the melting curve of He^4 see:

20. F. E. Simon and C. A. Swenson, *Nature*, **165**, 829, 1950.

For the surface-tension curve of He^3 see Reference 3 and

21. K. N. Zinov'eva, *J. Exptl. Theoret. Phys.* (*U.S.S.R.*), **29**, 899, 1955; *Soviet Phys.—JETP*, **2**, 774, 1956.

For a discussion of the phase separation of He^3 and He^4 see Reference 3 and

22. W. M. Fairbank and G. K. Walters, *Nuovo Cimento, Suppl.*, **9**, 297, 1958.
23. V. P. Peshkov and K. N. Zinov'eva, *J. Exptl. Theoret. Phys.* (*U.S.S.R.*), **32**, 1256, 1957; *Soviet Phys.—JETP*, **5**, 1025, 1957.

TABLE OF PHYSICAL CONSTANTS

In the table we give the values of a number of physical constants taken from a table by J. A. Bearden and H. M. Watts (*Phys. Rev.*, **81,** 73, 1951).* We have included a few constants that are not used in the text in order to increase the usefulness of the table.

Symbol	Constant	Value
R	absolute gas constant	$8.314 \cdot 10^{7}$ g cm^2 sec^{-2} mol^{-1} deg^{-1}
m_H	atomic weight unit	$1.660 \cdot 10^{-24}$ g
N_A	Avogadro's number†	$6.026 \cdot 10^{23}$ mol^{-1}
μ_B	Bohr magneton‡	$9.271 \cdot 10^{-21}$ g cm^2 sec^{-2} gauss^{-1}
a_B	Bohr radius	$5.291 \cdot 10^{-9}$ cm
k	Boltzmann's constant	$1.380 \cdot 10^{-16}$ g cm^2 sec^{-2} deg^{-1}
a	constant in the Stefan-Boltzmann law	$7.564 \cdot 10^{-15}$ g cm^{-1} sec^{-2} deg^{-4}
h	Planck's constant	$6.623\,6 \cdot 10^{-27}$ g cm^2 sec^{-1}
$\hbar$	Dirac's constant	$1.054\,2 \cdot 10^{-27}$ g cm^2 sec^{-1}
m	electron mass	$9.107 \cdot 10^{-28}$ g
e	electronic charge	$4.802 \cdot 10^{-10}$ esu
$\alpha\left(=\frac{e^2}{\hbar c}\right)$	fine structure constant	137.04^{-1}
μ_p	proton magnetic moment	$1.410 \cdot 10^{-23}$ g cm^2 sec^{-2} gauss^{-1}
m_p/m	ratio of proton to electron mass	1 836.09
R_∞	Rydberg constant for infinite mass	109737.3 cm^{-1}
c	velocity of light	$2.997\,90 \cdot 10^{10}$ cm sec^{-1}

* The author would like to express his thanks to J. A. Bearden and to the American Institute of Physics for their kind permission to quote from this table.

† We use the physical scale of atomic weights.

‡ There is a misprint in the exponent of 10 in the case of μ_B in Bearden and Watts' paper (see *Phys. Rev.*, **83,** 457, 1951).

GLOSSARY OF SYMBOLS

This glossary contains a list of the more important symbols occurring in the text. The page numbers indicate where these symbols are introduced.

a, a'	constants in van der Waals equation, 9
a	constant in Stefan-Boltzmann law, 73
A	collision term, 13
A	moment of inertia, 46
$\mathscr{A}$	cross section, 38
$\mathbf{A}$	vector potential, 58
a_i	external parameter, 50
A_i	generalized force, 51
a_{kl}	coefficient in kinetic energy, 40
A_{kl}	Einstein transition probability, 74
A_{ik}	generalized force, 103
$a_n^{\ k}$	probability amplitude, 209
$a_{12\to 1'2'}$	transition probability, 15
b	constant in van der Waals equation, 9
B	second virial coefficient, 10
B	collision term, 13
$\mathbf{B}$, B	magnetic (induction) field, 46
B	moment of inertia, 58
b_{kl}	coefficient in potential energy, 44
B_{kl}	Einstein transition probability, 74
$B_J(x)$	Brillouin function, 87
$\mathbf{c}$	velocity vector, 2
c	absolute magnitude of velocity, 4
c	velocity of light, 71

C	third virial coefficient,	10
C	moment of inertia,	57
$\mathfrak{C}$	Equation (4.616),	116
c_V, C_v	specific heat,	26, 45
c_{rel}	relative velocity,	15
d	length of dumbbell molecule,	40
D	atomic diameter,	15
$D(\omega)$	density of frequencies,	71
$D(p, q)$	density in phase,	152
e	electric charge,	56
E	total energy,	23
$\mathbf{E}$	electric field,	56
f	distribution function,	3
$\mathbf{F}$	force,	10
F	Helmholtz free energy,	53
g	gravitational acceleration,	38
g	degree of degeneracy,	70
G	Gibbs free energy,	106
g_n	degree of degeneracy,	75
$g(\omega)$	vibrational mode density,	85
$g(\alpha)$	Equation (4.710),	118
h	Planck's constant,	64
$\hbar$	Dirac's constant,	66
H	Boltzmann's H-function,	20
$\mathscr{H}$	Hamiltonian,	41
$\hat{\mathscr{H}}$	Hamiltonian operator,	94
$h(\alpha)$	Equation (4.805),	127
i	chemical constant,	295
I	phase integral,	65
$\mathbf{j}$	electrical current,	56
j	rotational quantum number,	81
J	Jacobian,	18
J_k	action variable,	77
J_j	flux,	275
k	Boltzmann's constant,	4
K, K'	normalization constants,	51, 64

l	one-dimensional volume, 51
L_{ij}	proportionality coefficient, 275
$\mathscr{L}(x)$	Langevin function, 46
m	mass, 4
M	molecular weight, 26
$\mathscr{M}$	magnetization, 87
m_{H}	molecular weight unit, 116
n	number density, 3
n	quantum number, 66
N	number of atoms in system, 2
N_{A}	Avogadro's number, 6
N_i	number of atoms in cell, 27
P	pressure, 6
$\mathbf{P}$	total momentum, 23
P	permutation operator, 95
p_i	generalized momenta, 40
q	q-potential, 106
q	grand potential, 183
q, Q, Q_m	heat, 26
q_i	generalized coordinates, 39
$\dot{q}_i$	generalized velocities, 39
$\mathbf{r}$	position coordinate vector, 2
r	ortho- to parahydrogen ratio, 84
R	gas constant, 6
R	Hall coefficient, 57
r_0	size of diatomic molecule, 81
s	velocity of sound, 85
s	number of degrees of freedom, 39
s, S, S_m	entropy, 26
S	nuclear spin, 82
t	time, 3
t	number of quadratic terms in potential energy, 44
T	temperature, 4
$\mathscr{T}$	kinetic energy, 4
T_0	condensation temperature of Bose-Einstein gas, 125

u, U, U_m	internal energy,	26
U	potential energy,	9
v	volume of 1 g of gas,	25
V	volume,	2
$\mathscr{V}$	virial,	10
V_m	volume of 1 mole,	26
$\mathbf{w}$	center of mass velocity,	13
$\mathbf{w}$	heat current,	57
$\mathscr{W}$	part of virial,	11
w_k	angle variables,	77
W_{Bo}	Boltzmann weight,	96
W_{BE}	Bose-Einstein weight,	97
W_{FD}	Fermi-Dirac weight,	98
W_j	probability for cell Z_j,	99
$W(n_i)$	probability for n_i-distribution,	92
$W(N_j)$	probability for N_j-distribution,	27
X, Y, Z	center of mass coordinates,	40
X_j	affinity,	275
Z_i	volume of cell,	27
Z_μ	partition function,	51
Z_Γ	partition function,	62
Z^{qu}	partition function,	64
dZ	number of energy levels in given range,	115
α	Lagrangian multiplier,	23
β	Lagrangian multiplier,	23
β	$= 1/kT$,	52
γ_k	multiplicity,	218
$\gamma_{\mathrm{Bo}}, \gamma_{\mathrm{FD}}, \gamma_{\mathrm{BE}}$	Equation (4.307),	101
$\delta(x)$	Dirac delta-function,	56
ε	energy,	39
ε_i	energy in cell Z_i,	28
ε_n	energy eigenvalue,	63
ε_P	$= (\pm)^P$,	97

ζ	Riemann zeta-function,	73
ζ	degeneracy parameter,	118
η	degeneracy parameter,	132
η	index of probability,	155
ϑ	Euler angle,	57
θ	Equation (7.118),	251
Θ	rotational temperature,	81
Θ	Debye temperature,	85
Θ	ensemble modulus,	158
κ	thermal conductivity,	57
κ	c_P/c_V,	120
κ	compressibility,	265
λ	mean free path,	56
λ	wavelength,	71
$\boldsymbol{\mu}$, μ	magnetic moment,	46
μ	partial free energy,	105
ν_i	number of particles of kind i,	182
Π	Peltier heat,	284
ρ	density,	38
ρ	ensemble density,	155
$\hat{\rho}$	density matrix,	210
ρ_{mn}	matrix elements of $\hat{\rho}$,	210
$\rho(\omega)$	radiation energy density,	72
φ	Euler angle,	57
φ	work function,	142
ϕ	Equation (5.0504),	167
φ_i	integrals of motion,	152
φ_n	complete orthonormal set,	209
ψ	Euler angle,	57
ψ	wave function,	78
χ	depth of potential well,	141
χ	sublimation heat,	294

$\boldsymbol{\omega}$	unit vector along line of centers,	14
ω	frequency of harmonic oscillator,	65
Ω	volume in phase space,	78
Ω	Equation (5.0502),	166
ω_{kl}	transition frequency,	74
$d\omega$	volume element in μ-space,	41
$d\Omega$	volume element in Γ-space,	153
∇	gradient operator,	49
∇_c	velocity gradient operator,	49
$\{\ldots,\ldots\}$	Poisson bracket,	154
$\wedge$	indicates operator,	94

INDEX